INTERNATIONAL ORDER AND THE POLITICS OF DISASTER

In this indispensable and comprehensive text, Scott D. Watson critically examines the current understanding of international order that underpins international disaster management and disaster diplomacy.

Based on empirical analysis of the three international disaster management regimes – disaster relief, disaster risk reduction, and disaster migration – and case studies of disaster diplomacy in the United States, Egypt and China, Watson argues that international disaster management and disaster diplomacy are not simply efforts to reduce the impact of disasters or to manage bilateral relations but to reinforce key beliefs about the larger international order. Challenging the conventional understandings of disasters as natural, as exogenous shocks, or as unintended and accidental outcomes of the current order, this text shows how the ideological foundations of the current heterogenous international order produce recurrent disasters.

International Order and the Politics of Disaster is a vital source for undergraduate or graduate students interested in international responses to disasters and complex humanitarian emergencies, forced migration and displacement, as well as climate change and development.

Scott D. Watson is associate professor of International Relations and chair of the Department of Political Science at the University of Victoria. He holds a PhD in Political Science from the University of British Columbia (2006) and is the author of *The Securitisation of Humanitarian Migration* (Routledge, 2009).

INTERNATIONAL ORDER AND THE POLITICS OF DISASTER

Scott D. Watson

Routledge
Taylor & Francis Group
NEW YORK AND LONDON

First published 2020
by Routledge
52 Vanderbilt Avenue, New York, NY 10017

and by Routledge
2 Park Square, Milton Park, Abingdon, Oxon, OX14 4RN

Routledge is an imprint of the Taylor & Francis Group, an informa business

Library of Congress Cataloging-in-Publication Data
A catalog record for this title has been requested

ISBN: 978-0-367-20036-7 (hbk)
ISBN: 978-0-367-20037-4 (pbk)
ISBN: 978-0-429-25927-2 (ebk)

Typeset in Bembo
by Lumina Datamatics Limited

CONTENTS

ACKNOWLEDGEMENTS

Since I began this project in 2013, I have accumulated a significant debt to all those who provided support, suggested additional readings and ideas, and challenged my claims. I would like to acknowledge the support of my colleagues and students at the University of Victoria, and express my gratitude to work with such a collegial and generous group. I would like to thank the Centre for Global Studies, where I spent a very fruitful and stimulating fellowship working on this project. The anonymous reviewers of this manuscript and audiences in various conferences where I presented parts of this work have provided insightful feedback, and I am appreciative of their constructive feedback.

I had the great pleasure to have Regan Burles and Can Zhao as research assistants on various parts of the project; thank you for your assistance.

Last, I would like to thank Alicia, Caleb, and Jacqueline for their love and support; completing this project would not have been possible without you.

1

INTRODUCTION

Humanity appears to be rapidly approaching an age of disasters that may well rival or even surpass the disruptions of previous periods of calamity.[1] Climate scientists expect the range of climate variability to expand, increasing the destructiveness of regular seasonal weather patterns and making once-rare climate events much more frequent. By most accounts, these changes are no longer preventable, even if the international community succeeded in reaching a binding resolution to curb greenhouse gas emissions. If the past decade is even a mild precursor to the near future, we should expect significant hardship and dislocation. Five of the hottest years on record have occurred in the 2010s. In the past 10 years, disasters have affected 1.5 billion people worldwide, killed over 700,000, and rendered another 23 million homeless (UNISDR, 2015a, 2015b). The costliest year on record for the United States in terms of natural disasters was 2017, with total damage exceeding $306 billion and with 16 different billion-dollar disasters (NOAA, 2018). According to the National Oceanic and Atmospheric Association (NOAA), government agencies and the insurance industry now consider losses exceeding $100 billion in damage due to natural disasters as the "new normal".

The near inevitability of an escalation in the number and severity of disasters necessitates further exploration of how political communities understand, prepare for, and respond to disasters. The deadly combination of climate change, environmental degradation, burgeoning world population, and expanding inequality will drastically boost human vulnerability to natural and human-made hazards (Chipper and Pelling, 2006; IPCC Working Group II, 2014). This book examines current practices for managing disasters, arguing that they are not simply inadequate but actually function to sustain an international order that has given rise to the social, political, and economic factors at the root of the looming period of persistent disaster-induced crises.

International Order and the Politics of Disaster draws and builds on the now well-established claim that disasters are socially produced – the outcome of complex interactions of natural hazards, social vulnerabilities, and intersubjectively established expectations about how the natural and social worlds operate. In contrast to accounts that depict disasters as completely natural, as accidents, or as "interruptions in a linear process of development" (Hannigan, 2012:51), the social view of disasters suggests that disasters are a function of social systems. Disasters disrupt important social routines and overwhelm the existing capacity to cope with these disruptions. In doing so, they reveal how the existing social capacity to handle disruptions has been (or has not been) developed and maintained (Stallings, 1988). Actors recognize events as disasters through intersubjective agreement that the event is in fact a disaster – revealing how some social actors enjoy a privileged position in defining events as disasters (and not others) and marshalling extra social resources.

The central contribution of this book to the critical disaster studies literature that informs this project is that disasters are *international* social events. To date, the critical disaster studies literature has paid insufficient attention to the global realm (Oliver-Smith, 1998), just as the international relations (IR) literature has largely neglected the field of disaster; the field of IR has integrated natural disasters either as a transboundary issue that can either foster cooperation or exacerbate conflict, or as a humanitarian issue largely independent from the politics of international relations (Hannigan, 2012:2). As Hannigan notes, disasters are not the disruption of normal politics at the global level; rather, they are the product of the "social, political and economic environments" (Hannigan, 2012:17). By treating disasters as international social events, I draw attention to the global processes that promote standard forms of knowledge and practices, produce new hazards while eliminating or reducing others, and increase or decrease human vulnerability. Major disasters disrupt important social routines across international borders, and the existing social capacity to prepare for and cope with disaster now includes extensive international disaster management regimes. In short, when thinking about disasters and their production and management, it is necessary to think globally, even as the effects of disasters are intensely local. As cliché as it may sound, the local is increasingly global (and vice versa) – local responses are governed by international frameworks, implemented through national legislation and the regulation of local practices, as well as by the practices of international nongovernmental organizations (NGOs) and corporations that operate at multiple sites and levels of disaster governance. Thus, the local, which is where disasters are experienced and managed (usually by community associations and neighbours who are often the first responders) becomes global.

The globalization of disaster occurs through a "re-articulation of scale" as local issues are understood as a matter of global concern and governance (such as risk reduction and building resilience) and as global developments are incorporated into local planning (such as climate change) (Arts, 2004). While disasters continue to be highly localized in their primary effects and in the initial response stage, local actors understand these developments through common frames of meaning and practices established through international processes. Recognizing the influence of international factors in the management of disaster reorients attention to the international order – how it creates hazards and vulnerability, how it shapes social capacity to handle unexpected events, and how it reproduces power relations in which some are privileged in defining disasters and shaping how they ought to be managed.

Historical comparison further demonstrates the importance of the international realm. The historical record shows that vulnerability to natural hazards that threaten the foundations of the existing social order is a recurrent issue across a wide range of human societies; some societies manage to rebuild and prosper in the aftermath of disaster, while others rebuild but never fully recover their previous level of prosperity and power. In other cases, communities have failed to recover from a major disaster or have migrated to other areas. The current era differs from historical precedents in a number of key ways associated with globalization; first and foremost is a changed worldview on the nature of disasters and the ability and desirability of human communities to "manage" the natural world. Other important changes include restrictive mobility regimes that reduce the viability of migration as a basic form of adaptation. At the same time, changes in information and transportation technology facilitate near simultaneous knowledge of disaster events and the rapid mobilization of international assistance. In the current era, governments and media networks know about and can respond to disasters in even the most remote human communities almost immediately. The United Nations network of humanitarian institutions often has information within minutes of a disaster, and many states have military disaster response teams whose response time is hours or days, rather than weeks and months. Facebook and Google have developed technology to track disaster displacement in real time. While the historical record of humanity and disasters should temper overgeneralizations about the novelty or unprecedented nature of the current era, recent technological developments and forms of global governance suggest that the current era differs markedly from earlier historical periods, which comes with profound implications for how societies understand and respond to disasters.

The modern ability to plan for and respond rapidly to disasters, however, is not value neutral. As with other forms of intervention, disaster preparation, relief, and recovery challenge basic beliefs about who is responsible for protecting people from calamity, and how best to do so. Just as disasters and disaster recovery highlight the social processes that produce inequality and vulnerability at the local level, so too do they demonstrate these same processes internationally. As I will demonstrate throughout this book, current international disaster management arrangements attempt to resolve a range of tensions inherent to the current international order: (1) the relative value of scientific and localized/traditional forms of knowledge; (2) the hierarchical division of responsibility between nation states and non-state actors; (3) the impact of power competition between states that prioritizes certain forms of insecurity over others; (4) the promotion of economic systems that justify and indeed necessitate certain forms of poverty, vulnerability, and degradation of nature; and (5) the nature of obligations to reduce or prevent suffering across international borders in a globalized context. The current system of disaster management attempts to provide contingent resolutions to these problems and in doing so privileges powerful actors who produce knowledge of disasters and who possess the resources to manage them (Pelling, 2003:6).

The establishment and functionality of the international disaster management system, combined with the failure to develop a broader regime for reducing disaster risk, combatting climate change, and facilitating migration demonstrates that the current regime serves to reproduce a specific set of beliefs about the source of international order. Disaster management inscribes particular values, interests, and practices, of which environmental degradation, climate change, economic inequality, and limited mobility are intrinsic and inevitable outcomes. The emphasis on emergency response reflects a concern with maintaining and implementing this international order and containing sources of disorder. I am concerned in this book with identifying the foundations of this order and with the discrepancies in that order exposed by disasters.

In this context, disasters expose cracks in the foundations of sociopolitical order by demonstrating the incongruity between expectations about how the world works and lived experiences. For Stallings, disasters are destabilizing because they "reduce certainty … and undermine faith" in existing institutions and bodies of knowledge (1988:129–31). Yet, as noted earlier, there are a wide range of belief systems about how the world works, and the foundations destabilized by disaster are matters of political contestation. Contrary to those who see international order emerging from a single overarching belief system (liberal order, for instance; see Sorenson, 2011), I suggest that international order is heterogenous, resting on an amalgam of interrelated belief systems (or *ideologies of international order*) that provide an account of social life and that direct action toward certain values and goals. International order does not revolve around a single objective; it entails the advancement of "a normatively thick and culturally and historically contingent vision of the good" as promoted by the dominant actors in the system (Phillips, 2011:5). For Phillips (2011), international order is "the constellation of constitutional norms and fundamental institutions" whose primary purposes are to "cultivate cooperation" and "contain conflict" by gaining consensus through "authoritative and coercive institutions" (Phillips, 2011). Where I depart from Philipps and others is that international order does not rest on a single conception of "the good"; rather, it is an amalgam of complementary and contradictory conceptions of "the good". Consequently, the "constellation of constitutional norms and fundamental institutions" that make up an international order may work toward different conceptions of the good and may rely on different mechanisms to achieve cooperation and limit conflict. This is the defining feature of a heterogenous international order.

This project employs a "neutral" conception of ideology that focuses on the "indispensable function of political belief systems" that are necessary to all human societies, regardless of the actual nature of those systems (Steger, 2008:4). It also draws on the multidisciplinary work that depicts ideology as consisting of "social, cognitive, and discursive components" (Van Dijk, 2006:115). Consequently, ideology is intersubjective, although it is rooted in individual cognitive processes profoundly shaped by prevailing discourses.[2] These ideologies, or belief systems essentially provide the foundation for a society's "identity, actions, aims, norms and values…as well as relations to other social groups" (Van Dijk, 2006). Ideologies simultaneously identify relevant actors, encourage those actors to act, and constrain their freedom of action (Steger, 2008:5). Buzan and Lawson use the concept of ideology to explain the international system as well, identifying how these belief systems "address how polities, economies and cultural orders relate to each other, how individuals and groups fit in these assemblages, and how communities should be governed" (Buzan and Lawson, 2015:99).[3] I argue that the current international order relies on multiple ideologies that address and govern the complex and uncertain interactions between the natural world, specific political communities, and humanity as a whole. These *ideologies of international order* include (but are not limited to) modernism (addresses the value of scientific and non-scientific knowledge), nation-statism (addresses the relation between state and society in providing public goods), realism (addresses the relations between states), economic liberalism (addresses the relationship between economic activity and vulnerability), and humanitarianism (addresses the obligations between societies).

This typology of ideologies expands on those provided by Simon Hollis who wrote specifically of disaster management and identified three "legitimizing discourses" of disaster management: humanitarian ethics of care, scientific ontology, and sovereignty (Hollis, 2014). In addition to disaggregating sovereignty to its "domestic" and "international" components (which I describe as nation-statism and realism, respectively), I also add economic liberalism as the dominant ideology of economic relations internationally. The following chapter develops these ideologies of order more thoroughly, although it bears emphasizing at this point that the aforementioned ideologies of order are not an exhaustive list of the belief systems and relations foundational to the existing international order. Rather, I argue that they are the belief systems most deeply implicated in the management of disaster and of international disaster relief in the post–World War II era – a claim I develop and support in subsequent chapters.

Periods of disaster are a particularly appropriate time to examine ideologies of order – as Geertz notes, ideologies are themselves a "response to strain" and function to orient societies facing disruption (Geertz, 1973). The first section of this book, which includes Chapters 3 through 5, argues powerful actors developed the various disaster management regimes (DMRs) as key mechanisms to shore up the ideological foundations of international order; however, I show that this practice itself exposes further ambiguities within, and between, these belief systems. Like other international order-sustaining practices, such as war and humanitarian intervention (indeed, disaster relief often bears a striking similarity to these practices), disaster management opens up the possibility of altering the existing order that it seeks to maintain, often through incremental change. Whether societies alter or replace disaster management practices, which Stallings refers to as "exception routines", depends on "organizational and professional power" (Stallings, 1988:140).

The creation of international disaster management regimes, as well as post-disaster political contestation at the national and international level, illustrates how "organizational and professional power" is brought to bear on maintaining, altering, or replacing existing "exception routines". The post-disaster period is a particularly useful time to examine the underlying ideologies of an existing order. It is an "unsettled" period, in which political actors rely on ideology to guide action and to maintain order (see Swidler, 1986). Disasters are unsettling because, beyond the death toll and economic impact, disasters profoundly disrupt the routine functioning of existing systems (Kreps, 1989; Wisner et al., 2004). Consequently, disasters reveal the limits of our current knowledge and challenge claims of stability and security provided by the current social, political, and economic order (Hewitt, 1983:9; Stallings, 1988).

In this sense, disasters are a form or subcategory of crisis (Quarantelli, 1998), a concept that has come to occupy a central position in recent IR scholarship explaining change and international order (e.g. Hay, 1999; Blyth, 2002; Phillips, 2011; Jordheim and Wigen, 2018).[4] Hay defines a crisis as a moment when "systematic contradictions require decisive intervention" that will either stabilize the system or "exacerbate its latent contradictions" (1999:325). Guzzini, applying the concept of crisis to identity, defines crisis as a "moment in which debates cannot update or combine repertoires to fix tensions, either internally or externally induced", thus requiring new interpretations and identities (Guzzini, 2017:7). At the root of this literature on crisis is an understanding of an existing order beset with tensions and a theory of change that emerges from these internal contradictions. This project shares their view of the international system as comprised of multiple foundations beset with internal tensions and contradictions that require ongoing maintenance and periodic substantial intervention and reinvention. Disasters are a form of crisis because they demonstrate the failure of existing systems to provide goods or outcomes central to their own narrative of legitimacy and authority. The expectation of increased instances of disaster not only threatens to kill and displace larger numbers of people and to stall or reverse economic development; it will also present increasing opportunities to challenge and/or alter the ideological foundations of political and economic systems, at the national and international levels.

To explain how states maintain international order through the management of disasters, I examine (1) efforts to create international DMRs and (2) diplomacy in the aftermath of disaster. In Section 1 of this book, I focus on three DMRs: the disaster relief regime, the risk reduction regime, and the migration regime, all of which attempt to manage the inconsistencies and failures of the existing international order by separating these three interrelated issues, and by reinscribing the authority of key beliefs about international order.

In Section 2, I focus on disaster diplomacy, more specifically on how disaster-affected states respond to disaster assistance. In this section, I focus in depth on three cases, the United States, Egypt, and China, to show how political elites within recipient states make decisions on disaster assistance to maintain or challenge existing social, political, and economic relations. I include the United States as one of the key authors of the international disaster management system and the current international order more generally. Earlier chapters on DMRs (Chapters 3 through 5) demonstrate that the United States was instrumental in shaping the development of these regimes. By examining the United States as an aid recipient after Hurricane Katrina, this analysis demonstrates (1) how disasters disrupt belief systems even in dominant states, (2) how ideologies of order profoundly shape disaster management, and (3) how political actors creatively reinforce existing beliefs through disaster management practices.

The remaining two cases of Egypt and China examine non-Western states where the relative influence of the dominant ideologies of international order may be expected to be less pervasive, a topic that is understudied in the field of disaster management (Quarantelli, 1998). As a rising global power that has both a long history of disaster management as well as opposition to "Western" forms of humanitarian intervention and global governance, China is an important case. This analysis of disaster management in China helps illuminate (1) how shifts in the distribution of power may induce international ideological change, (2) how rising powers incorporate and alter foundational international belief systems, and (3) how China's rise may (or may not) affect existing practices of international disaster management.

I include a chapter on Egypt as an example of how a developing, ideologically divided, post-colonial state responds to disaster and international disaster assistance. As with many other such states, Egypt has, at times, challenged key aspects of the American-led international order and has faced significant ideological contestation, internally and regionally. Egypt is included to demonstrate (1) the relative influence of dominant ideologies of international order in shaping disaster response in developing states, and (2) how political elites may use international disaster management to reduce domestic opposition and ideological contestation.

Together, these three cases assess the global spread of these ideologies of international order and provide a relatively robust overview of when and how international ideologies of order shape disaster management practices, as well as when disasters may trigger change in ideological frameworks.

Areas of Significance

This book begins with the claim that disasters are not exogenous developments that states, civil society actors, and the international community must manage; they are products of the current international order and are central to the dominant understandings of international relations. While the book builds on what might broadly be conceived as a constructivist position, its primary aim is not to defend or develop a particular theoretical perspective in IR. Rather, the aim of this book is to contribute to several important areas and approaches at the forefront of international relations and disaster studies scholarship. The first concerns the concept of international order, which is central to several different schools of thought. This project sees the current international order as heterogenous and constituted by material and ideational power. Neither are overly controversial positions as they follow several recent interventions on the question of international order (see Keene, 2002; Phillips, 2011; Buzan and Lawson, 2015; Acharya, 2018), that provide a corrective for the prevalence of writing on order that relies on a single overriding principle or set of actors around which the system operates and that stresses either material or ideational factors. Typical examples of "homogenous" conceptions of international order include Westphalian order, liberal order, or capitalist order, and are most clearly illustrated in cases where a particular set of policies is attributed to a single theoretical or ideological tradition, such as recent debates over whether the authors of the Iraq War were acting on liberal or realist beliefs (Deudney and Ikenberry, 2017). This project contends that such depictions of international order tend to overstate the coherence and unity of the international order and policy makers' ideological positions and thus provide a narrow understanding of how actors maintain and alter that order.

For example, the depiction of the current order as a liberal international order (see Sorenson, 2011; Ikenberry et al., 2018); stretches the concept of liberalism to encompass such a wide range of institutions and behaviours (American dominance, democratic institutions, human rights, free trade, collective security, secularism), that the concept loses its coherence and does not account for the tensions inherent between several of these principles. Such a depiction ignores the prevalence of a wide range of non-liberal values, goals, and behaviours – the undermining of democratically elected governments by democratic states, restrictive migration policies, ongoing economic protectionism, arms competitions, human rights violations, persistence of nationalism, and so on. An additional problem with depicting the current order as a Western liberal order is that it neglects how non-Western states contribute to the principles and practices often identified as

Western (Acharya, 2018) and that many aspects of the current order are not exclusively liberal in any meaningful sense of that term. Such a critique applies to all those who define international order in terms of a single overriding principle.

To be sure, not all scholars view international order through a single ideological prism: Haass (2017) sees a tension between liberal democratic order and Cold War competition; Hurrell (2007), between pluralism and solidarism; Keene, between toleration and civilization; while Buzan and Lawson (2015) see competing foundations that include industrialization, rational states, and several ideologies of progress. There is growing recognition of the heterogeneity of international order, including in terms of the variety of polity forms (Phillips and Sharman, 2015). Charles Kupchan similarly suggests understanding international order in terms of heterogeneity, as a "packages of ideas and rules that inform the nature of a given order and govern social relations within that order" (2014:20). Building on recent interventions that highlight both the crucial role of ideas and the heteronomy of the current international order (Kupchan, 2014; Buzan and Lawson, 2015), this book suggests that international order should be understood as a heterogenous order: a composite of several *ideologies of order* that interact to sustain existing practices, behaviours, and institutions toward a varied set of goals and values. This heterogeneity both facilitates and constrains actor agency; it allows political actors to frame developments through a relatively prescribed set of frameworks. However, this heterogeneity is fundamentally unstable as it contains both internal ideological contradictions as well as tensions between these potentially competing belief systems.

Internal contradictions within individual ideologies of order arise due to the simplifying logic characteristic of all ideologies and their reliance on binaries that break down when applied to the material world. Inter-ideological tension emerges because ideologies of order offer competing or conflicting claims about similar material developments. In short, actors may interpret the same observed development in multiple ways depending on the normative framework employed. The available repertoire of frames are not fixed, however; an international order may face "ideological schism", in which actors articulate an "anti-systemic ideology" that fundamentally rejects the concepts of "the good" that constitutes the existing order (Phillips, 2011:8).

A heterogenous account of international order provides a more coherent and robust account of change (and continuity) in international order, although obviously at the expense of parsimony and a commitment to prediction. This contrasts with the dominant accounts of change in international order. Although Iaddress this at greater length in subsequent chapters, it is worth highlighting here that this account challenges the dominant account in IR – in which distribution of power is the primary determinant of international order (Gilpin, 1983; see Ikenberry, 2014; Lake, 2014). For those working in this tradition, the ideas and institutions that characterize an international order reflect the ideas and interests of the most powerful state(s). Continuity occurs due to the persistence of the existing power distribution, while instability emerges through a changing distribution of power and the appearance of a new hegemonic power. Gilpin's account is useful in highlighting the role of power in sustaining and changing international order and stresses the importance of alignment between material factors and the ideas and institutions that comprise the international order. Change is the product of contradictions or tensions that arise when material conditions (distribution of power) diverge from ideas and institutions.

The heterogenous account of international order necessarily departs from those who see order strictly in terms of hegemonic power, on several fronts. The first is by highlighting the importance of incremental change. The hegemonic stability model of change implies war is the prominent mechanism of change, and that post-hegemonic war involves a complete alteration of the existing order – change in this account is catastrophic and the result of a single event or series of direct confrontations. In contrast, I argue that incremental change within institutions is the more common mechanism of change, as states respond incrementally to the instability caused by the contradictions and tensions in the existing order. The added advantage of this approach is that it better explains the possibility of continuity across changes in distribution of power. States are not either revisionist or status quo – they can occupy a wide range of positions on this spectrum (see Chan et al., 2018).

A second point of departure is that instability (and thus change) may arise from mechanisms other than changing distribution of power and war. Philipps emphasizes that international orders change due to "ideational and material transformations" that include "ideological dissent", "popular dissatisfaction with existing governance structures", and the rulers' declining "ability to manage violent conflict" (Phillips, 2011:7). I propose that the first two factors potentially arise in any instance where material conditions do not match existing ideas about how the world works, such as the persistence and growth of inequality in a liberal economic system that promises a reduction in inequality. In short, the distribution of power is not the only material development that can destabilize an order (although it remains important), and war is not a necessary process to produce change nor is a new hegemonic power.

This approach is more in keeping with liberal internationalist explanations for continuity and change that focus on the authority and legitimacy of the rules rather than on the power of the dominant actor. For Ikenberry and other liberal internationalists, the existing order derives its authority and legitimacy from "solving collective action problems" and "providing services and benefits to states within the order" (2014:84). Where the analysis in this book diverges from the liberal internationalist account is in rejecting the claim that the existing order is fundamentally liberal. Ikenberry admits that the international order is "a layer cake of ideas, norms, rule and institutions – most of which are complementary but some of which conflict" (2014:94); that there is a tension between a "Westphalian" order and a "liberal" order (2014:95), and that the liberal order has incorporated a significant dose of realism (2014:96), but he does not view these tensions and contradictions as key factors in producing change. This book explores the implications of taking seriously the heterogeneity of beliefs that underpin international order.

The second area of contribution is to critical studies of disaster, management, and international relations. The departure point of this project is to call into question "existing social and power relations" by constructing a broader account of contemporary politics of which "the initially contemplated part is just one component" (Cox, 1981:129). The main component under scrutiny here – disaster management – must be understood within the broader relations articulated through dominant ideologies of order. This book approaches management in a way that is consistent with the growing "critical management studies" literature, which is to say that its primary goal is to "denaturalize" identities, relationships, and practices central to "management" (Grey and Willmott, 2005) and to show how these identities and practices sustain ideologies about how the world works (Tadajewski et al., 2011:2).

This book draws on the considerable body of critical work on disasters, a field of study largely ignored in international relations. This, I suggest, is a significant shortcoming, as disaster management reproduces foundational beliefs and institutions by attempting to manage – rather than eliminate, prevent, or accept – disaster. As Calhoun observes, "management" as a response to disruptions depoliticizes the processes that lead to emergencies and the actors involved (Calhoun, 2004). The project of disaster management is fundamentally depoliticizing, rooted in the belief that disasters are endogenous to the social order, and that disaster management as a project should sustain current configurations of power while limiting challenges to this order when it erupts into periodic crises.

I argue in this book that the practice of disaster management is tied to important historical developments associated with the shift to a modern understanding of disasters, the emergence of the nation state and belief in a competitive international environment, the global expansion of capitalist economic systems, and the growth of humanitarianism to ameliorate the negative outcomes of these belief systems and associated practices. These developments form the background against which the project of disaster management operates. They are not static or completed processes, however, and ongoing shifts in belief systems and distribution of power have an impact on practices of disaster management. Some of the more notable recent shifts include the promotion of sustainable development, the spread of neoliberalism, and an attendant shift toward resilience in the management of disaster, as well as the renewed resistance to "universal" solutions as signified by China's growing role in development aid. I address the subject of resilience, and the growing body of critical work on this subject, in Chapter 4 and the impact of China in Chapter 8.

To explain changes in the field of global disaster governance, this book engages with the work on international norms and regimes most often associated with constructivist or liberal IR scholarship. For scholars concerned with the spread of ideas and norms, the creation of disaster management regimes and the selective or conditional acceptance of disaster assistance is important because it points to aspects of norm diffusion and institutionalization that remain overlooked. One contribution to the norm diffusion literature is to draw attention to normative splintering, in which interrelated issues are separated and subject to distinct normative contestation. The separation of international treaty making on disaster relief, climate change, sustainable development, disaster risk reduction, and migration illustrates this process. While these issues are clearly interrelated, they are subject to discrete processes of norm diffusion and treaty making as though they were separate issues. While there is pragmatic and strategic purposes to separate complex issues (Sebenius, 1983), doing so has important effects on norm diffusion and international agreements. Haas, in his work on issue linkage in international regimes, shows that linking issues requires a consensus on shared knowledge and social goals – something I suggest is missing (but persistently pursued) in the case of disaster management – and international order more generally. Haas also notes that stronger states are able to dictate the normative framework around which separated issues are to be addressed (Haas, 1980:367). While some scholars have begun to examine norm clusters, in which norms are tied together based on conceptual interlinkage (Winston, 2018), few have examined norm splintering, which is conceptual delinkage to prevent subsuming issues under a common normative framework.

A second contribution to constructivist IR literature is to draw attention to the importance of material developments to norm failure or success. The articulation of normative principles does not rely only on a logic of appropriateness (Finnemore and Sikkink, 1998; March and Olsen, 2005) but also on a logic of consequences, which are subject to empirical or material assessment. Normative claims inevitably include an if-then prediction: if the relevant actors behave in the appropriate way, then certain outcomes are expected. Disasters offer a window of opportunity for political actors to highlight and interpret the failure of the promised outcome in one of three ways: (1) the development is truly exogenous and unrelated to norm compliance, (2) the development is due to normative noncompliance, or (3) the development occurred even though (or potentially because) actors complied with the norm. The interpretation of "norm failures", which is in essence what disasters are, is crucial to the development and maintenance of international norms.

The third contribution is to draw greater attention to the production of identity and difference in the diffusion and institutionalization of norms. Norms are "collective expectations for the proper behavior of actors *with a given identity*" (Katzenstein, 1996:5). Norms attach meaning to these identities by articulating what is "good or bad" or "what ought to be" (Schmidt, 2008:306). Much of the critical constructivist work in IR focuses then on identity discourses, particularly the politics of creating and sustaining identity and difference. The norm diffusion literature generally approaches identity less critically, focusing on "positive" identity constructs and how changing norms may in fact encourage states to alter their behaviour and identity, producing a universal identity. Yet this project suggests that the universalized identities at the root of international disaster management cannot be universalized, rooted as they are in producing and sustaining difference. The carrying capacity of the earth could not sustain all human communities achieving the identity (and consumption necessary to sustain that identity) at the root of the current international order. Even if this were physically achievable, there are incompatible conceptions of identity and difference foundational to the current international order. As the cases in this book demonstrate, disasters present opportunities to challenge and/or reassert identity constructs. The identities and behavioural expectations created by different normative frameworks lead to normative instability and recurrent crises.

Selective acceptance of disaster management norms, and assistance, draws our attention to how international norms constitute the identity of actors and influence interests and behaviour (Wendt, 1999). A frequent criticism of the constructivist literature is its overemphasis on the positive aspect of producing identity and difference inducing positive international change (Kowert and Legro, 1996; Price, 2008). Yet a growing number of "critical" constructivist scholars are turning to the less-positive aspects of identity/difference produced through international norms (Diez, 2005; Rumelili, 2008). The assertion of international norms bifurcate international society through the production of binary identities: universal/particularistic (Walker, 1993), civilized/uncivilized (Gong, 1984; Todorov, 2010); masculine/feminine (Peterson, 1992), and so on. Disaster management is deeply implicated in the operation of several international binaries – donors/recipients, humanitarian/political, Western/non-Western, modern/traditional, worthy/unworthy recipients – binaries that are produced through the dialectical operation of ideology.

The project also highlights normative ambiguity. Variation in practical implementation and contestation over basic normative principles demonstrate that norms of humanitarian assistance are ambiguous and subject to varying interpretations (Betts, 2010; Betts and Orchard, 2014). The analysis in this book suggests that, while the practice of disaster management may be widespread, its normative foundations are sufficiently ambiguous that they are subject to interpretation and political contestation. In short, offering disaster assistance, either as relief or as risk reduction, is not understood unambiguously as a humanitarian act; it has multiple, contested meanings. To take one example, the frequently selective provision and acceptance of aid, coupled with non–Development Assistance Committee (DAC) donors providing assistance outside the mechanisms developed by the DAC donors (Binder et al., 2010), suggests a certain level of ambiguity and a fragmentation of the normative foundation for disaster aid. This is true in terms of development aid as well; nontraditional donors emphasize different normative foundations for aid, notably sovereignty, equality, and mutual respect, rather than externally imposed conditionality (Woods, 2008). One goal of this book is to expose how political actors interpret and contest the norms of disaster management through international negotiations on framework agreements as well as through the practice of accepting and rejecting aid.

The fourth set of concerns in IR that this project addresses is the small existing IR literature on disaster diplomacy, whose primary focus is to examine how disasters and the management of disasters reflect and affect bilateral relations between states. For the most part, this literature suggests that decisions on international disaster assistance reflect a rational assessment of the benefits and costs of offering, receiving, and/or rejecting assistance. For instance, Ilan Kelman's extensive work on disaster diplomacy suggests post-disaster aid has limited independent effect on bilateral relations and that aid acceptance and rejection reflect the level of historical hostility between states (Kelman, 2011). Others working in this area argue that disaster assistance is unlikely when states perceive the offering state(s) as a threat (Selth, 2008; Nelson, 2010; Allan and O'Donnell, 2012).

While I address this literature in greater detail in the second section of this book, I want to highlight here how the analysis offered in this book contributes to this literature. I do so, first and foremost, by analyzing the logic behind disaster assistance and assessment of threat and hostility. In doing so, I focus on how interests and perceptions of threat are themselves the product of beliefs about international order. For instance, is not immediately clear why states should view aid as threatening or be less likely to accept aid from a historical rival or enemy. From a strictly materialist perspective, accepting aid would appear to be a relatively straightforward case of positive relative gains (see Snidal, 1991; Baldwin, 1993). Humanitarian assistance, if it is untied and if there is little economic interdependence between the two states, is a direct transfer of material resources from one state to another. Consequently, offers of aid, and rejection of such offers, are inconsistent with a materialist concern for relative gains. To understand how a material transfer of goods can be perceived as threatening or hostile requires understanding how the meaning of aid is constructed and how it fits within broader understandings of the international order.

Moving away from a strictly materialist conception of power, humanitarian assistance might be understood as a form of soft power (Nye, 2004) and contributing to international prestige or reputation (Wang, 2006; Mercer, 2010). If an offer of humanitarian assistance is understood in terms of soft power, acceptance of aid becomes puzzling, because it increases another state's soft power. By examining the ideological foundations of disaster assistance, this book explores how certain frames evoke international competition and concern with material gains or soft power. The analysis here also identifies competing frameworks for understanding offers of aid, in which power considerations are largely immaterial to the decision on aid.

The issue of relief acceptance addresses a second concern raised in the disaster diplomacy literature that is central to realist IR scholarship, namely, how a changing distribution of power in the system alters state behaviour (see Waltz, 1979; Gilpin, 1983; Kennedy, 1987). Although this literature is predominantly concerned with how changing distribution of power produces violent conflict or gives rise to revisionist powers, the literature on great power status suggests that states will attempt to expand their influence as they rise in power (Nayar and Paul, 2003; Zheng, 2005). Crafting international regimes and contributing to humanitarian assistance are key methods of expanding influence. As noted earlier, the existing literature strongly demonstrates that, as states aspire to regional or great power status, they often seek to alter their identity from relief recipient to donor. The cases explored in this book demonstrate how power, manifest in knowledge production as well as economic and military capability, strongly shapes the process of norm diffusion and institutionalization. Similarly, in Section 2 where I examine cases of disaster assistance acceptance and rejection, this book examines how political elites understand offers of aid through the lens of power politics and how they justify the selective acceptance of disaster relief; this topic should be of interest to scholars concerned with changing power status.

The last body of work in international relations for whom this project is significant is the literature on humanitarian assistance. For advocates and practitioners, the management of disasters and the timely and effective provision of disaster relief can reduce casualties, contribute to reconstruction efforts, and allow long-term cooperation between states. From a humanitarian perspective, disaster management has weighty implications. The introduction of nonhumanitarian principles into DMRs can reduce their efficacy or even subsume their humanitarian objectives. Failed management practices and the refusal or delay of disaster relief can lead to large-scale loss of life, prolonged public health crises, increased poverty, and stalled development and community rebuilding (Cohen and Werker, 2008; Pelling and Dill, 2010). Insufficient or unequal distribution of aid can lead to increased violence and terrorist incidents within divided communities (Berrebi and Ostwald, 2011), while denying disaster aid may even lead to more antagonistic relations between states (Kelman, 2005, 2007, 2011; Streich and Mislan, 2014). Following the Indian Ocean tsunami in 2004, Le Billon and Waizenegger (2007) found that delayed and unequal distribution of disaster relief contributed to renewed conflict in Sri Lanka, while prompt, equitable distribution in Indonesia contributed to reduced conflict in Aceh (Gaillard et al., 2008; Hyndman, 2009).

While the prioritization of humanitarian considerations in dealing with disasters can reduce the negative effects of these events, it is not a panacea for managing disasters and reducing human vulnerability. Indeed, the existing literature suggests otherwise. Seminal works by authors such as Klein and Mutter on disaster capitalism have shown how disaster management operates to further empower and enrich national and international actors over local actors (Klein, 2007; Mutter, 2015). Practitioners and recipients have shown that humanitarian assistance is neither "neutral" nor wholly beneficial and can cause further harm to donor recipients, increase inequality, and in some situations prolong crises (Rieff, 1997; Anderson, 1998; Macrae, 1998). Humanitarian relief operations often prioritize the economic needs of relief institutions and their donors above those of relief recipients (Cooley and Ron, 2002); contribute to the spread of neoliberal economic policies (Middleton and O'Keefe, 1997); and are increasingly tied to the promotion of "Western" values, such as human rights and liberal democracy (Weiss and Campbell, 1991; Chandler, 2001; Barnett, 2005), and "Western" forms of knowledge and social organization (Hewitt, 1983; Kent, 1987; Donini, 2010).

Given some of the well-documented shortcomings of the existing disaster management system, it should not be overly surprising that the expansion of DMRs faces resistance despite some of its positive effects. In international negotiations over disaster management frameworks, there has been significant contestation over technology transfers and lack of issue linkage to climate change and sustainable development. Questions about consent and state sovereignty, maintenance of regulatory frameworks that limit the movement of emergency personnel and equipment, and outright refusal of disaster assistance continue to plague international efforts to implement existing agreements. Identifying the ideological foundations of disaster management that shape decisions on both norm diffusion and aid acceptance amply demonstrates (as if this needed further evidence) the highly politicized nature of disaster management and humanitarian assistance. Thus, these ongoing disagreements could be construed as a humanitarian crisis and a crisis of humanitarianism. This book adds to the body of work on the crisis facing humanitarianism (Chandler, 2001; Rieff, 2003; Barnett and Weiss, 2011) by showing how humanitarianism is ideologically nested with other belief systems, belief systems that produce and legitimize human vulnerability. This project also shows how humanitarianism operates as a powerful frame in reproducing binary identities that legitimize the civilizing project.

Framing Disaster

This book employs two common methods in critical IR scholarship: interpretive process tracing (IPT) (see Guzzini, 2012; Norman, 2015) and framing analysis. IPT examines the longer-term constitutive and causal dynamics that contribute to specific outcomes – in this case, the creation of international DMRs and disaster diplomacy. As Guzzini notes, IPT explains outcomes such as policy decisions as the result of historically contingent processes rather than a mixture of variables that combine to produce an outcome regardless of context.

IPT is "interpretive" in that it focuses on actors' understanding of developments rather than on the intrinsic qualities of those developments (Guzzini, 2012:4). IPT examines the intersubjective and context-specific meanings held by actors that in turn are necessary to understand and explain political effects as well as changes in social structure (Norman, 2015:8).

The project combines IPT with framing analysis, which explores communicative acts that highlight and prioritize "some aspect of a perceived reality to promote a particular problem definition, causal interpretation, moral evaluation, and/or treatment recommendation" (Entman, 1993:52). In this context, framing analysis is less attentive to long-term processes compared to the short-term discursive politics used to explain how particular understandings and policies emerge. Rather than simply employ framing analysis to explain policy choice, I use framing analysis to expose the ideological foundations of disaster management that shape how actors understand disasters and the production of social order. In this book, I analyze how frames draw on and reproduce belief systems about the source of societal, political, and economic order. The purpose then is not to predict the negotiating positions of specific actors or when actors will accept or reject aid, or to explain whether particular disaster narratives are more or less successful in legitimizing emergency aid, but to explain the relative strength of certain ideologies and to highlight ideological foundations to those working to improve or alter the international system of disaster management.

Reintroducing ideology to the study of discursive framing highlights the existence of multiple, overlapping ideologies shaping our understanding of social order. Because disasters unsettle a number of beliefs about how the world works; decision makers engage with multiple, often contesting frames in their response to disaster and offers of aid. Some offers of aid are understood in terms of modernization, others in humanitarian terms, still others in terms of power politics or in terms of economic development and property rights. Understanding how actors respond to disaster requires understanding the belief systems that underpin the existing sociopolitical order.

The collective focus on discursive and visual modes of meaning making in framing analysis places framing on familiar analytical terrain with the concept of ideology, yet this body of work tends to neglect the study of ideology. Indeed, some have suggested that the concept of frames displaces the role that ideology has played in previous work on the construction of meaning and its relationship to policy outcomes (Oliver and Johnston, 2000). The problem with this substitution is that the concept of ideology entails a more explicitly political interpretation and offers a deeper and more complex understanding of how meaning is constructed through various processes, including but not limited to framing (Oliver and Johnston, 2000). The use of the word *frame* instead of *ideology* frames the analysis of meaning making in a particular way that reinforces ideological positions on "social scientific" inquiry and on social-psychological explanations over political (Oliver and Johnston, 2000).

Employing the concept of ideology brings with it its own set of challenges, not least of which is analytical confusion. As with most concepts employed in the social sciences, ideology is contested. Ideology has been variously described as a belief system or mental framework "aspiring to offer a unified answer to problems of social action" (Swidler, 1986:279), that "make sense of" and "render intelligible the way society works" (Hall, 1996:26), or that "make existing social arrangements appear natural or inevitable" (Billig, 1995:15). To this I would add that ideology constructs categories and makes sense of specific social relations, as between men and women, workers and capitalists, politicians and citizens, and so on. In so doing, ideologies justify certain relations of domination (Thompson, 2013). Based on these definitions, the analytical overlap between frames and ideology is obvious – they both refer to structures of thought that provide a relatively coherent narrative of (self-selected) events and the possibility of meaningful choices within this narrative. Master frames in particular, understood as a "central interpretive framework" that contain "general conceptions of injustice and general visions of socio-political alternatives" (Carroll and Ratner, 1996:411), seem to capture much of what ideology is meant to capture.

For Oliver and Johnston (2000), the key distinction between frames and ideology is that ideology consists of (1) a social theory and (2) normative commitments. Frames, in contrast, are problem-oriented; they focus on a particular problem, establishing cause and effect, and necessary solutions. Although they are rooted in an ideological belief system, frames rarely entail an explicit and fully elaborated theory of society and moral relations within them. Because of this, actors from opposing ideological camps can draw on the same frame, such as a rights frame or a security frame, for different ends – ends defined as valuable according to a larger belief system. Consequently, framing performs "remedial ideological work" in that actors must interpret events that reveal the disjuncture between ideological belief systems and real-world experience, though often without explicit articulation of the ideological framework (Snow and Benford, 2005:209). Frames allude to but do not extensively articulate ideological foundations. Focusing exclusively on identifying different frames, or on framing effects, obscures the political effect of framing, namely, the ideologies they sustain. This book analyzes how political actors frame disasters and offers of disaster assistance to uncover the ideologies that make particular frames possible/resonant and to show how frames function to challenge ideologies or to restore belief in ideologies whose core claims are questioned. From this perspective, framing is the process through which actors discursively respond to developments that are ideologically unsettling. Actors frame these developments in ways that legitimate actions as logically coherent with existing beliefs or that delegitimize these ideologies. Thus, frames allow actors to interpret the world in ways that reconcile material developments and concrete action taken in response to these developments with belief systems about how the world works.

The disaster frame itself reveals the distinction between frames and ideology. The disaster frame refers to a disruption to normal societal routines (Tobin and Montz, 1997:6; Wisner et al., 2004:10) and in doing so reinforces beliefs about what is normal. The distinction between normal and disaster is not objectively or materially established but is ideologically established. While a credible disaster frame may need to appeal to materially observable developments, such as widespread physical destruction associated with a specific event (Kreps, 1989:2; Fischer, 1998:3; UNISDR, 2005:1), it is a shared, intersubjectively established interpretation of events. For this reason, social and psychological disruption is often more significant in the establishment of something as a disaster than is sheer physical destruction (Dynes, 1993:178; Aldrich, 2012). Furthermore, the disaster frame relies on a strict distinction between normal social routines and disrupted social routines, in turn privileging certain groups of people and their social routines. This perspective neglects that "for many people, "normal" daily life is largely indistinguishable from the social upheaval and disruption associated with disaster events" (Varley, 1994:4–5; see also: Middleton and O'Keefe, 1997:3; Wisner et al., 2004:10).

A second ideological move implied in disaster framing is to focus on an "event" that exposed vulnerable populations to existing hazards. Typical events include natural events such as earthquakes, tsunamis, droughts, or storms, or a technological event such as train derailment, nuclear meltdown, or blackout. Events-focused narratives exclude consideration of social structures and processes that produce human vulnerability. The negative effects of events are not solely a product of the events themselves, but of social processes and economic practices that render some people more vulnerable than others (Blaikie, 1994; Heijmans, 2001; Kahn, 2005; Bolin and Kurtz, 2007; Neumayer and Plümper, 2007). Events that disrupt social routines but that produce minimal damage to property, such as heat waves or epidemics, are rarely even understood as disasters warranting inclusion in disaster management (Klinenberg, 2002). Klinenberg suggests that political actors seldomly frame heat waves as disasters due to their low impact on property and infrastructure, reinforcing a liberal economic view of disaster. The disaster frame, by focusing on events, and on particular types of events with particular types of effects, reinforces belief systems about social order; in short, it is ideological.

Actors from various and even incommensurate ideological backgrounds may employ the disaster frame, although what qualifies as a disaster is profoundly shaped by ideology. Developments considered disastrous by nationalists or humanists may not be considered disastrous by cosmopolitans or deep ecologists. The specific ideological work performed by a disaster frame appears only through the examination of specific instances of disaster framing. Regardless of how and by whom it is used, disaster performs ideological work, even as its primary goal may be to mobilize a particular policy response.

In examining the discourse surrounding disaster events, institutionalization of disaster management regimes and the provision of emergency relief, I am focusing on specific frames, starting with the policy prescription, namely, what to do in the aftermath of disaster. I then work back to see how political elites defined the problem, identified relevant causal mechanisms, and marshaled moral arguments in coming to their policy recommendation. In each of these areas, I explore their discursive foundation by identifying a number of binary oppositions that underpin the classification of problems, causes, and moral implications. I related a number of these earlier in the chapter, such as humanitarian/political, modern/traditional, private/public, order/chaos, donor/recipient, and so on. The textual material examined throughout the book are wide-ranging; they include policy, speeches, working papers, governmental and nongovernmental reports, legislation and regulatory frameworks, and media coverage of disasters and disaster response. Individual chapters explain the specific texts under analysis in that case.

Outline of the Book

This book is divided into two sections: the first section explores DMR, while the second examines cases of disaster diplomacy. Prior to that, chapter two focuses on the concept of international order and several of the belief systems foundational to the current order. This chapter will be of interest to those studying international order and/or the specific belief systems related to nation-statism, modernism, realism, economic liberalism, and humanitarianism. In Chapter 2, I introduce the theoretical framework of how multiple and potentially contrasting ideologies interact to (1) sustain order and (2) produce ambiguities and tensions that periodically erupt in the form of disaster. I then turn to how these belief systems shape the current international order and influence how communities understand and respond to natural hazards and vulnerability.

The first section of this book examines the development of international regimes of disaster management by tracing the emergence and implementation of the three DMRs. This section will be of interest to those studying the diffusion and institutionalization of international norms, as well as those interested in international disaster management. In focusing on the three international regimes from the most institutionalized (disaster relief) to the least institutionalized (disaster migration), this section highlights how material and ideational developments lead to crises of existing regimes and the beliefs that sustain them, necessitating within-regime modifications as well as regime proliferation. Chapter 3 examines the historical development of the disaster relief regime, tracing the practice of disaster relief from Lisbon earthquake of 1755 to current efforts by governments, NGOs, and social media firms to enhance disaster relief.

Chapter 4 addresses the disaster risk reduction regime. Beginning with the declaration of a decade of disaster risk reduction in the 1990s to the contentious negotiations that resulted in the 2015 Sendai Framework for Disaster Risk Reduction, this chapter addresses the pressures that produced a (largely rhetorical) shift from disaster relief to risk reduction in terms of sustaining beliefs about the sources and purposes of international order. In Chapter 5, I explore the emerging disaster migration regime, tracing the exclusion of disaster-induced migration from other migration regimes, such as the refugee regime, to current efforts encapsulated in the Nansen initiative and Global Compacts, to produce a protection regime for disaster or climate migrants. Drawing on the themes developed to this point, Chapter 5 demonstrates the ideological crisis necessitating the development of a governance framework for disaster migration.

The second section examines three case studies of disaster diplomacy and post-disaster response. Chapter 6 discusses a detailed case study of the response of the United States to offers of aid following Hurricane Katrina; Chapter 7 addresses the Egyptian response to the Cairo earthquake; the final chapter examines China's rise and participation in international disaster management practices. This chapter focuses on existing accounts of change in international order, with an extensive examination of China's rise and its approach to foreign aid, broadly conceived. This chapter addresses the extent to which the rise of China, and other nontraditional donor states, may affect the existing international order and the potential implications for disaster management. The three case studies demonstrate the common concerns in the aftermath of disaster and provide a wide range of variability in terms of relative power internationally, domestic regime type, culture, and economic prosperity. Those concerned with how individual states manage disasters will find this section to be most relevant.

Notes

1 This phrase is borrowed from Barbara Tuchman's book *A Distant Mirror*, which examines the upheaval of the fourteenth century.

2 A number of scholars are engaging in inter- and multidisciplinary approaches to ideology. Maynard (2013) suggests that methodological differences divide this field of study but argues that there are areas of convergence between "conceptual, discursive, and quantitative approaches". Jost and Amodio (2011) call for the integration of cognitive, neural, and motivational process that would produce a "political neuroscience". *International Order and the Politics of Disaster* focuses primarily on the discursive but notes that ideology operates through various processes, including the cognitive that are not systematically analyzed in this book.

3 Juergensmeyer prefers the term *ideology of order* to capture "frameworks of thought that [that] conceive the world in coherent, manageable ways" (Juergensmeyer, 1993:31), largely to avoid the "loaded" connotations of the term *ideology*.

4 Jordheim and Wigen see "crisis" as replacing "progress" as the temporal ordering aspect of the current international order, an aspect of order not addressed in *International Order and the Politics of Disaster*.

References

Aldrich, D.P., 2012. *Building Resilience: Social Capital in Post-Disaster Recovery*, Chicago: University of Chicago Press.

Allan, C., & O'Donnell, T., 2012. A Call to Alms?: Natural Disasters, R2P, Duties of Cooperation and Uncharted Consequences. *Journal of Conflict and Security Law*, 17(3), pp. 337–71.

Anderson, M.B., 1998. You Save My Life Today, but for What Tomorrow? In Moore, J. (ed.). *Hard Choices: Moral Dilemmas in Humanitarian Intervention*, Lanham: Rowman & Littlefield.
Arts, B., 2004. The Global-Local Nexus: NGOs and the Articulation of Scale. *Journal of Economic and Social Geography*, 95(5), pp. 498–510.
Baldwin, D.A., 1993. *Neorealism and Neoliberalism: The Contemporary Debate*, New York: Columbia University Press.
Barnett, M., 2005. Humanitarianism transformed. *Perspectives on Politics*, 3(4), p. 723.
Barnett, M.N., & Weiss, T.G., 2011. *Humanitarianism Contested: Where Angels Fear to Tread*, London: Routledge.
Berrebi, C., & Ostwald, J., 2011. Earthquakes, hurricanes, and terrorism: Do natural disasters incite terror? *Public Choice*, 149(3–4), pp. 383–403.
Betts, A., 2010. The Refugee Regime Complex. *Refugee Survey Quarterly*, 29(1), pp. 12–37.
Betts, A., & Orchard, P. (eds.), 2014. *Implementation and World Politics: How International Norms Change Practice*, Oxford: Oxford University Press.
Billig, M., 1995. *Banal Nationalism*, London: Sage Publications Limited.
Binder, A., Meier, C., & Steets, J., 2010. Humanitarian Assistance: Truly Universal? A Mapping Study of Non-Western Donors, Berlin. https://www.k4health.org/sites/default/files/Binder_Meier_Steets__2010__Truly_Universal_-_Mapping_Study._GPPi_RP_12.pdf.
Blaikie, P.M., 1994. *At Risk: Natural Hazards, People's Vulnerability, and Disasters*, London: Routledge.
Blyth, M., 2002. *Great Transformations: Economic Ideas and Institutional Change in the Twentieth Century*, Cambridge: Cambridge University Press.
Bolin, B., & L. Kurtz, 2007. Race, class, ethnicity, and disaster vulnerability. In Rodriguez, E. Donner, W., & Trainor, J. (eds.). *Handbook of Disaster Research*, Albany, NY, Springer, pp. 113–29.
Buzan, B., & Lawson, G., 2015. *The Global Transformation: History, Modernity and the Making of International Relations*, Cambridge: Cambridge University Press.
Calhoun, C., 2004. A world of emergencies: Fear, intervention, and the limits of cosmopolitan order. *Canadian Review of Sociology*, 41(4), pp. 373–95.
Carroll, W.K., & Ratner, R.S., 1996. Master framing and cross-movement networking in contemporary social movements. *The Sociological Quarterly*, 37(4), pp. 601–25.
Chan, S., W. Hu, & K. He, 2018. Discerning states' revisionist and status-quo orientations: Comparing China and the US. *European Journal of International Relations*. doi:10.1177/1354066118804622.
Chandler, D., 2001. The road to military humanitarianism: How the human rights NGOs shaped a new humanitarian agenda. *Human Rights Quarterly*, 23(3), pp. 678–700.
Chipper, L., & Pelling, M., 2006. Disaster risk, climate change and international development: Scope for, and challenges to, integration. *Disasters*, 30(1), pp. 19–38.
Cohen, C., & Werker, E.D., 2008. The political economy of 'Natural' disasters. *Journal of Conflict Resolution*, 52(6), pp. 795–819.
Cooley, A., & Ron, J., 2002. The NGO scramble: Organizational insecurity and the political economy of transnational action. *International Security*, 27(1), pp. 5–39.
Cox, R.W., 1981. Social forces, states and world orders: Beyond international relations theory. *Millennium: Journal of International Studies*, 10(2), pp. 126–55.
Deudney, D., & Ikenberry, G.J., 2017. Realism, liberalism and the Iraq war. *Survival*, 59(4), pp. 7–26.
Diez, T., 2005. Constructing the self and changing others: Reconsidering 'normative power Europe'. *Millennium-Journal of International Studies*, 33(3), pp. 613–36.

Donini, A., 2010. The far side: The meta functions of humanitarianism in a globalised world. *Disasters*, 34(s2), pp. S220–37.
Dynes, R.R., 1993. Disaster reduction: The importance of adequate assumptions about Social Organization. *Sociological Spectrum*, 13(1), pp. 175–92.
Entman, R., 1993. Framing: Toward clarification of a fractured paradigm. *Journal of Communication*, 43(4), pp. 51–8.
Finnemore, M., & Sikkink, K., 1998. International norm dynamics and political change. *International Organization*, 52(4), pp. 887–917.
Fischer, H.W., 1998. *Response to Disaster: Fact Versus Fiction & Its Perpetuation: The Sociology of Disaster*, Lanham: University Press of America.
Gaillard, J.-C., Clavé, E., & Kelman, I., 2008. Wave of peace? Tsunami disaster diplomacy in Aceh, Indonesia. *Geoforum*, 39(1), pp. 511–26.
Geertz, C., 1973. *Ideology as Cultural Artifact: In the Interpretation of Cultures*, New York: Basic Books.
Gilpin, R., 1983. *War and Change in World Politics*, Cambridge: Cambridge University Press.
Gong, G.W., 1984. *The Standard of "Civilization" in International Society*, Oxford: Clarendon Press.
Grey, C., & H. Willmott, 2005. *Critical Management Studies: A Reader*, Oxford: Oxford University Press.
Guzzini, S., 2012. *The Return of Geopolitics in Europe? Social Mechanisms and Foreign Policy Identity Crises*, Cambridge: Cambridge University Press.
Guzzini, S., 2017. Militarizing politics, essentializing identities: Interpretivist process tracing and the power of geopolitics. *Cooperation and Conflict*, 52(3), pp. 423–45.
Haas, E.B., 1980. Why collaborate? Issue-linkage and international regimes. *World Politics*, 32(3), pp. 357–405.
Haass, R., 2017. *A World in Disarray: American Foreign Policy and the Crisis of the Old Order*, London: Penguin.
Hall, S., 1996. The problem of ideology: Marxism without guarantees. In D. Morley & K.-H. Chen (eds.). *Stuart Hall: Critical Dialogues in Cultural Studies*, London: Routledge.
Hannigan, J., 2012. *Disasters Without Borders*, Cambridge: Polity Press.
Hay, C., 1999. Crisis and the structural transformation of the state: Interrogating the process of change. *The British Journal of Politics and International Relations*, 1(3), pp. 317–44.
Heijmans, A., 2001. *Vulnerability: A Matter of Perception*, London: Benfield Greig Hazard Research Centre London.
Hewitt, K., 1983. *Interpretation of Calamities: From the Viewpoint of Human Ecology*, Boston: Allen and Unwin.
Hollis, S., 2014. Competing and complimentary discourses in global disaster risk management. *Risk, Hazards & Crisis in Public Policy*, 5(3), 342–63.
Hurrell, A., 2007. *On Global Order*, Oxford: Oxford University Press.
Hyndman, J., 2009. Siting conflict and peace in post-tsunami Sri Lanka and Aceh, Indonesia. *Norsk Geografisk Tidsskrift-Norwegian Journal of Geography*, 63(1), pp. 89–96.
Ikenberry, J., 2014a. Introduction: Power, order and change in world politics. In J.I. Ikenberry (ed.). *Power, Order and Change in World Politics*, Cambridge: Cambridge University Press.
Ikenberry, J., Parmar, I., & Stokes, D., 2018. Introduction: Ordering the world? Liberal internationalism in theory and practice. *International Affairs*, 94(1), pp. 1–5.
IPCC Working Group II, 2014. *IPCC Climate Change 2014 Summary for Policymakers*, Cambridge: Cambridge University Press.

Jordheim, H., & Wigen, E., 2018. Conceptual synchronisation: From progress to crisis. *Millennium: Journal of International Studies*. doi:10.1177/0305829818774781.
Jost, J.T., & Amodio, D.M., 2011. Political ideology as motivated social cognition: Behavioral and neuroscientific evidence. *Motivation and Emotion*, 36(1), pp. 55–64.
Juergensmeyer, M., 1993. *The New Cold War? Religious Nationalism Confronts the Secular State*, Oakland: University of California Press.
Kahn, M.E., 2005. The death toll from natural disasters: The role of income, geography, and institutions. *Review of Economics and Statistics*, 87(2), pp. 271–84.
Katzenstein, P.J., 1996. *The Culture of National Security: Norms and Identity in World Politics*, New York: Columbia University Press.
Keene, E., 2002. *Beyond the Anarchical Society: Grotius, Colonialism and Order in World Politics*, Cambridge: Cambridge University Press.
Kelman, I., 2005. Tsunami diplomacy: Will the 26 December, 2004 tsunami bring peace to the affected countries? *Sociological Research Online*, 10(1), pp. 1–6.
Kelman, I., 2007. Hurricane Katrina disaster diplomacy. *Disasters*, 31(3), pp. 288–309.
Kelman, I., 2011. *Disaster Diplomacy*, London: Routledge.
Kennedy, P., 1987. *The Rise and Fall of the Great Powers: Economic and Military Conflict from 1500 to 2000*, New York: Random House.
Kent, R.C., 1987. *Anatomy of Disaster Relief: The International Network in Action*, Sussex: Pinter Publishers.
Klein, N., 2007. *The Shock Doctrine: The Rise of Disaster Capitalism*, New York: Metropolitan Books/Henry Holt.
Klinenberg, E., 2002. *Heat Wave: A Social Autopsy of Disaster in Chicago*, Chicago: University of Chicago Press.
Kowert, P., & Legro, J., 1996. Norms, identity, and their limits: A theoretical reprise. In P. Katzenstein (ed.). *The Culture of National Security: Norms and Identity in World Politics*, New York: Columbia University Press.
Kreps, G.A., 1989. Future directions in disaster research. *International Journal of Mass Emergencies and Disasters*, 7(3), pp. 215–41.
Kupchan, C., 2014. Unpacking Hegemony: The social foundations of hierarchical order. In J. Ikenberry (ed.). *Power, Order and Change in World Politics*, Cambridge: Cambridge University Press.
Lake, D.A., 2014. Dominance and subordination in world politics: Authority, liberalism, and stability in the modern international order. In J. Ikenberry (ed.). *Power, Order and Change in World Politics*, Cambridge: Cambridge University Press.
Le Billon, P., & Waizenegger, A., 2007. Peace in the wake of disaster? Secessionist conflicts and the 2004 Indian Ocean tsunami. *Transactions of the Institute of British Geographers*, 32(3), pp. 411–27.
Macrae, J., 1998. The death of humanitarianism? An anatomy of the attack. *Disasters*, 22(4), pp. 309–17.
March, J.G., & Olsen, J.P., 2005. The logic of appropriateness. In M. Rein, M. Moran, & and R.E. Godin (eds.). *The Oxford Handbook of Public Policy*, Oxford: Oxford University Press.
Maynard, J.L., 2013. A map of the field of ideological analysis. *Journal of Political Ideologies*, 18(3), pp. 299–327.
Mercer, J., 2010b. *Reputation and International Politics*, Ithaca: Cornell University Press.
Middleton, N., & O'Keefe, P., 1997. *Disaster and Development: The Politics of Humanitarian Aid*, London: Pluto Press.

Mutter, J., 2015. *The Disaster Profiteers: How Natural Disasters Make the Rich Richer and the Poor Even Poorer*, Toronto: Macmillan.

Nayar, B.R., & T.V. Paul, 2003. *India in the World Order: Searching for Major-Power Status*, Cambridge: Cambridge University Press.

Nelson, T., 2010. Rejecting the gift horse: International politics of disaster aid refusal. *Conflict, Security & Development*, 10(3), pp. 379–402.

Neumayer, E., & Plümper, T., 2007. The gendered nature of natural disasters: The impact of catastrophic events on the gender gap in life expectancy, 1981–2002. *Annals of the Association of American Geographers*, 97(3), pp. 551–66.

NOAA, 2018. Billion dollar weather and climate disasters 2018, Washington, DC. https://www.ncdc.noaa.gov/billions/.

Norman, L., 2015. Interpretive process tracing and causal explanations. *Qualitative and Multi-Method Research*, 13(2), pp. 4–9.

Nye, J.S., 2004. Soft power and American foreign policy. *Political Science Quarterly*, 119(2), pp. 255–70.

Oliver, P., & Johnston, H., 2000. What a good idea! Ideologies and frames in social movement research. *Mobilization: An International Quarterly*, 5(1), pp. 37–54.

Oliver-Smith, A. 1998. Global changes and the definition of disaster, Chapter 15. In E.L. Quarantelli, (ed.). *What Is a Disaster? A Dozen Perspectives on the Question*, London: Routledge.

Pelling, M., 2003. *Paradigms of Risk: In Natural Disasters and Development in the Developing World*, London: Routledge.

Pelling, M., & Dill, K., 2010. Disaster politics: Tipping points for change in the adaptation of sociopolitical regimes. *Progress in Human Geography*, 34(1), pp. 21–37.

Peterson, V.S., 1992. Transgressing boundaries: Theories of knowledge, gender and International Relations. *Millennium: Journal of International Studies*, 21(2), pp. 183–206.

Phillips, A., 2011. *War, Religion and Empire*, Cambridge: Cambridge University Press.

Phillips, A., & Sharman, J., 2015. *International Order in Diversity: War, Trade and Rule in the Indian Ocean*, Cambridge: Cambridge University Press.

Price, R., 2008. Moral limit and possibility in world politics. *International Organization*, 62(2), pp. 191–220.

Quarantelli, E.L. (ed.), 1998. *What Is a Disaster? A Dozen Perspectives on the Question*, London: Routledge.

Rieff, D., 1997. Charity on the rampage: The business of foreign aid. *Foreign Affairs*, 76(1), pp. 132–38.

Rieff, D., 2003. *A Bed for the Night: Humanitarianism in Crisis*, New York: Simon and Schuster.

Rumelili, B., 2008. Interstate community-building and the identity/difference predicament. *Cambridge Studies in International Relations*, 107(1), pp. 253–80.

Schmidt, V.A., 2008a. Discursive Institutionalism: The explanatory power of ideas and discourse. *Annual Review of Political Science*, 11(1), pp. 303–26.

Sebenius, J.K., 1983. Negotiation arithmetic: adding and subtracting issues and parties. *International Organization*, 37(2), p. 281.

Selth, A., 2008. Even paranoids have enemies: Cyclone Nargis and Myanmar's fears of invasion. *Contemporary Southeast Asia: A Journal of International and Strategic Affairs*, 30(3), pp. 379–402.

Snow, D., & Benford, R., 2005. Clarifying the relationship between framing and ideology. In H. Johnston & J. Noakes (eds.). *Frames of Protest: Social Movements and the Framing Perspective*, New York: Rowman & Littlefield.

Snidal, D., 1991. Relative gains and the pattern of international cooperation. *The American Political Science Review*, 85(3), pp. 701–26.
Sorenson, G., 2011. *A Liberal World Order in Crisis: Choosing between Imposition and Restraint*, Ithaca: Cornell University Press.
Stallings, R., 1988. Conflict in natural disasters: A codification of consensus and conflict theories. *Social Science Quarterly*, 69(3), pp. 569–86.
Steger, M.B., 2008. *The Rise of the Global Imaginary: Political Ideologies from the French Revolution to the Global War on Terror*, Oxford: Oxford University Press.
Streich, P.A., & Mislan, D.B., 2014. What follows the storm? Research on the effect of disasters on conflict and cooperation. *Global Change, Peace & Security*, 26(1), pp. 55–70.
Swidler, A., 1986. Culture in action: Symbols and strategies. *American Sociological Review*, 51(2), pp. 273–86.
Tadajewski, M., et al., 2011. *Key Concepts in Critical Management Studies*, London: Sage.
Thompson, J.B., 2013. *Ideology and Modern Culture: Critical Social Theory in the Era of Mass Communication*, Hoboken: John Wiley & Sons.
Tobin, G.A., & Montz, B.E., 1997. *Natural Hazards: Integration and Explanation*, New York: Guilford Press.
Todorov, T., 2010. *The Fear of Barbarians: Beyond the Clash of Civilizations*, Chicago: University of Chicago Press.
Tuchman, B.W., 1978. *A Distant Mirror: The Calamitous 14th Century*, New York: Knopf.
UNISDR, 2005. *Hyogo Framework for Action 2005–2015*, Geneva. https://www.unisdr.org/we/coordinate/hfa.
UNISDR, 2015a. *Indicators to Monitor Global Targets of the Sendai Framework for Disaster Risk Reduction*, http://www.preventionweb.net/files/45466_indicatorspaperaugust-2015final.pdf.
UNISDR, 2015b. *Sendai Framework for Disaster Risk Reduction 2015–2030*, Sendai. https://www.unisdr.org/we/inform/publications/43291.
Van Dijk, T.A., 2006. Ideology and discourse analysis. *Journal of Political Ideologies*, 11(2), pp. 115–40.
Varley, A., 1994. *Disasters, Development and Environment*, New Haven: John Wiley & Sons.
Walker, R.B.J., 1993. *Inside/Outside: International Relations As Political Theory*, Cambridge: Cambridge University Press.
Waltz, K., 1979. *Theory of International Politics*, Boston: McGraw-Hill.
Wang, J., 2006. Managing national reputation and international relations in the global era: Public diplomacy revisited. *Public Relations Review*, 32(2), pp. 91–6.
Weiss, T., & Campbell, K., 1991a. Military humanitarianism. *Survival*, 33(5), pp. 451–65.
Wendt, A., 1999. *Social Theory of International Politics*, Cambridge: Cambridge University Press.
Winston, C., 2018. Norm structure, diffusion, and evolution: A conceptual approach. *European Journal of International Relations*, 24(3), pp. 638–61.
Wisner, B., et al., 2004. *At Risk Natural Hazards, People's Vulnerability, and Disasters*, London: Routledge.
Woods, N., 2008. Whose aid? Whose influence? China, emerging donors and the silent revolution in development assistance. *International Affairs*, 84(6), pp. 1205–21.
Zheng, B., 2005. China's "Peaceful Rise" to great-power status. *Foreign Affairs*, 84, pp. 18–25.

2

THE IDEOLOGICAL FOUNDATIONS OF INTERNATIONAL ORDER

In Chapter 1, I argued that social order is maintained through ideologies that provide a historical narrative to explain the origin and future of that order, legitimize institutions, proscribe a moral framework for action, and identify material developments that should result if relevant actors implement and follow these moral frameworks. These ideologies are relatively stable systems of belief that identify relevant social actors and proscribe proper conduct; however, they contain internal inconsistencies and are in tension with other ideologies needed to sustain a complex social order. Analyzing multiple overlapping belief systems is necessary, because, as Michael Freeden notes, "[P]olitical thinking engaged in by people across this planet is less neat and far more diverse and confused" than most political analysts recognize (Freeden, 2006:9). Internal contradictions and external tensions produce possibility for change in light of lived experience and societal disruptions. These disruptions can lead to change in the relative prominence of certain belief systems as well as the emergence of multiple positions within a single ideological system (e.g. classic liberalism, neoliberalism). Disasters expose internal inconsistencies that present opportunities for actors to frame events by referencing a particular ideology, to either reinforce, challenge, or amend the social order. Through this process of framing events and developing new policy responses, political actors maintain, and alter, ideological foundations of social order.

As it pertains to the management of natural disasters, five ideologies of order are most prevalent: modernism, nation-statism, realism, economic liberalism, and humanitarianism. These five ideologies are institutionalized in various disaster management regimes (DMRs) and through practices of international disaster diplomacy. State actors participating in international disaster assistance generally understand these offers through the lens of one or more of these ideological frameworks – what does an offer of aid suggest about their status in the international system, their relative level of technological sophistication, their state capacity, and their level of economic development, and the intentions of the donating or recipient actor. Individually, these ideological frameworks do not determine how actors will respond to disasters as they each contain contradictions and ambiguities; rarely are disaster events understood through the lens of a single ideological frame. There may be a number of distinct issue-specific frames and problem solutions consistent with each ideological framework. For instance, recipients of aid may understand disaster assistance as signalling the need for technological development and/or market reforms. Or they may view offers of aid as politically motivated rather than humanitarian, and/or as recognition of their power/status. The relative influence of individual belief systems vary substantially in different contexts. These ideological frameworks are interrelated and historically connected; however, they are not harmonious and rarely produce synchronous pressures or legitimize identical solutions. As governments attempt to manage disasters, they may prioritize one frame, problem definition, or set of considerations over others, potentially producing tensions with the other beliefs.

This chapter lays the foundation for the chapters that follow by demonstrating that disasters and disaster management expose inconsistencies within a number of belief systems that sustain the current international order. In this chapter, I examine five ideologies of order in detail: their ideological foundation, their historical connection to the phenomenon of disaster, as well as the variety of issue frames that operate within these belief systems. In the empirical chapters that follow, I will show how political elites frame disasters and the impact that these frames have on the production of international disaster management regimes as well as on decisions to accept or reject assistance.

Nation-Statism and Sovereignty

It is not particularly contentious to depict nation-state[1] sovereignty as a foundational ideology of international order, by which I mean the belief that the nation state should hold ultimate authority within a given territory. Yet this understanding of state sovereignty is insufficiently attentive to the larger ideological system that underpins this principle, and it obfuscates the tensions at the root of the belief systems that justify state authority. The legitimization of state authority is rooted in functional as well as representational expectations of the state, evidenced in the field of disaster management and the practices of international aid. For instance, the state has claimed the leading role in emergency planning/implementation and has insisted on consent to international assistance. Yet this is a relatively recent development; the central organizing structure of international politics throughout long periods of history was imperial rather than statist, and only with the processes of colonization and decolonization did we have a period of institutional homogeneity centred around the sovereign nation state (Phillips and Sharman, 2015:204;

Chowdhury, 2018). Necessarily incomplete, and largely deficient in many states, the prevalence of the statist form of political authority nonetheless creates "expectations of order" and shapes the understanding and response to disasters.

Political elites invoke and perform state sovereignty in numerous ways in the management of disasters: developing emergency services; participating in international negotiations; and offering, accepting, and rejecting foreign aid are all assertions of sovereignty. The principle of state consent, which is a core component of international disaster management, requires affected states to agree to outside intervention, whether in the form of an international treaty, the creation of intergovernmental organizations, or the dispersal of aid. All of these acts perform state sovereignty and reinforce the belief that the state has ultimate authority within a given territory and ultimate responsibility for the wellbeing of its population. The ability of governments to refuse, delay, co-opt, or mismanage international humanitarian assistance as part of their sovereign right has led many analysts to see state sovereignty as the primary obstacle to the provision of international humanitarian assistance (Stephens and Green, 1979; Barnett and Weiss, 2011).

While state sovereignty permits certain practices, it is also important to keep in mind that it is a malleable and transferable concept that applies to both domestic and international processes (Krasner, 1995; Biersteker and Weber, 1996; Cooley and Spruyt, 2009). The growing body of work on state sovereignty demonstrates that, rather than the term having a fixed meaning, it has varied and changed over time in response to political developments and crises of political authority. This necessitates understanding what sovereignty does. Rather than approaching state sovereignty as a fixed principle or set of rights and obligations, it is best understood as a belief system that authorizes ultimate authority within a specified territory (James, 1986; Biersteker and Weber, 1996; Camilleri, 2008). As historically attuned scholars observe, it is a relatively recent ideology of international order, reaching perhaps its zenith only in the post-colonial period.

The authorization of the state as sovereign rests on a rather specific set of beliefs that the state (1) represents a particular national community and (2) is best suited, and morally obliged, to perform certain basic functions, including emergency services, for its nationals. Early theorists of state sovereignty demonstrated that locating the source of state authority is highly problematic (Wilks, 1955; Kelsen, 1959) and thus becomes a practice of continuous authorization of state authority by "ultimately unauthorized" authorities (Suganami, 2007:522). The ideology of nation-state sovereignty hides its problematic foundations (by which I mean its violent, recent, and ultimately unauthorized authority) by appealing to some authorial moment or source, such as a social contract, a declaration of independence, revolutionary war, popular consent, the Peace of Westphalia, a constitution, and/or international law (Walker, 1993:8; Biersteker and Weber, 1996:3). Under scrutiny, the mythical foundations of state authority are primarily symbolic and provide a weak foundation. In the absence of an uncontested foundation that validates claims to state authority, the state must continuously authorize itself not only by appealing to foundational myths but also through the provision of various public goods (Suganami, 2007:522). As Suganami notes, belief in the authority of the state is maintained through everyday practices, such as functioning roads and traffic systems and collecting trash.

It is worth reflecting here on the role of nationalism in authorizing state authority, as it emerges as a source of tension within the ideology of nation-state sovereignty. As Billig notes, "[N]ationalism is an international ideology" that makes "national identities... appear as natural" (1995:9). Nationalism is the leading justification narrative for state sovereignty and roots state authority in the principle of national self-determination based on the will of the people. According to Anthony Smith, the "political program of nationalism" entailed "the conquest of the state by the nation and the fusing of a popular-national with a territorial–political identity" (Smith, 1995:112). For Smith and other scholars of nationalism, state authority is legitimized almost exclusively through appeals to nationalism, or (in the parlance of international relations [IR]) the national interest. The state grounds its authority in a nation that it simultaneously produces. The sovereignty of the nation state is thus crucial to the ability of nations to exercise their self-determination.

Grounding state authority as an expression of national self-determination required the ongoing imagination of the nation, to borrow Anderson's terminology (1983). Despite variation among states in the forms of nationalism central to each state's legitimation, as an account of state authority, nation-statism introduced a novel set of contradictions due to the misalignment between the territorial basis of the state and the people-centric concept of the nation, which seldom overlapped perfectly. If not all people falling within a state's territory occupy the same position in the authorization of the state's authority due to their exclusion from the nation, then failure of the state to provide basic public goods to these populations are less likely to undermine the state's self-authorization.

These tensions would become particularly problematic with the spread of statist ideology that contributed to the division of the globe into distinct territorial units expected to perform similar functions, to adopt particular institutional forms, and to generate and promote certain types of knowledge (Meyer et al., 1997:159). Some of the routine practices expected of the state include the regulation of border movement, treaty making, participation in diplomacy, and routine domestic provision of basic public goods (Jackson, 1990:29; Krasner, 1999), but these took place in areas where other forms of identity had been cultivated and a strong sense of nationalist identity was not fully developed.

Thus, the state came to rely on two modes of authorization – bureaucratic and nationalist. Central to the bureaucratic state's authorization is the provision of emergency services, including disaster management. This has led to the establishment of similar institutional forms around the globe – military officers, law enforcement officers, firefighters, ambulances, and disaster managers. At the current time, citizens in most states not only expect states to respond to emergencies, manage hazards, and provide disaster assistance; they also expect timely, "fair and sufficient" levels of assistance (Olson and Drury, 1997). So central is emergency response to belief in state authority that an inadequate or unfair response can undermine a government's legitimacy and contribute to electoral defeat, protests, and even violence against the state (Pelling and Dill, 2006; Nel and Righarts, 2008; Berrebi and Ostwald, 2011).

In the European context, the expectation of state-provided emergency services arose as the state displaced local, often voluntary, actors or placed them into a network of responsibility alongside or subservient to state actors. This occurred as state areas of interest expanded, from "concern with numbers of people", "their necessities of life", "their health", their "activity", and the "circulation of goods" (Foucault, 2007:324–5). The modern state became increasingly concerned with "productivity, health, sanitation, education, transportation, mineral resources, grain productivity and investment" in addition to "keep[ing] order, extract[ing] tax and rais[ing] armies" (Scott, 1998:51–2). The state came to view more and more issues as a threat to its people and to their health, wealth, and productivity and came to regulate those areas, in turn creating subjects who believed states could and should do so (Foucault, 1978). As part of this move toward increased concern with productivity and health, disaster prevention, management, and assistance, the state invested in social services to lessen the shocks that natural disasters represented to the economic system and to assist with the "accumulation of capital" (Jones, 1981:41).

The state's attempts to manage all of these new areas of interest involved reordering existing patterns of social responsibility through the authoritative allocation of responsibility for protection. Rather than establishing a monopoly, the state produced a hierarchical network of voluntary, private, and public actors, with the national state at the top (Dupont, 2004). Most of these emergency services initiatives emerged in major urban centres, although this gradually spread to non-urban centres, and the national state often took on the role of actor of last resort when local capacity was insufficient.

The networked approach to emergency response has not culminated in state monopoly on the provision of emergency services; rather, the authorization of the state entailed granting it the authority to authorize the provision of services and goods. In some cases this meant circumscribing the authority of voluntary organizations and private corporate actors that had previously provided these goods or sought to develop this capacity (Scott, 1998; Jones and Newburn, 1998), while in others it meant creating and empowering civil society actors. The incremental authorization of state institutions to provide emergency services, such as policing and firefighting, involved public agencies co-existing with private actors rather than complete displacement (Shearing and Stenning, 1981, 1983; Robert, 1989; Jones and Newburn, 1998; Johnston, 1992, 1996, 1999; Hensler, 2011:16).

The primary outcome of the nation-state ideology of order was the placement of centralized state actors at the top of a hierarchical allocation of authority for the provision of public goods. The actual allocation of responsibility for emergency response and disaster relief between the national state and local state actors and private actors varies greatly between states and is often a matter of contestation and ongoing negotiation. Despite this variation, the ideology of nation-statism serves two functions: (1) to authorize the state to regulate other actors empowered to provide emergency assistance and (2) to identify the state as ultimately responsible for emergency services within a given territory. As noted, the national state regulates various state agencies through constitutionally demarcated responsibility, while private actors may be regulated through compulsory registration requirements (Breen et al., 2009) and tax law (Irvin, 2005; Fremont-Smith, 2009). States regulate international actors that provide public goods through international treaty and the enforcement of state sovereignty and the principle of consent.

It is once again imperative here to consider the impact of nationalism within the ideological framework authorizing state authority. While the ideology of state sovereignty implies authority over a territory, nationalism implies authority of, and over, people. Not all people within a state's territory may have similar expectations regarding the provision of emergency services and other public goods; to use Foucault's insights – their health, education, and labour are less crucial to the state's self-authorization. This is amply evident in the types of developments considered emergencies or disasters – and whose normal routine lives operate as the standard establishing a condition of emergency or disaster (Varley, 1994; Middleton and O'Keefe, 1997; Wisner et al, 2004). Consequently, the management of disaster is implicated in the ongoing project of reproducing the nation – as well as authorizing the state.

Of course, the ideology of nation-statism does not incorporate a belief that the state can eliminate emergencies but that it is essential to manage them. Indeed, the periodic disruption of social life – what we might term *emergency* – is built in to the ontology of nation-state sovereignty. This ideology reinforces the belief that disruptions are external or exogenous to the nation state and that national citizens will periodically face disruptions. Emergency situations themselves do not represent a fundamental challenge to nation-statist ideology; rather they further reinforce the necessity of the state and in doing so highlight the performance of the state's emergency services. In this respect, it is the failure of the state's provision of emergency services and other public goods in the aftermath of disaster that defines disaster and that challenges the ideology of nation-state sovereignty.

The need for external assistance exposes a basic tension: ultimate responsibility for the protection of national citizens rests in the hands of the state. The seminal United Nations General Assembly (UNGA) Resolution 46/182 that forms the basis of current disaster relief reiterates the principle of state sovereignty by asserting that affected states have the "responsibility first and foremost to take care of the victims of natural disasters and other emergencies occurring on [their] territory". Similarly, the recently finalized Sendai Framework for Disaster Risk Reduction states as its first guideline: "Each state has the primary responsibility to prevent and reduce disaster risk" (United Nations International strategy for Disaster Reduction, 2015:13).

At the same time, international law identifies the international community as the actor of last resort for states overwhelmed by natural disaster. UNGA Resolution 46/182 states that international cooperation occurs in cases where the extent of the disaster is "beyond the response capacity" of the affected state. This is not dissimilar to the language employed in federal states to articulate when the federal government is responsible to assist sub-federal units. To resolve this tension, international humanitarian law stresses assistance occurs with "the consent of the affected country" and that the "sovereignty, territorial integrity… of the receiving state must be fully respected" (UNGA Resolution 46/182).

The aftermath of a disaster then is a seminal moment, because it represents an empirical test of the ability of the national state to fulfil a functional expectation foundational to its own authorization. The need for the international community to manage disasters and to provide assistance both bolsters and undermines a key aspect in the ideology of nation-state sovereignty. The principle and practice of consent is meant to preserve this ideology at the moment in which nation-state sovereignty, and its claim to provide protection for its citizens, is most demonstratively called into question. The act of negotiating and signing an international treaty, or of requesting, consenting, or denying international assistance, performs remedial ideological work by framing these acts as an exercise of sovereignty and self-determination.[2]

Beyond signing treaties, and offering and/or accepting aid, political elites may further draw on the sovereignty frame to legitimize the decision to reject aid. Disaster aid may be declined based on the claim that the state is indeed capable of responding on its own and does not need international assistance or that certain offers of aid would undermine the state's ultimate authority within its territory over local actors or international actors. As the cases in the second section of this book demonstrate, belief in nation-state sovereignty has a strong impact on the practice of disaster relief and on decision makers in societies affected by disaster.

Modernism

Understanding modernism as an ideology of order requires separating it from the concept of modernity itself. In its traditional usage, the word *modernity* applies quite broadly to a particular historical period marked by a set of ideological, political, and economic systems that differed in substantive ways from the period that preceded it. The shift to modernity brought about profound, wide-ranging changes, a process that is ongoing as disruptions within the modern period have led to various modifiers such as high-, late-, or post-modernity (see Lyotard, 1984; Giddens, 1991; Habermas, 2012). The various ideologies of order I address in this chapter are all clearly linked to and products of the historical shift to modernity and in that regard are all symptomatic of modernity. I suggest, however, that these belief systems are not equivalent to modernist ideology, which offers a particular set of beliefs about societal order and a particular way of framing disasters. Framing disaster management through a modernist belief system establishes different priorities from those emphasized through the lens of nation-statism, although the nation state is clearly a product of modernity.

Numerous scholars depict modernism as an ideology that elevates scientific thought and technological solutions and endorses human mastery of nature. James Scott depicts modernist ideology as "a strong version of the self-confidence about scientific and technical progress, the expansion of production, the growing satisfaction of human needs, the mastery of nature ... the rational design of social order commensurate with the scientific understanding of natural laws" (1998:4). Modernist ideology elevates "science, technical fields and [rationalized] economic activity" (Meyer et al., 1997:165), and asserts that human societies can plan and effectively respond to "geophysical processes and their human impacts" (Duffield, 2012:480). Modernism is often set in contrast with non-modernist modes of thought, requiring the "rejection of tradition and superstition" (Toulmin, 1990:13) and purporting to offer "liberation from the irrationalities of myth, religion and superstition" (Harvey, 1990:12). Given the inherent critique of religion in modernist ideology, it also contains a strong link to secularization. Kedourie summarizes modernist belief well, as "faith in progress ... that history will not let us down, that no catastrophe is final, no disaster irremediable. This is the prevailing note in modern culture" (1971:93).

These various writers identify the central feature of modernist ideology – human interventions based on scientific assessment count as progress and thus operate as the standard by which different polities and policies should be assessed, notwithstanding the existing literature that demonstrate that many modern innovations to control nature and prevent disasters have produced highly negative and unexpected consequences (Wisner et al., 1994; Platt, 1999; Bankoff, 2003; Collier, 2014). Increasingly modernized humanitarian efforts to improve the human condition rely on a belief in "the possibility of using social institutions to bring progress" and not necessarily on objectively identifiable or scientific evidence of progress (Barnett and Weiss, 2011:21).

Modernist ideology and disasters are closely linked because modern modes of thinking altered how societies understood geophysical processes and how to respond to them. Some of the key Enlightenment thinkers integral to the modernist project, such as Kant and Rousseau (among many others), were the first to frame disasters in terms of scientific explanations and solutions rather than theological explanations, as I will show in the following chapter. The modern framing of disasters was an unqualified success in terms of altering how people understood disasters and led to the establishment of new scientific fields, such as seismology, meteorology, volcanology, and geology; new technologies, ranging from relief efforts to building codes, early warning systems, and disaster insurance; and new institutions, from university departments to disaster assistance response teams.

International disaster management is thoroughly modernist. Hannigan notes that scientific and technical solutions have formed the bedrock of international governance of disasters in the postwar period – demonstrated in key institutions such as the World Meteorological Organization (WMO)and the International Decade for Natural Disaster Reduction (IDNDR) (2012:61). The establishment of the IDNDR called for "scientific and engineering endeavours", "disseminat[ing] existing and new information... for assessment, prediction, prevention and mitigation", programs of "technical assistance and technology transfer... and education and training" (UNGA Resolution 42/169, 1987). Quoting Terry Canon's critique, Hannigan observes that international disaster management "is overwhelmingly oriented to scientific and engineering approaches" (2012:66).

As modernism rests on the claim to be able to control or master nature, disasters expose ambiguities and tensions in modernist thought by demonstrating that the modernist belief in human progress and increasing human security and control over nature has not developed as claimed. Disasters demonstrate that humans have not solved the issue of natural hazards and that many of their interventions have exacerbated human vulnerability or introduced new forms of vulnerability as it has reduced others (Beck, 1992). Even as advanced industrial societies have reduced the number of deaths caused by natural disasters, the economic costs and number of people affected by disasters have increased dramatically. Many disasters are the partial product of modernist advances themselves, with technological disasters and climate change the most obvious examples.

Modernist ideology is sufficiently hegemonic in many societies that disasters rarely require an explicit appeal to legitimize modernist interpretations. Technical experts and scientists enjoy an almost exclusive monopoly on explaining disasters and do so through the language of scientific processes and objective measurements, which obscures the contribution of human interventions in creating risk and vulnerability. For example, in the case of earthquakes, this usually entails discussion of fault lines, epicentres, and the Richter scale, familiar terms in the scientific language of explaining earthquakes that affirm the possibility of knowing, and conquering, the vagaries of nature. Similarly, actors may frame disasters in terms of failed or insufficient modernization, and recommended solutions typically reaffirm faith in modernist ideology by seeking to perfect modernization – such as updating outdated structures or technology, establishing early warning systems, increasing funding for research and preventative systems, and promoting sustainable development.

As a near hegemonic ideology in the discussion of disaster, modernism plays an important role in the framing of disaster management and in shaping decisions on aid by operating as a standard of civilization. The ability to tame nature, which is central to modernist belief, operates as a standard of civilization used to legitimize and delegitimize certain forms of authority (Yao, 2018). The international DMRs reinforce this binary and actively encourage all states to meet a basic civilizational standard. Similarly, donors and recipients often understand disaster assistance through the lens of modernism, in which the level of technological advancement operates as a standard of civilization to differentiate actors in the system. By employing a modernist frame, actors construct other actors or types of assistance as sufficiently or insufficiently modern. Used in this way, the modernist frame sustains the belief in mastery of nature, scientific progress, and technological solutions at the core of modernist ideology, but it does so by framing actors or policies as modern/non-modern. States may decline assistance that is perceived to be insufficiently technologically advanced, including both equipment and personnel. Political elites and technical experts may also ban or reject offers of aid from groups that promote non-scientific explanations of disaster or that interpret disaster as divine retribution.

Alternatively, states may actively pursue assistance from actors perceived to be technologically sophisticated. States do this both to acquire technologically sophisticated assistance and equipment but also to have their disaster management units observe, train, and cooperate with disaster experts from advanced states. The modernist frame of disaster relief is also tied to international reputation, in that having one's disaster experts accepted by another state signals a recognition of one's modernity/expertise. This also raises potential diplomatic difficulties, when the refusal of disaster assistance is understood as a statement of one's level of modernity and technological sophistication.

While modernism is nearly hegemonic in disaster narratives, it does not determine disaster response. As I explore in subsequent chapters, international conventions on disaster management increasingly recognize the importance of incorporating "traditional" sources of knowledge on disaster. Hannigan (2012) traces the international reaction against overly modernist approaches to disaster management that have led to greater acceptance of the vulnerability approach and the importance of community-based disaster management (CBDM) (69). Furthermore, modernism does not exclusively shape how states respond to offers of aid. States often accept aid from actors less technologically advanced, although modernist ideology is still on display in the types of assistance that these states can supply, often basic aid (cash) and moral support. However, not all aid from technologically advanced states is accepted, suggesting that there are other ways decision makers think about and understand offers of assistance. In nearly all societies, including the most technologically advanced, political elites rely on and encourage traditional beliefs and mechanisms for coping with disaster. This is often evoked through condolences and offers of prayer for those affected by disaster or by emphasizing local resilience. Modernism, though often presented in terms of its opposite – tradition, religion, or superstition – has not and cannot completely displace these systems of meaning making.

Recent developments in disaster management have opened up space for forms of knowledge that are often constructed in opposition to modernist ways of knowing (local, traditional, indigenous). Yet as the empirical cases in subsequent chapters demonstrate, this move has introduced a number of tensions into the modernist ideology of order that is foundational in the management of disasters – particularly when these forms of knowledge stand in direct contrast with one another in attributing causation and identifying policy recommendations. The primary mechanism to resolve this tension has been to severely restrict how, when, and where non-modernist knowledge is incorporated. Beyond this, decision makers have continued to employ the standard of modernity to marginalize particular actors in the management of disaster.

Economic Liberalism

Liberalism is a large tent, with various distinct strands pertaining to political and economic order, although there is broad agreement that it coalesces around three foundational themes: rational individuals, market exchange, and representative democracy (Parsons, 1968:91; Poggi, 1972:137; Wrong, 1994:5; Buzan and Lawson, 2015:102). Given the highly varied nature of liberal thought, I separate economic liberalism from political liberalism. I justify this separation on two accounts. First, the empirical study of disaster management suggests that actors understand disasters and disaster recovery almost exclusively in economic terms: (re)building to facilitate economic exchange and better use of natural resources for economic development. Conversely, DMRs exhibit little concern over principles of democratic government and in many cases may be explicitly non-democratic, endorsing technocratic governance. Second, the political and economic components of liberal ideology are not intrinsically related – politically illiberal states may adopt a market economy, and politically liberal states may reject basic principles of economic liberalism and adopt socialist, communist, or mercantilist economic policies.

Consequently, I focus on economic liberalism as the most significant component of liberal thought in relation to managing disaster, and I largely exclude analysis of political liberalism. While many scholars of IR see political liberalism (understood as democracy promotion) as a core, if not *the* core, aspect of a liberal international order, the various international practices associated with disaster management suggest this is not the case. One aspect of political liberalism – primacy of the individual and shared humanity – has been very influential in disaster management, and I take this up at length in the discussion on humanitarianism later in this chapter.

Despite this separation for analytical purposes, the two realms of liberal thought interact, primarily around the question of state authority. At its core, liberal ideology emphasizes the common values of freedom, popular sovereignty, and limited government, thus placing limits on state power (Deudney, 2007:13). And yet a coercive state is essential to liberal thought, which relies on a coercive resolution to the problem of order (see Neocleous, 2008). However, in liberal thought, a coercive state on its own is insufficient for the emergence of order and may become a threat to it. Thus, the nation-statist resolution to the problem of order, which requires a capable state to provide security and other public goods, produces tension with liberal ideology that warns against too much state interference. On this question, liberal ideology has proven to tolerate a fair amount of fluidity in how much the state intervenes to provide security, enforce borders, maintain and promote the free market, and protect private property ownership, although it is heavily contested and divides different strands of liberal ideology (e.g. classical liberalism, social liberalism, New Deal liberalism, neoliberalism).

Economic liberalism emphasizes the importance of private property and market exchange for the establishment of a good and prosperous society (Wyatt-Walter, 1996). In this context, individuals should be free to own and develop property, exchange goods, and move freely in pursuit of market exchange. Given the centrality of property ownership, development, and exchange, economic liberalism must differentiate between legitimate and illegitimate forms of property accumulation and exchange, which in turn requires a role for the state to prevent market manipulations and failures. State intervention, however, introduces the potential to overregulate the market (thus limiting freedom) or undermine the disciplining aspect of the market (thus producing too much freedom). Legitimate changes in property ownership are to occur through market mechanisms enshrined in codes of law and enforced by the state – in turn legitimizing some capital destruction, property dispossession, and poverty as a necessary function of the market.

Disasters that damage or destroy property and that disrupt market exchange introduce and expose tensions in economic liberal thought because they reveal the disparity between ideological beliefs about economic exchange, property, and poverty, and lived experience (Hewitt, 1983:9; Stallings, 1988). Disasters demonstrate that, under some conditions, a market-based system of exchange is insufficient to sustain social order and that the market system requires a significant level of state capacity and coercion. Consequently, one of the most prominent, and disproportionately reported, frames in the aftermath of disaster is the "theft and looting" frame (Fischer, 1998; Heide, 2004; Tierney et al., 2006). The theft and looting frame depicts non-market and non-state-controlled redistribution of basic (and luxury) goods after disaster as illegitimate and as a fundamental breakdown in social order necessitating action. In this frame, increased security and the controlled delivery of disaster relief is a necessary solution to the problem of order because it increases the supply of basic life necessities and reduces the need for survivors to steal private property.

Many international offers of assistance include security personnel, and state militaries are often the key institutions employed in the delivery of disaster relief because they have the capacity to both deliver supplies and use coercion to control supplies. In many cases, international offers of security personnel are rejected, not on the grounds that a coercive state is not needed but because it is a threat to sovereignty or on the grounds that the state has the coercive capacity necessary to distribute relief materials and restore/enforce the market.

In contrast, economic liberalism does not deem all forms of wealth accumulation in the aftermath of disaster as illegitimate. While price gouging is represented as an illegitimate market distortion, disasters function as productive destructive events that free up and "recycle capital" and present new opportunities to develop damaged areas and invest in new projects (Rozario, 2001:81). Ozerdem and Jacoby show that disaster often affects low-income areas most profoundly but that post-disaster reconstruction often increases property prices in these areas, contributing to further marginalization (2006:5). In most cases, the opportunity to accumulate wealth after disaster accrues to those already enriched by the liberal economic system, resulting in the further consolidation of wealth and property in the hands of wealthy capitalists (Klein, 2007). This is often regarded as legitimate wealth accumulation, unlike the theft and looting narrative that is disproportionately employed during disasters and that profoundly shapes national and international disaster management. Disaster may also strengthen economic liberalism by further incorporating marginalized groups into the market economy in the form of wage labour (Reid, 2012:75). International disaster management projects now include consideration of poverty alleviation, understood within an economically liberal framework that emphasizes market-based solutions (Buchanan-Smith and Maxwell, 1994; Longhurst, 1994; Vatsa and Krimgold, 2000).

Even as many aspects of disaster management are meant to sustain or expand economic liberalism, it also creates additional tensions and ambiguities by potentially weakening domestic/local market forces, creating dependence, or encouraging risky economic behaviour. A number of aid scholars have noted that international humanitarian assistance contributes to weakened local markets and aid dependency, leading to calls for alternatives to aid (Bourguignon and Sundberg, 2007; Easterly, 2007; Moyo, 2009; Tadesse and Shively, 2009). Poverty alleviation is in tension with key facets of liberal ideology that legitimize economic inequality and even poverty (Wade, 2004).

At its core, liberalism establishes a normative justification for impoverishment on the basis of individual responsibility. In a liberal economic system, property is viewed primarily in terms of development – how human populations can alter the natural landscape to increase capital (Platt and Dawson, 1999:151). To facilitate this, property owners may alter, dispose, and/or invest their property and capital as they see fit, but they also bear the risk of these actions. Impoverishment is a legitimate market mechanism to discipline those who have not used their capital or property wisely. Impoverishment due to events for which the property owner bears no responsibility, however, muddies the normative basis of economic liberalism and the story it tells about the current and future distribution of property. Even staunch proponents of neoliberalism, such as Hayek, argued that collective action should be taken to ameliorate the effects of events, such as natural disasters, that individuals cannot protect against or prepare for (Hayek, 1944:148). Consequently, a key liberal response to the problem of disaster assistance is moral assessment of responsibility for loss of property (Landis, 1997). Disaster assistance exists to assist the "deserving poor", meaning those who (1) owned property and (2) suffered loss of property from natural disaster rather than personal shortcomings – such as risky economic behaviour or indolence. In previous cases of natural disaster, differentiating the poor has involved requiring the victims to show proof of property ownership and denying assistance to those who were homeless prior to the disaster (Phillips, 1998). Thus, disaster assistance is intricately tied up with restoring property ownership. Indeed, the concept of disaster itself is understood first and foremost in terms of loss to property (Tierney, 2007:508). As noted earlier, events that disrupt social routines but that produce minimal damage to property, such as heat waves or epidemics,[3] are rarely even understood as disasters warranting disaster relief programs (Klinenberg, 2002).

Disaster assistance also raises the related problem of moral hazard, which refers to the negative effect that collective assistance for individual loss has on property owners. According to Platt, collective assistance may reduce the prudence that property owners ordinarily practice in dealing with "natural hazards" in decisions to develop, invest, or locate their property (Platt, 1999:9). As noted earlier, liberal ideology is built on the foundation that property owners should lose or incur the cost of damage to their property if they take unnecessary risks, and disaster assistance may unjustly subsidize risk and may even encourage irresponsible, risky behaviour (Collier, 2014).[4] Collective assistance in the aftermath of disaster may have unintended consequences that increase the costs of disaster by encouraging people to build in disaster-prone areas (Collier, 2014). The problem of moral hazard is also a central concern in international disaster relief programs (Raschky and Shwindt, 2009). Economic liberalism contains a number of potential ambiguities and tensions that are exposed in disaster situations. Consequently, disasters have been instrumental in bringing about changes within liberal ideology: the perceived failure of different liberal schemes to preserve key aspects of liberal ideology in the aftermath of disaster produced new liberal technologies of governance, including relief, prevention, and insurance (Collier, 2014). More recently, the resilience frame has emerged as the primary (neo)liberal solution to the problem of order in the aftermath of disaster.

To further complicate matters, economic liberalism also informs understandings of international order. Extending its basic principles to the international realm, economic liberalism postulates that free trade and movement between states increases wealth, reduces poverty, and contributes to stable and peaceful economic relations between states (Kiely, 2005). To reap the benefits of international trade, states must open up to the authority of the market through the reduction of trade barriers, currency manipulation, and discriminatory practices (Gilpin and Gilpin, 1987:188; Sorenson, 2011:117).[5] In the same way that the liberal commitment to free markets domestically stands in tension with the need for a coercive power to provide security and the functioning of the market, commitment to international trade is in tension with the role of the state in preserving domestic industries and providing security in an anarchic international structure.

International economic liberalism emerged in contrast with economic nationalism, which endorsed protectionist measures ranging from complete economic self-sufficiency to the manipulation and/or restriction of international trade and exchange (Helleiner, 2002). In contrast, early liberal theorists promoted the concept of comparative advantage, in which states could specialize, producing gains for both partners in a trade relationship (Ruffin, 2002). Rather than completely abandon protection of domestic industries and workers, powerful states implemented a compromise, what Ruggie (1982) refers to as embedded liberalism. This compromise sought to balance the importance of international free trade with domestic economic interests in maintaining certain core industries, high levels of employment, and protection for workers. The embedded liberal compromise encouraged states to adopt international exchange controls and free trade policies while encouraging them to insulate their citizens from the global market through social security insurance and job creation programs rather than protectionist trade policies (Hays, 2009:3). This represented a departure from the protection of national industries typical of a mercantilist economic order in favour of the protection of individuals displaced as a consequence of the loss of nationally protected industries, or other impacts of globalization. The drastic increase in social security after World War II, including disaster management, occurred alongside a decrease in protectionist trade policies for numerous sectors of the economy.

Disasters expose various fault lines in the ideological foundation of embedded liberalism. First, disasters destroy infrastructure crucial to international trade, reduce export commodity production, and alter a state's import/export balance (Gassebner et al., 2010; Cavallo and Noy, 2011). In doing so, they force decision makers to make difficult trade-offs in restoring the domestic economy or abiding by international trade agreements. In a number of cases, disasters are exacerbated by decisions to maintain and expand international trade relations rather than alleviate domestic economic hardship, such as when states continue to export goods that could be used to meet domestic demand after disaster.

Disaster assistance can also alter trade relations if previously exported goods are redirected for domestic purposes and may involve the importation of types of aid that undermine the ability of domestic industries to compete. Consequently, states often approach disasters as an opportunity to prioritize the domestic economy above international trade commitments, to maintain and strengthen trade ties, or as the impetus to diversify and reduce reliance on a specific trade partner.

Disasters may also represent an opportunity to challenge the dominant liberal international economic order (LIEO). Over the past 50 years, there have been many changes in the key institutions of LIEO as actors attempt to manage the competing objectives that characterize this compromise (Lang, 2006:93). In addition, numerous states reject the basic tenets of LIEO and have promoted a new international economic order with improved/preferential trade conditions for less developed states (Cox, 1979; Ruggie, 1982). One of the key points of contestation emerged from the exclusion of the question of development within embedded liberalism, and of certain states from shaping the international economic order (Dillon, 2002; Lang, 2006:100). These tensions were evident in the international negotiations over the Sendai Framework for Disaster Risk Reduction, with developing states emphasizing the importance of development and technology transfers to reduce disaster risk and developed states resisting such commitments.

States may frame disasters in terms of contesting the LIEO, from those that critique the failure of LIEO to redistribute wealth globally and from economic nationalists who endorse protectionist policies. In this context, disasters may represent a threat to the stability of economic liberalism. Disaster assistance may be viewed as a means of propping up the LIEO, as is evident from the overwhelming focus on protecting, rebuilding, or expanding infrastructure essential to the functioning of the "economy, the state and international system" (Hewitt, 1983:28).

Because governments must consider both the basis of domestic economic order as well as their international economic relationships, we should expect states with different economic systems, and with different positions vis-à-vis the international economic order, to frame international disaster management treaties and offers of assistance differently. We certainly see this in regard to negotiating positions on DMRs, in which states' position in the international economic order strongly influenced their preferred disaster management policies. However, this relationship is not determinative in that the Group of Eight (G8) or the Group of 77 will exhibit similar patterns of participation in DMR or aid refusal or acceptance based on their support for a new international economic order. What it means is that all political elites responding to offers of disaster assistance have to consider the potential impact that disaster and disaster assistance will have on their domestic economic arrangements and, to a lesser extent, their international trade relationships. In more specific terms, political elites not only assess the impact that accepting aid will have on trade relations; they also assess what types of aid they will accept based on domestic economic factors, and how to distribute international offers of aid.

Realism

In IR scholarship, realism is often (problematically) depicted in opposition to liberalism, at least among the more passionate defenders of each of these perspectives. What the ongoing contestation over the relative influence of realist or liberal beliefs in shaping the Iraq War demonstrate, at the very least, is that there are instances in which both realist and liberal ideology influence decision makers and that both are used to justify similar policy positions, although they may vary in the logic of argumentation. This book argues that this is the case because, in addition to sharing a number of central tenets, realism and economic liberalism operate in tandem (as well as tension) in the maintenance of the current international order, alongside several other ideologies of order. Rather than ask whether a specific policy outcome is specifically liberal or realist, the analysis in this book suggests that we need to understand how both liberal and realist beliefs shape the current international order and to what extent and in what contexts these belief systems influence foreign policy decisions.

As an ideology of international order, realism strongly overlaps with, and indeed rests upon, the ideology of nation-state sovereignty. As Hyde-Price (2009) observes, realist ethics are grounded in communitarian political theorizing and take the nation as the foundational political community. While nation-statist ideology addresses the authorization of nation-state authority, realism provides a social theory and normative guidelines for the interaction among these units. As an outward-oriented ideology, realism encourages states actors into "geopolitical thinking" among states (see Guzzini, 2017). It asserts an anarchic international realm in which states ought to behave as though power is the primary determinant of outcomes; nation-statism asserts a hierarchical realm in which states and groups within states ought to behave as though state authority, derived from nationalism and the provision of public goods, shapes outcomes.

Realism operates as an ideology because it is "an attempt to explain the supposedly immutable patterns of international politics…and to formulate the best possible way to live within these ineluctable constraints" (Bell, 2002:222). The core beliefs associated with realism are that the state is the central actor in human affairs, its primary goal is survival, and there is no higher authority in the international realm than the state (Behr and Heath, 2009). In such a world, the primary determinant of systemic outcomes is power, and therefore the appropriate behaviour for states is that of self-help through the accumulation of power (Herz, 1950; Waltz, 1979:91–2). At the same time, realists tend to ameliorate the stronger ideological aspects by restricting unrestrained power accumulation and competition by calling for "prudence, skepticism, reciprocity" (Hyde-Price, 2009), or "prudence and practicality" (Rafshoon, 2001).[6] Thus, realists do not reject, on principle, all cooperation between states, often endorsing balancing and bandwagoning strategies. This moderating aspect of realism produces several tensions and inconsistencies about how states ought to behave: whether they should balance or maximize power, balance against power or threat, balance or bandwagon, or even whether to pursue hard or soft power (Morganthau, 1948:219; Waltz, 1979:126; Snidal, 1985; Walt, 1985; Grieco, 1988; Nye Jr, 1990; Schweller, 1997). In short, realism, as do other ideologies, contains a number of distinct strands that offer different guidelines on how states ought to act in a world characterized by sovereign states operating in an anarchic system, although it does contain strong anti-ideological components. At its core, realists assert that interests should be understood in terms of power – rather than moralistic pursuit – and generally promote the belief that the pursuit of power (in its various forms) is the best strategy to ensure the security of the state and its population, which is it highest priority (Bell, 2002:227).

This has two further implications that are central to this book. The first is that the nature of the system makes international cooperation difficult to sustain, although not inconsistent with an environment encouraging self-help and power politics (Glaser, 1996; Abbott and Snidal, 1998). From this perspective, cooperation is understood as self-help and power politics, as a form of balancing or enhancing soft power. The second related implication is that states are defined largely in terms of how much power they possess (super powers, major powers, middle powers, minor powers, regional powers, etc.) (Cooper et al., 1993; Huntington, 1999; Jordaan, 2003), their orientation toward the current distribution of power (status quo or revisionist powers) (Gilpin, 1983; Kennedy, 1987; Mead, 2014), and their orientation toward the other (ally, enemy). Realism endorses the belief that the rules of international institutions that enable cooperation are themselves a function of power and self-help behaviour designed to reflect the interests of the powerful states in the system (Mearsheimer, 1995; Glaser, 1996); consequently, rising powers may seek greater influence (and rewards) from the existing system or may challenge the status quo when it no longer reflect the distribution of power in the system (Gilpin, 1983).

To the extent that realism offers an account of disaster and how to respond (which it largely regards as the realm of low politics or as an uncontrollable externality) realism encourages actors to see disasters and disaster response through the lens of its impact on state power, or perceptions of state power, internationally. Disaster itself is not necessarily a reflection of weakness; disasters can happen to any state. Disaster management and response, however, can affect a state's power by reducing perceptions of competent leadership (see Walt, 2009), which figures in classic realist conceptions of state power (Morganthau, 1948). Great powers are expected to have competent leadership and sufficient capability to handle disasters without external assistance, while weaker powers lack this capacity.

Disaster aid, on the other hand, is a bit more problematic for realist ideology. On the one hand, it may be understood as an important source of soft power and influence; on the other, it may be seen, along with most foreign aid, as creeping liberal influence that sees aid as an obligation of powerful states to weaker states – an understanding realist ideology strongly rejects (see Morganthau, 1962). Morganthau, one of the leading figures of realism, rejected the representation of foreign aid "extended to nations which are victims of natural disasters" as strictly "nonpolitical"; he notes that, if used effectively, it should have useful political effects for the donating state (301–2). Morganthau was not opposed in principle to humanitarian aid or to other types of foreign aid; however, he sought to situate foreign aid within a realist ideology that required the analyst to ask whether the provision of aid concretely contributed to a clearly defined foreign policy goal (p. 308). Morganthau stressed the need to recognize the political efficacy of foreign aid as distinct from its economic or humanitarian efficacy, and he argued that, for aid to be politically effective, the donating state must ensure that the receiving state understands and accepts a "connection between the aid and the politics of the giver" (Morganthau, 1962:309). For Morganthau, foreign aid is "no different from diplomatic or military policy or propaganda. They are all weapons in the armory of the nation" (Morganthau, 1962:309).

On the other hand, international disaster management and international cooperation in the aftermath of disaster exposes tensions within realist ideology, since it seems to violate some of its basic tenets – namely, the centrality of the state in providing for its citizens' security as well as the importance of accumulating military and economic power for this purpose. Reducing vulnerability to disaster is often at odds with the forms of insecurity that realists prioritize for the state and requires investments in areas not overly conducive to increasing power internationally. One potential resolution to this tension has been to promote military reform toward a multi-purpose army that can fight wars and provide disaster relief. At present, the military often undertakes disaster relief duties due to its ability to transport goods and people quickly and over difficult terrain – duties essential to both combat and disaster relief operations. In addition to complicating the mission, purpose, training, and procurement policies of military forces, relying on military forces to provide disaster relief militarizes aid (see Walker, 1992; Chandler, 2001). Even if a military could act as the lead agency in providing disaster assistance without causing any problems, it would not reduce the tension produced by realist beliefs about the primary source of insecurity in the world and the primary purpose of military force.

Further tensions in realist belief arise as disasters have the potential to weaken the affected state; providing assistance, particularly to a rival or enemy state, goes against the injunction to maximize power. Furthermore, and as Morganthau notes, expectations that powerful states should provide disaster aid may result in providing aid that does not achieve a foreign policy objective and may even shore up an opposing regime.

Accepting aid also poses a challenge to realist ideology, particularly for major powers and for states in antagonistic relationships. First and foremost, it demonstrates that self-help is not the only appropriate behaviour for states and that the accumulation of power does not necessarily increase the security of their citizens. It may also demonstrate a source of authority beyond the state in the form of universal values of humanitarianism (which I address in the next section). Consequently, the decision to accept or reject aid may be framed in a way that maintains belief in realism, usually by emphasizing the effect of one's power in the system, or by distinguishing between the types of actors providing aid. Offers of cooperation may be presented as confirmation of one's power (hard or soft) in the international system or as the desperate attempt of other states to increase their own power. Again, both frames may be consistent with realist ideology, although they endorse diametrically opposing courses of action, further accentuating the need to study frames and ideology together.

As with the other ideologies explored in this chapter, disasters and disaster management is potentially highly disruptive to realist ideology. Disasters and the response to them may be viewed as key moments in maintaining a state's position in the international system or in reinforcing existing relations with other states in the system, either in maintaining friendly or hostile relations, or in demonstrating one's position in an international hierarchy. The need for disaster assistance demonstrates dependence on other states, which stands in contrast to the realist belief in accumulating sufficient power to act independently of other states. This is not primarily a concern with loss of sovereignty but with diminishing the state's place in an international hierarchy of power. Conversely, providing disaster assistance is seen as a demonstration of having power and of expanding influence and soft power. Many rising states use disaster assistance to demonstrate their place among the more powerful states in the system, even as they reject or revise the rules established by the traditional donor states in distributing disaster aid (Manning, 2006; Kragelund, 2008; Woods, 2008).

Humanitarianism

The last ideology of international order I examine is that of humanitarianism, which is intricately intertwined with the other ideologies explored in this chapter. In its purest form, humanitarianism asserts a belief that all humans have moral standing and that there is a moral obligation to alleviate human suffering. This emerged from a modernist orientation that sees suffering as the product of human activity (Greenfeld, 2005:329), creating an obligation for human intervention. The most common articulation of the humanitarian obligation is for the neutral and impartial provision of humanitarian assistance, a stance mostly closely associated with the International Committee of the Red Cross (ICRC), which has been referred to as "the high priests of humanitarianism" (Barnett and Weiss, 2011:9). According to the ICRC, humanitarianism does not discriminate between victims, take sides in a conflict, or depend on state governments, and it is guided by altruistic motives rather than personal or national gain. A key tenet of humanitarian ideology is that humanitarianism is free of politics, political interests, or the application of power (Parekh, 1997:54; Warner, 1999; Edkins, 2003:255). The apolitical status of humanitarian action is further reinforced through a variety of international resolutions and professional charters (Office for the Coordination of Humanitarian Affairs (UNOCHA), 1991; Minear and Weiss, 1995:18–20;

Sphere, 2004; Humanitarian Accountability Project (HAP), 2007). Since its inception, the ideology of humanitarianism and its practitioners has contributed to the alleviation of suffering of millions of people around the globe and transformed thinking about universalist moral obligations.

At the same time, humanitarianism is deeply political and operates "to sustain global relations of domination" (Chimni, 2000:245). Rather than endorse a radical reordering of international society and the elimination of the full spectrum of human vulnerabilities that arise from the modern international order, humanitarianism promotes a set of beliefs that operates within and alongside existing beliefs about a good and just international order (Barnett, 2010). Humanitarianism is highly modernist, promoting scientific solutions with Western societies as the model of development, in the process justifying various forms of intervention by Western states to assist the modernization process globally (Sending and Neuman, 2006:659). From the outset, one of the key agendas of humanitarianism has been containment of disaster in an effort to maintain the political and economic status quo (Walker and Maxwell, 2009:20–1). In its early stages, it did so by ameliorating a more fundamental challenge to the international order from a growing pacifist tradition (Slim, 2001) that challenged key aspects of realist beliefs, namely, the necessity of increasing power and using coercion to produce security. Humanitarianism also operated to alleviate the threat posed by the newly impoverished to law and order necessary for the functioning of a global capitalist system (Walker and Maxwell, 2009:18) and to alleviate the worst effects of market economies (Barnett and Weiss, 2011:19). Through the promotion of global civil society actors, embodied in this case in the Red Cross movement, humanitarianism helped resolve the problem of alleviating suffering outside the state and the market mechanisms, without displacing the state or/and market as the foundational institutions of global order (Amoore and Langely, 2004:100; Bartelson, 2006:374).

In providing a resolution to some of the tensions produced by modernist, statist, realist, and liberal beliefs, and thus shoring up this order, humanitarianism has also altered international order in important ways – giving rise to new tensions and ambiguities exposed in instances of disaster. To understand these processes, it is necessary to further unpack humanitarian ideology. Although there are many articulations of humanitarianism – notably by the ICRC and the SPHERE project on humanitarian assistance – perhaps the clearest enunciation of humanitarian ideology is actually Peter Singer's seminal 1972 piece "Famine, Affluence and Morality". In arguing for a humanitarian response to famine,[7] Singer sets out the basic beliefs of humanitarianism: (1) human suffering is bad, (2) everyone who can has a moral obligation to prevent suffering from happening, and (3) this moral obligation obtains regardless of distance or particular attachments. Of course, Singer's article involves much more logical elucidation of his position and is far more demanding than most humanitarians suggest, yet it stands as a paradigmatic expression of the humanitarian belief system.

The first belief, that suffering is bad, is evident in the ICRC's statement of principles and is taken as so obvious and true by Singer as to require no further justification or argumentation in its favour (p. 231). While it seems an uncontroversial point, it is still a belief that excludes other perspectives on suffering. It excludes from consideration any claim that suffering may serve some purpose, either in a religious/transcendental sense or in a pragmatic/political sense, and consequently, there are political implications of this position. It encourages a representation of those suffering as innocent and passive victims, which in turn may lead to poor policy response (Malkki, 1996; Rieff, 2003; Finnemore, 2008). It also overlooks structural causes or political processes that cause suffering and to accept that suffering is inevitable. From the humanitarian perspective, the only appropriate response to suffering is to alleviate suffering (Rieff, 2003).

This stands in tension with other foundational belief systems, such economic liberalism, which ascribes responsibility for and legitimizes some forms of poverty. The market serves a crucial role in economic liberalism and produces suffering in a variety of forms: failed businesses, unemployment, ecological damage, loss of property, and poverty. That society has a responsibility to alleviate suffering stands in tension with the economically liberal belief that individuals (and potentially societies) bear some responsibility for their own economic lot in life – hence the concern with moral hazard. Consequently, just as disaster assistance within states required the discrimination of the deserving from the undeserving poor in the allocation of aid, so too does this pressure exist at the international level.

One example of this tension is evident in contestation over technology transfers and levels of economic development assistance in the 2015 Sendai Framework for Disaster Risk Reduction. Economic liberalism asserts that individuals and communities that produce valuable commodities, including knowledge and technology, should benefit from the investment and risk associated with producing those commodities. Conversely, humanitarianism asserts that free or subsidized transfer of these goods would alleviate human suffering and is morally good (perhaps required). At the heart of these questions and the Sendai negotiations was the tension between belief systems that offer contrasting answers to the question of who deserves to benefit from technological advancements and material resources – and, by necessity, who deserves to suffer by not having access to them.

Singer's second and third assertions construct the universal category of "humanity" against the particularism at the heart of nation-statist and realist ideology. For Singer, moral obligations pertain to all humans, regardless of particular attachments. This challenge to particularism has contributed to the expansion of who qualifies as human, and thus worthy of action to alleviate suffering, by contesting the meaning of humanity and the limits of moral obligations (Finnemore, 2004). The assertion of a common humanity has also gained strength due to the growing global interconnections brought about by globalization and transnational capitalist practices (Chimni, 2000:245).

The category of the human often appears to be self-evident; yet, the human subject constructed in humanitarian ideology has political implications; it constructs humans as particular kinds of subjects (Campbell, 1998; Edkins, 2003). Humanitarianism produces a human-subject that stands in stark contrast to the citizen-subject, which privileges and reifies the state/citizen nexus at the heart of statist and realist thought and empowers certain types of assistance/governance. Nonetheless, it also challenges realist assertions that moral considerations of justice are unique to political communities and that moral obligations cease at the border (Kennan, 1985). Consequently, normative guidance about how to respond to suffering outside one's own political community is a key aspect of humanitarian ideology. By promoting solutions such as global economic redistribution and military intervention, humanitarianism provides a social theory on "how the world works" with a clear set of principles that instruct individuals and collective entities on how they ought to act in the world when faced with suffering.

Given its commitment to a universal humanity over particular attachments, humanitarianism appears to stand in opposition to the ideology and practice of nation-statism and/or realism; however, this relationship is far more complex than one of simple opposition. Humanitarianism both challenges and reaffirms nation-state sovereignty, most obviously by ignoring the structural causes of suffering and by alleviating some of the worst effects of nation states exercising their sovereignty. Furthermore, humanitarianism is contested, malleable, and ambiguous, and actions undertaken in the name of humanitarianism are often difficult to separate from political or self-interested action. Humanitarian actors are often motivated by geopolitical calculations, media attention, and professional and self-interested considerations (Olsen et al., 2003; Barnett and Weiss, 2011), and humanitarian interventions often privilege donor interests over recipients, leading to further victimization, prolonged crises, and the manipulation and abuse of humanitarian principles to justify self-interested and violent interventions (Minear and Weiss, 1995:18; Rieff, 1997, 2003; Anderson, 1998; Chimni, 2000:244; Weiss, 2001; Kennedy, 2005).

Disasters expose tensions in humanitarian ideology first by demonstrating the limitations of humanitarianism in resolving the problem of disasters and human vulnerability. By eschewing structural reform or political advocacy, humanitarianism provides no normative guideline for how to construct a good society in terms of preventing disaster or suffering. Many studies demonstrate that humanitarian assistance does not alleviate suffering and in some cases may prolong or deepen a crisis, thus contributing to ongoing suffering. At its roots, humanitarianism takes suffering as inevitable and offers a normative guideline for how to respond to suffering once it has occurred – a criticism, not surprisingly, levelled at Singer, and one ultimately embraced by Rieff. The limitations of humanitarianism have resulted in a number of developments, such as the call for a human rights–based approach that goes further than humanitarian assistance. This important shift is evident in recent efforts to enshrine a human right to assistance and to reduce the risk of disaster rather than just provide relief. The promotion of human rights is now a significant aspect of the current international order, although it remains quite limited in the management of disaster. The challenge, which I outline in the next chapter, has been that more substantial systemic changes necessary to promote human rights and to reduce disaster risk have yet to gain traction, while support for humanitarian assistance increases. Given humanitarianism's contribution to sustaining the existing order, the continued preference for humanitarian solutions over structural changes is not terribly perplexing. However, repeated demonstrations of the failure of a humanitarian approach to resolve the problem of suffering and disasters has led to calls for alternative approaches.

This leads to the second way that disasters expose tensions in humanitarian ideology, which is by demonstrating as problematic the belief that one can act non-politically. Because humanitarianism is intertwined in supporting the existing international order, it is difficult to separate humanitarian acts from political acts that preserve beliefs and serve interests beyond alleviating human suffering. Disaster assistance, often framed as a humanitarian response, is not unambiguously humanitarian and must constantly be interpreted or framed as such. Many donors do seem to act out of a genuine concern to help others, demonstrating the strength of humanitarian ideology in shaping state identity and policies. However, the process of offering and accepting aid is a site of political struggle over who qualifies, and what types of acts qualify, as genuinely humanitarian (Hilhorst and Jansen, 2010). The ambiguity of the humanitarian/political distinction means that acts of assistance must be framed as humanitarian and become a primary site in the contestation over aid. Because humanitarianism operates to sustain standards of civilization, actors may be depicted as acting politically rather than humanely and thus call into question their status as a civilized state.

Some forms of aid are more credibly represented as political rather than humanitarian by virtue of the offer itself: discriminatory aid, tied aid, or conditional assistance is more likely to be construed as political. Military humanitarian intervention is one of the key dividing lines within the humanitarian community, with purists suggesting military involvement is fundamentally anathema to humanitarian activity (Hannigan, 2012:57). Certain actors may frame others as having political intentions – by supporting one side in a conflict, appealing to a domestic constituency, or increasing prestige/soft power. By examining the humanitarian frame, the project explores how disaster management regimes – ranging from relief and risk reduction to migration and diplomacy – reinforces or challenges humanitarian ideology and, by extension, how humanitarianism helps mend some of the cracks in the foundations of international order exposed by disasters.

Conclusion

In this chapter, I have identified some of the ideological foundations of international order that are most directly implicated in the management of disaster. I have also shown how disasters expose inconsistencies in these beliefs and how DMRs and disaster assistance attempt to preserve these ideological foundations. As I have stressed throughout, these ideological frameworks are neither static nor determinative, but they do structure how actors think about and respond to disasters. It is important to understand these ideological foundations because they shape decisions on how to prepare for and respond to disasters. Frames and policy options are not simply invented; they emerge from an existing reservoir of cultural resources. What I have tried to do in this chapter is explore the range of cultural resources readily available to actors who interpret and frame disasters and disaster assistance.

Ideological analysis moves the analysis beyond a simple description of the process of framing disasters and disaster assistance to an analysis of the belief systems that underpin systems of disaster management. Critiquing an actor for employing a particular frame or implementing a specific policy is meaningless without understanding the belief system that the particular frame or specific policy is meant to uphold. The only means of critique available to those focusing on process is either in terms of process itself (actors have violated legitimized processes) or in terms of the empirical accuracy of their claims (actors have misled or misperceived reality). I find both sorts of critique insufficient because they limit critical analysis by accepting the legitimization of particular processes and claims to objective measurement and empirical accuracy. Exposing ideological foundations permits a deeper form of critique.

In subsequent chapters, I focus on how decision makers have understood and framed international disaster management, beginning with the construction of the international disaster relief regime. In Chapter 3, I explore how particular ideologies, backed by powerful actors, shaped the system that currently sets the parameters for disaster response. Again, this is not to suggest that the regime is static and unchanging; the international disaster regime has undergone numerous changes since its inception, largely in response to recurrent disasters. To understand and explain these changes, as well as the changes that have not taken place, requires ideological analysis and a deeper understanding of the ideologies of international order that I have attempted to provide in this chapter.

Notes

1 I use the terms *nation-state* and *state* interchangeably, while emphasizing the tension between nationalist and bureaucratic legitimization of state authority.

2 In contrast, Chowdhury contends that war making is the primary self-authorizing practice of the state and that in the near absence of war making, citizens are largely unwilling to pay the state to provide public goods.

3 Contagious disease is governed through an entirely different institutional structure than other forms of disaster and is not covered in consolidated appeals for assistance.
4 Though less relevant to the framing of disaster assistance, advocacy groups have also framed disaster assistance and prevention programs in terms of property rights. In this frame, building codes, zoning laws, insurance programs, and environmental protection laws implemented to reduce the likelihood and cost of disasters represent an illiberal regulation and restriction on how property owners can use their property (Platt, 1999), violating a core tenet of liberal ideology, and a key strand in its neoliberal variant.
5 Over the past 30 years, there is considerable debate about whether this belief system has been successful; see Dollar and Kraay, 2002. For an opposing interpretation, see Lubker et al., 2002 and Wade, 2004.
6 These are clearly not exclusively realist values. Moderates of all ideological stripes endorse these principles. However, realism is among the few belief systems to ideologize these principles – by this I mean, to turn these principles into defining characteristic of the belief system and to use them to identify the self and other.
7 I should note that Singer's humanitarianism is utilitarian, unlike the deontological humanitarianism of the ICRC.

References

Abbott, K.W., & Snidal, D., 1998. Why states act through formal international organizations. *Journal of Conflict Resolution*, 42(1), pp. 3–32.

Amoore, L., & Langley, P., 2004. Ambiguities of global civil society. *Review of International Studies*, 30(1), pp. 89–110.

Anderson, B., 1983. *Imagined Communities: Reflections on the Origin and Spread of Nationalism*, London: Verso.

Anderson, M.B., 1998. You save my life today, but for what tomorrow? In Moore, J. (ed.). *Hard Choices: Moral Dilemmas in Humanitarian Intervention*, Lanham: Rowman & Littlefield.

Bankoff, G., 2003. *Cultures of Disaster: Society and Natural History in the Philippines*, London: Routledge.

Barnett, M.N., 2010. *The International Humanitarian Order*, London: Routledge.

Barnett, M.N., & Weiss, T.G., 2011. *Humanitarianism Contested: Where Angels Fear to Tread*, London: Routledge.

Bartelson, J., 2006. Making sense of global civil society. *European Journal of International Relations*, 12(3), pp. 371–95.

Beck, U., 1992. From industrial society to the risk society: Questions of survival, social structure and ecological enlightenment. *Theory, Culture & Society*, 9(1), p. 97.

Behr, H., & Heath, A., 2009. Misreading in IR theory and ideology critique: Morgenthau, Waltz and neo-realism. *Review of International Studies*, 35(2), p. 327.

Bell, D., 2002. Anarchy, power and death: Contemporary political realism as ideology. *Journal of Political Ideologies*, 7(2), pp. 221–39.

Berrebi, C., & Ostwald, J., 2011. Earthquakes, hurricanes, and terrorism: Do natural disasters incite terror? *Public Choice*, 149(3–4), pp. 383–403.

Biersteker, T.J., & Weber, C., 1996. *State Sovereignty as Social Construct*, Cambridge: Cambridge University Press.

Billig, M., 1995. *Banal Nationalism*, London: Sage Publications.

Bourguignon, F., & Sundberg, M., 2007. Aid effectiveness: Opening the black box. *American Economic Review*, 97(2), pp. 316–21.

Breen, O.B., Ford, P., & Morgan, G.G., 2009. Cross-border issues in the regulation of charities: Experiences from the UK and Ireland. *International Journal of Not-for-Profit Law*, 11(3), pp. 5–41.

Buchanan-Smith, M., & Maxwell, S., 1994. Linking relief and development: An introduction and overview. *IDS Bulletin*, 25(4), pp. 2–16.

Buzan, B., & Lawson, G., 2015. *The Global Transformation: History, Modernity and the Making of International Relations*, Cambridge: Cambridge University Press.

Camilleri, J., 2008. Sovereignty discourse and practice. In T. Jacobsen (ed.). *Re-envisioning Sovereignty: The End of Westphalia?* London: Routledge.

Campbell, D., 1998. Why fight: Humanitarianism, principles, and post-structuralism. *Millennium: Journal of International Studies*, 27(3), pp. 497–521.

Cavallo, E., & Noy, I., 2011. Natural disasters and the economy: A survey. *International Review of Environmental and Resource Economics*, 5(1), pp. 63–102.

Chandler, D., 2001. The road to military humanitarianism: How the human rights NGOs shaped a new humanitarian agenda. *Human Rights Quarterly*, 23(3), pp. 678–700.

Chimni, B.S., 2000. Globalization, humanitarianism and the erosion of refugee protection. *Journal of Refugee Studies*, 13(3), pp. 243–63.

Chowdhury, A., 2018. *The Myth of International Order*, Oxford: Oxford University Press.

Collier, S., 2014. Neoliberalism and natural disaster. *Journal of Cultural Economy*, 7(3), pp. 273–90.

Cooley, A., & Spruyt, H., 2009. *Contracting States: Sovereign Transfers in International Relations*, Princeton, NJ: Princeton University Press.

Cooper, A.F., Higgott, R.A., & Nossal, K.R., 1993. *Relocating Middle Powers: Australia and Canada in Changing World Order*, Vancouver, Canada: UBC Press.

Cox, R.W., 1979. Ideologies and the new international economic order: Reflections on some recent literature. *International Organization*, 33(2), pp. 257–302.

Deudney, D., 2007. *Bounding Power: Republican Security Theory from the Polis to the Global Village*, Princeton, NJ: Princeton University Press.

Dillon, S., 2002. Farewell to linkage: International trade law and global sustainability indicators. *Rutgers Law Review*, 55, p. 87.

Dollar, D., & Kraay, A., 2002. Growth is good for the poor. *Journal of Economic Growth*, 7(3), pp. 195–225.

Duffield, M., 2012. Challenging environments: Danger, resilience and the aid industry. *Security Dialogue*, 43(5), pp. 475–92.

Dupont, B., 2004. Security in the age of networks. *Policing and Society*, 14(1), pp. 76–91.

Easterly, W., 2007. *The White Man's Burden: Why the West's Efforts to Aid the Rest Have Done So Much Ill and So Little Good*, Oxford: Oxford University Press.

Edkins, J., 2003. Humanitarianism, humanity, human. *Journal of Human Rights*, 2(2), pp. 253–8.

Finnemore, M., 2004. *The Purpose of Intervention: Changing Beliefs about the Use of Force*, Ithaca, NY: Cornell University Press.

Finnemore, M., 2008. *Paradoxes in Humanitarian Intervention*, Cambridge: Cambridge University Press.

Fischer, H.W., 1998. *Response to Disaster: Fact versus Fiction & Its Perpetuation: The Sociology of Disaster*. Lanham, MD: University Press of America.

Foucault, M., 2007. *Security, Territory, Population: Lectures at the College de France*, translated by G. Burchell, New York: Palgrave Macmillan.

Freeden, M., 2006. Ideology and political theory. *Journal of Political Ideologies*, 11(1), pp. 3–22.

Fremont-Smith, M.R., 2009. *Governing Nonprofit Organizations: Federal and State Law and Regulation*, Cambridge, MA: Harvard University Press.
Gassebner, M., Keck, A., & Teh, R., 2010. Shaken, not stirred: The impact of disasters on international trade. *Review of International Economics*, 18(2), pp. 351–68.
Giddens, A., 1991. *Modernity and Self-Identity: Self and Society in the Late Modern Age*, Stanford, CA: Stanford University Press.
Gilpin, R., 1983. *War and Change in World Politics*, Cambridge: Cambridge University Press.
Gilpin, R., & Gilpin, J.M., 1987. *The Political Economy of International Relations*, Princeton, NJ: Princeton University Press.
Glaser, C.L., 1996. Realists as optimists: Cooperation as self-help. *Security Studies*, 5(3), pp. 122–63.
Greenfeld, L., 2005. Nationalism and the mind. *Nations and Nationalism*, 11(3), pp. 325–41.
Grieco, J., 1988. Anarchy and the limits of cooperation: A realist critique of the newest liberal insitutionalism. *International Organization*, 42(3), pp. 485–507.
Guzzini, S., 2017. Militarizing politics, essentializing identities: Interpretivist process tracing and the power of geopolitics. *Cooperation and Conflict*, 52(3), pp. 423–45.
Habermas, J., 2012. Modernity: An unfinished project [1980]. In C. Calhoun, J. Gerteis, J. Moody, S. Pfaff, & I. Virk (eds.). *Contemporary Sociological Theory*, New York: John Wiley & Sons.
Hannigan, J., 2012. *Disasters without Borders*, Cambridge: Polity Press.
HAP, 2007. HAP 2007 *Standard in Humanitarian Accountability and Quality Management*, Adopted by HAP on January 30, 2007.
Harvey, D., 1990. *The Condition of Postmodernity: An Enquiry into the Origins of Cultural Change*, Cambridge, MA: Blackwell.
Hayek, F.A., 1944. *The Road to Serfdom*, London: Routledge.
Hays, J.C., 2009. *Globalization and the New Politics of Embedded Liberalism*, Oxford: Oxford University Press.
Heide, E.A., 2004. Common misconceptions about disasters: Panic, the "disaster syndrome," and looting. Chapter 27. In M. O'Leary (ed.). *The First 72 Hours*, New York: iUniverse.
Helleiner, E., 2002. Economic nationalism as a challenge to economic liberalism? Lessons from the 19th century. *International Studies Quarterly*, 46(3), pp. 307–29.
Hensler, B., 2011. *Crucible of Fire: 19th Century Urban Fires and the Making of the Modern Fire Service*, Washington, DC: Potomac Books.
Herz, J., 1950. Idealist internationalism and the security dilemma. *World Politics*, 2(2), pp. 157–80.
Hewitt, K., 1983. *Interpretation of Calamities: From the Viewpoint of Human Ecology*, Boston: Allen and Unwin.
Hilhorst, D., & Jansen, B., 2010. Humanitarian space as arena: A perspective on the everyday politics of aid. *Development and Change*, 41(6), pp. 1117–39.
Huntington, S.P., 1999. The lonely superpower. *Foreign Affairs*, 78(2), p. 35.
Hyde-Price, A., 2009. Realist ethics and the war on terror. *Globalizations*, 6(1), pp. 23–40.
Irvin, R., 2005. State regulation of nonprofit organizations: Accountability regardless of outcome. *Nonprofit and Voluntary Sector Quarterly*, 34, p. 2.
Jackson, R.H., 1990. *Quasi-States: Sovereignty, International Relations and the Third World*, New York: Cambridge University Press.
James, A., 1986. *Sovereign Statehood: The Basis of International Society*, Milton Park: Taylor & Francis Group.

Johnston, L., 1992. *The Rebirth of Private Policing*, London: Routledge.
Johnston, L., 1996. Policing diversity: The impact of the public-private complex in policing. In F. Leishman, B. Loveday, & S, Savage (eds.). *Core Issues in Policing*, Harrow: Longman, pp. 54–70.
Johnston, L., 1999. Private policing in context. *European Journal on Criminal Policy and Research*, 7(2), pp. 175–96.
Jones, E., 1981. *The European Miracle: Environments, Economies and Geopolitics in the History of Europe and Asia*, 1st ed., Cambridge: Cambridge University Press.
Jones, T., & Newburn, T., 1998. *Private Security and Public Policing*, Oxford: Clarendon Press.
Jordaan, E., 2003. The concept of a middle power in international relations: Distinguishing between emerging and traditional middle powers. *Politikon*, 30(1), pp. 165–81.
Kedourie, E., 1971. *Nationalism in Asia and Africa*, London: Weidenfeld & Nicholson.
Kelsen, H., 1959. Sovereignty and international law. *Georgetown Law Journal*, 48(4), pp. 627–40.
Kennan, G., 1985. Morality and foreign policy. *Foreign Affairs*, 64(2), pp. 205–18.
Kennedy, D., 2005. *The Dark Sides of Virtue: Reassessing International Humanitarianism*, Princeton, NJ: Princeton University Press.
Kennedy, P., 1987. *The Rise and Fall of the Great Powers: Economic and Military Conflict from 1500 to 2000*, New York: Random House.
Kiely, R., 2005. Globalization and poverty, and the poverty of globalization theory. *Current Sociology*, 53(6), pp. 895–914.
Klein, N., 2007. *The Shock Doctrine: The Rise of Disaster Capitalism*, New York: Metropolitan Books/Henry Holt.
Klinenberg, E., 2002. *Heat Wave: A Social Autopsy of Disaster in Chicago*, Chicago, IL: University of Chicago Press.
Kragelund, P., 2008. The return of non-DAC donors to Africa: New prospects for African development? *Development Policy Review*, 26(5), pp. 555–84.
Krasner, S.D., 1995. Compromising Westphalia. *International Security*, 20(3), pp. 115–51.
Krasner, S.D., 1999. *Sovereignty: Organized Hypocrisy*, Princeton, NJ: Princeton University Press.
Landis, M.L., 1997. Let me next time be tried by fire: Disaster relief and the origins of the American welfare state 1789–1874. *Northwestern University Law Review*, 92, pp. 967–1034.
Lang, A.T.F., 2006. Reconstructing embedded liberalism: John Gerard Ruggie and constructivist approaches to the study of the international trade regime. *Journal of International Economic Law*, 9(1), pp. 81–116.
Longhurst, R., 1994. Conceptual frameworks for linking relief and development. *IDS Bulletin*, 25(4), pp. 17–23.
Lubker G., Smith, G., & Weeks, J., 2002. Growth and the poor: A comment on Dollar and Kraay. *Journal of International Development*, 14(5), pp. 555–71.
Lyotard, J.-F., 1984. *The Postmodern Condition*, translated by G. B. and B. Massumi, Manchester: Manchester University Press.
Malkki, L.H., 1996. Speechless emissaries: Refugees, humanitarianism, and dehistoricization. *Cultural Anthropology*, 11(3), pp. 377–404.
Manning, R., 2006. Will "emerging donors" change the face of international co-operation? *Development Policy Review*, 24(4), pp. 371–85.
Mead, W.R., 2014. The return of geopolitics: The revenge of the revisionist powers. *Foreign Affairs*, XXXIII(2), pp. 81–7.

Mearsheimer, J., 1995. The false promise of international institutions. *International Security*, 19(3), pp. 5–49.
Meyer, J., Boli, J., Thomas, G., & Ramirez, F., 1997. World society and the nation-state. *American Journal of Sociology*, 103(1), pp. 144–81.
Middleton, N., & O'Keefe, P., 1997. *Disaster and Development: The Politics of Humanitarian Aid*, London: Pluto Press.
Minear, L., & Weiss, T.G., 1995. *Mercy under Fire: War and the Global Humanitarian Community*, Boulder, CO: Westview Press.
Morganthau, H., 1948. *Politics among Nations: The Struggle for Power and Peace*, New York: Knopf.
Morganthau, H., 1962. A political theory of foreign aid. *American Political Science Review*, 56(2), pp. 301–9.
Moyo, D., 2009. *Dead Aid: Why Aid Is Not Working and How There Is a Better Way for Africa*, New York: Farrar, Straus and Giroux.
Nel, P., & Righarts, M., 2008. Natural disasters and the risk of violent civil conflict. *International Studies Quarterly*, 52(1), pp. 159–85.
Neocleous, M., 2008. *Critique of Security*, Edinburgh: Edinburgh University Press.
Nye, J.S., Jr, 1990. Soft power. *Foreign Policy*, 80, pp. 153–71.
UNOCHA, 1991. *What Is OCHA? United Nations Office for the Coordination of Humanitarian Affairs*, New York: United Nations Secretariat.
Olsen, G.R., Carstensen, N., & Høyen, K., 2003. Humanitarian crises: What determines the level of emergency assistance? Media coverage, donor interests and the aid business. *Disasters*, 27(2), pp. 109–26.
Olson, R., & Drury, A., 1997. Un-therapeutic communities: A cross-national analysis of post-disaster political unrest. *International Journal of Mass Emergencies and Disasters*, 15(2), pp. 221–38.
Özerdem, A., & Jacoby, T., 2006. *Disaster Management and Civil Society: Earthquake Relief in Japan, Turkey and India*, London: IB Tauris.
Parekh, B., 1997. Rethinking humanitarian intervention. *International Political Science Review*, 18(1), pp. 49–69.
Parsons, T., 1968. *The Structure of Social Action*, New York: Free Press.
Pelling, M., & Dill, K., 2006. "Natural" disasters as catalysts of political action. *Media Development*, 53(4), p. 7.
Phillips, B.D., 1998. Sheltering and housing of low-income and minority groups in Santa Cruz County after the Loma Prieta Earthquake. *U.S. Geological Survey Professional Paper*, (1553).
Phillips, A., 2011. *War, Religion and Empire*, Cambridge: Cambridge University Press.
Phillips, A., & Sharman, J., 2015. *International Order in Diversity: War, Trade and Rule in the Indian Ocean*, Cambridge: Cambridge University Press.
Platt, R., 1999. *Disasters and Democracy: The Politics of Extreme Natural Events*, Washington, DC: Island Press.
Platt, R., & Dawson, A., 1999. The takings issue and the regulation of hazardous areas. In R. Platt (ed.). *Disasters and Democracy: The Politics of Extreme Natural Events*, Washington, DC: Island Press.
Poggi, G., 1972. *Images of Society: Essays on the Sociological Theories of Tocqueville, Marx, and Durkheim*, Stanford, CA: Stanford University Press.
Rafshoon, E.G., 2001. A realist's moral opposition to war: Han J. Morgenthau and Vietnam. *Peace and Change*, 26(1), pp. 55–77.

Raschky, P.A., & Schwindt, M., 2009. *Aid, Natural Disasters and the Samaritan's Dilemma.* World Bank Policy Research Working Paper 4952.
Reid, J., 2012. The disastrous and politically debased subject of resilience. *Development Dialogue*, 58, pp. 67–79.
Rieff, D., 2003. *A Bed for the Night: Humanitarianism in Crisis*, New York: Simon and Schuster.
Robert, P., 1989. The privatisation of social control. In *Crime and Criminal Policy in Europe*, Proceedings of a European Colloquium, Oxford, pp. 104–20.
Rozario, K., 2001. What comes down must go up: Why disasters have been good for American capitalism. In S. Biel (ed.). *American Disasters*, New York: New York University Press.
Ruffin, R., 2002. David Ricardo's discovery of comparative advantage. *History of Political Economy*, 34(4), pp. 727–48.
Ruggie, J.G., 1982. International regimes, transactions, and change: Embedded liberalism in the postwar economic order. *International Organization*, 36(2), pp. 379–415.
Schweller, R.L., 1997. New realist research on alliances: Refining, not refuting, Waltz's balancing proposition. *The American Political Science Review*, 91(4), p. 927.
Scott, J.C., 1998. *Seeing Like a State: How Certain Schemes to Improve the Human Condition Have Failed*, New Haven, CT: Yale University Press.
Sending, O.J., & Neumann, I.B., 2006. Governance to governmentality: Analyzing NGOs, states, and power. *International Studies Quarterly*, 50(3), pp. 651–72.
Shearing, C.D., & Stenning, P.C., 1981. Modern private security: Its growth and implications. *Crime and Justice*, 3(1), pp. 193–245.
Shearing, C.D., & Stenning, P.C., 1983. Private security: Implications for social control. *Social Problems*, 30(5), pp. 493–506.
Singer, P., 1972. Famine, affluence, and morality. *Philosophy and Public Affairs*, 1(3), pp. 229–43.
Slim, H., 2001. Violence and humanitarianism: Moral paradox and the protection of civilians. *Security Dialogue*, 32(3), pp. 325–39.
Smith, Anthony D., 1995. *Nations and Nationalism in a Global Era*, Cambridge: Polity Press.
Snidal, D., 1985. The limits of hegemonic stability theory. *International Organization*, 39(4), pp. 579–614.
Sorenson, G., 2011. *A Liberal World Order in Crisis: Choosing between Imposition and Restraint*, Ithaca, NY: Cornell University Press.
SPHERE, 2004. The Humanitarian Charter: The Sphere Project: Humanitarian Charter and Minimum Standards in Disaster Response, 2004 Edition. http://www.sphereproject.org/dmdocuments/handbook/hdbkpdf/hdbk_hc.pdf.
Stallings, R., 1988. Conflict in natural disasters: A codification of consensus and conflict theories. *Social Science Quarterly*, 69(3), p. 569.
Stephens, L., & Green, S., 1979. *Disaster Assistance: Appraisal, Reform, and New Approaches*, New York: Macmillan Press.
Suganami, H., 2007. Understanding sovereignty through Kelsen/Schmitt. *Review of International Studies*, 33(3), pp. 511–30.
Tadesse, G., & Shively, G., 2009. Food aid, food prices, and producer disincentives in Ethiopia. *American Journal of Agricultural Economics*, 91(4), pp. 942–55.
Tierney, K.J., 2007. From the margins to the mainstream? Disaster research at the crossroads. *Annual Review of Sociology*, 33, pp. 503–25.
Tierney, K., Bevc, C., & Kuligowski, E., 2006. Metaphors matter: Disaster myths, media frames, and their consequences in Hurricane Katrina. *The Annals of the American Academy of Political and Social Science*, 604(1), pp. 57–81.

Toulmin, S., 1990. *Cosmopolis: The Hidden Agenda of Modernity*, New York: Free Press.
UNISDR, 2015. *Sendai Framework for Disaster Risk Reduction 2015–2030*, Sendai. https://www.unisdr.org/we/inform/publications/43291.
United Nations General Assembly Resolution 42/169, 1987. International decade for natural disaster reduction. *96th Plenary Meeting*, December 11, 1987.
United Nations General Assembly Resolution 46/182, 1991. Strengthening of the coordination of humanitarian emergency assistance of the United Nations. *78th Plenary Meeting*, December 19, 1991.
Varley, A., 1994. The exceptional and the everyday: Vulnerability analysis in the international decade for natural disaster reduction. In A. Varley (ed.). *Disasters, Development and Environment*, New Haven, CT: John Wiley & Sons.
Vatsa, K., & Krimgold, F., 2000. Financing disaster mitigation for the poor. In A. Kreimer & M. Arnold (eds.). *Managing Disaster Risk in Emerging Economies*, Chapter 12, Washington, DC, World Bank, pp. 129–36.
Wade, R.H., 2004. Is globalization reducing poverty and inequality? *World Development*, 32(4), pp. 567–89.
Walker, P., 1992. Foreign military resources for disaster relief: An NGO perspective. *Disasters*, 16(2), pp. 152–9.
Walker, P., & Maxwell, D.G., 2009. *Shaping the Humanitarian World*, London: Routledge.
Walker, R.B.J., 1993. *Inside/Outside: International Relations as Political Theory*, Cambridge: Cambridge University Press.
Walt, S.M., 1985. Alliance formation and the balance of world power. *International Security*, 9(4), pp. 3–43.
Walt, S., 2009. Alliances in a unipolar world. *World Politics*, 61(1), pp. 86–120.
Waltz, K., 1979. *Theory of International Politics*, Boston: McGraw-Hill.
Warner, D., 1999. The politics of the political/humanitarian divide. *Revue Internationale de la Croix-Rouge/International Review of the Red Cross*, 81(833), pp. 109–18.
Weiss, T., 2001. Researching humanitarian intervention: Some lessons. *Journal of Peace Research*, 38(4), pp. 419–28.
Wilks, I., 1955. A note on sovereignty. *The Philosophical Quarterly*, 5(21), pp. 342–7.
Wisner, B., Blaikie, P., Cannon, T., & Davis, I., 2004. *At Risk: Natural Hazards, People's Vulnerability, and Disasters*, New York: Routledge.
Woods, N., 2008. Whose aid? Whose influence? China, emerging donors and the silent revolution in development assistance. *International Affairs*, 84(6), pp. 1205–21.
Wrong, D., 1994. *Problem of Order*, New York: Simon and Schuster.
Wyatt-Walter, A., 1996. Adam Smith and the liberal tradition in international relations. *Review of International Studies*, 22, pp. 5–28.
Yao, J., 2018. 'Conquest from barbarism': The Danube Commission, international order and the control of nature as a standard of civilization. *European Journal of International Relations*. doi:10.1177/1354066118768379.

SECTION 1

International Disaster Management Regimes

In Chapters 1 and 2, I have explored the ideological foundations of the current international order and how disasters expose the inconsistencies within these belief systems as well as the tensions that exist between them. In periods of disaster, political actors attempt to reinforce these belief systems through slight alterations and the establishment of new concepts and institutions. While not fully resolving these inconsistencies and tensions, new concepts and institutions provide a contingent resolution that permits continuity of the international order. In the chapters of this section, I examine how the international community has attempted to manage disasters to preserve or to implement changes to aspects of the existing order.

In the next three chapters, I examine three international disaster management regimes (DMRs) through which the international community has sought to deal with disasters and preserve international order: disaster relief, disaster risk reduction, and disaster migration.

Chapter 3 focuses on the historical development of disaster relief, with particular attention to three distinct phases, or episodes. Beginning with the advent of disaster relief in the early modern period, I examine the 1755 Lisbon earthquake and tsunami to show how changing beliefs about science, economics, and political authority gave rise to the practice of disaster relief – in short, disaster relief became a mechanism to reinforce novel developments in the international system. I then turn to the early twentieth century, when political actors sought to institutionalize the practice of disaster relief, which had (since the Lisbon earthquake) remained ad hoc and bilateral. The development of the International Federation of the Red Cross/Crescent Societies and the International Relief Union (IRU) brought to the fore a number of tensions exposed by the international management of disasters in a world marked by nation states (and empires) and a growing liberal economic order. In this case, humanitarianism emerged as the core conceptual resolution to these tensions; however, humanitarianism created new ideological contestation. This chapter demonstrates how powerful actors have had a profound impact on the nature of international disaster relief.

The last development covered in Chapter 3 consists of an examination of the UN system of disaster relief to show how it strongly mirrored the IRU system implemented by the League of Nations. Although incremental changes during the UN period have resulted in a drastic increase in the amount of disaster relief and its more centralized coordination by UN agencies, I show the continuity of its ideological foundations with the period that preceded it. The failure of the relief system, along with the growth of new decolonized states, gave rise to new pressures within the UN, namely, for an alternative understanding of disasters and a new approach in the form of disaster risk reduction.

Chapter 4 explores the three frameworks for disaster risk reduction that followed from the UN's declaration of the International Decade for Disaster Risk Reduction (IDDRR). This chapter demonstrates the ongoing contestation over key beliefs about international order and the management of disasters. The chapter highlights the development of new frameworks, such as "sustainable development" and "resilience", that attempt to resolve the ideological inconsistencies within the foundations of international order. I also show how the institutional separation of disaster risk reduction from climate change and development serves to preserve a particular understanding of disaster that separates the question of responsibility from mitigation and adaptation. The chapter concludes by demonstrating how these contingent resolutions themselves introduce new ambiguities that give rise to reformist pressures and ongoing struggles at the international level.

In the final chapter of this section, I examine recent efforts to manage disaster migration. The impetus for this regime emerged from the failures of the existing DMRs to address the problem of human migration as a key mechanism of adaptation to disasters, as well as the failure of existing global migration regimes, mostly the refugee regime, to deal with displacement caused by environmental degradation, climate change, and disasters. In many respects, disaster relief and risk reduction functioned to contain one problem associated with disaster, namely, large-scale displacement. By reifying the principle of state sovereignty, in which states bear the primary responsibility for the protection of those affected by disaster, international DMRs offered tools that should minimize the attractiveness of migration as a solution to displacement and should encourage states to fulfil their protection responsibility. Focusing on the Nansen initiative and the Protection Agenda that resulted from this initiative, Chapter 5 concludes by demonstrating how the Protection Agenda operates similarly to the Sendai Framework by offering modest changes to the global migration regime to preserve the fundamental beliefs about international order.

3

THE INTERNATIONAL DISASTER RELIEF REGIME

Kent depicts the current system of international disaster relief as a network (complex and loosely connected) of public, private, and civil society actors organized locally, nationally, and transnationally (Kent, 1987). One of the prevailing beliefs about disaster relief is that the state is the central node in this network. Despite the near hegemonic status of this belief, there is great deal of variation in the capability of states to organize disaster relief. This variation reflects different historical experiences and exposure to disasters, levels of economic development, as well as processes of state formation. Despite this variation, most states possess some form of public emergency response mechanism that can be mobilized in an emergency, such as policing, ambulance corps, firefighting, national guard, and military forces, as well as a Red Cross/Crescent Society that provides relief and shelter after disasters.[1] In addition, private corporations, local nongovernmental organizations (NGOs), and community groups, both organized and ad hoc, play a tremendous role in the provision of disaster relief.

A large number of international organizations engage in disaster relief, including state-based aid agencies, military response teams, humanitarian NGOs, and intergovernmental organizations associated with the UN. Barnett and Weiss estimate that over 100 states participate in bilateral or multilateral forms of disaster relief, while over 2,500 NGOs may be involved – with 250 of those being "serious players" (Barnett and Weiss, 2011:27). At present, there are also close to ten separate UN agencies that mobilize in the aftermath of disasters; some participate directly in the provision of disaster relief, although the primary role for the UN is in coordinating the large number of actors that tend to converge on disaster-affected communities.[2] The response to a major disaster in any given state is likely to include a mixture of local community respondents;

emergency responders of the affected state; a host of national and international NGOs; foreign state aid organizations and military response teams; private corporations (both national and multinational); and close to ten separate UN agencies who, in addition to providing relief services, attempt to coordinate, along with the affected state, these various actors.[3]

The current system of disaster relief is a product of struggles to resolve ambiguities and inconsistencies in a complex normative environment. I suggest that the practice of disaster relief today reflects relatively stable resolutions to the problems and ideological tensions raised by disasters and international assistance. The current disaster relief system represents a historically contingent resolution to the ambiguities associated with the civilizing mission of modernism, the preservation of state sovereignty, the promotion of liberal economic order, and the maintenance of a realist international system – largely through the development of a highly circumscribed concept of humanitarianism. As I show in the next chapter, however, failure of this system has contributed to the development of alternative international frameworks for managing disasters, namely, disaster risk reduction and disaster migration.

More specifically, this chapter reveals that the practice of disaster relief was driven and shaped by norm entrepreneurs motivated by specific ideological beliefs, backed by powerful actors, and in contestation with alternative visions. In this chapter, I focus on a number of historical episodes that highlight the ideological contestation around the practice of disaster relief, and I show how the resolutions that emerged from these intense moments of contestation excluded alternative approaches to disaster and international assistance. While the examination of "episodes" may draw attention to short-term causes of change, it can also focus attention on how periodic contestation about the foundations of social order emerges from sources of tension and conflict during long-term processes.

In the cases examined in this chapter, I try to show that these seminal episodes arose due to long-term processes but that they also arose due to the interjection of disastrous events, such as the 1755 Lisbon earthquake and the first and second world wars. In the case of the UN-based system of disaster relief, it was not a single disaster but rather a series of disasters situated in the ideologically disruptive process of decolonization. These episodes reveal how disasters disrupt existing social orders and expose ideological inconsistencies and how they were incredibly influential in shaping the trajectory of Western political thought, the foundations of international order, and the management of disaster. As moments that capture the contingent resolution of major ideological conflicts, these episodes are highly instructive. Readers interested primarily in the major development of the post–World War II disaster management regime (DMR) – disaster risk reduction – can move to the next chapter, but they will find in this chapter transformative historical episodes that have a profound impact on current disaster management practices.

The history of international disaster relief reveals a number of distinct phases, all of which had a lasting impact on modern practices of disaster relief: ad hoc and irregular bilateral state assistance that has persisted since the Lisbon earthquake of 1755, a semi-autonomous civil society stream associated with the Red Cross movement and other humanitarian organizations that arose around the turn of the nineteenth century, and a disaster relief regime nested in formal international institutions such as the League of Nations and United Nations. Novel forms of disaster relief emerged from contestation over the ideological foundations of existing practices, and subsequent developments signal processes of ideological transitions that altered perspectives on disaster relief. Current international efforts to improve disaster relief through international institutions attempt to reconcile these models, which themselves represent ideological compromise and imposition.

Lisbon and the Birth of Disaster Relief

The Lisbon earthquake, tsunami, and fire of 1755 is an important watershed in the development of disaster relief and is often referred to as the "first modern disaster" because of the success that modernizing forces had in shaping the response to this disaster and subsequent disasters. Susan Neiman (2004) contends that Lisbon occupied the same place in eighteenth century thought that Auschwitz does in the current era: as a signifier of civilizational transformation. Although it is important to avoid overemphasizing a single historical episode, this particular disaster aptly demonstrates the foundations of current disaster relief practices and the intense ideological contestation that brought about a practice that is more or less now taken to be self-evident and common sense. It also shows the profound role that disaster can play in disrupting existing ideological belief systems.

The Lisbon disaster contributed immensely to the spread of modernist beliefs in Europe. Although modern explanations for earthquakes existed at the time of the Lisbon earthquake, they were marginal in much of Europe and not systematically studied. This was especially true in Portugal, which lagged behind other European states in terms of economic and intellectual modernization (Dynes, 1999:4). Key modernizing thinkers associated with the Enlightenment, such as Kant, Vattel, Rousseau, Smith, and Voltaire, all wrote on the Lisbon disaster to promote modernist understandings and to promote international disaster relief.

In Portugal, the earthquake played a decisive role in the contestation between modernizing and traditional sources of authority. In Portugal at this time, influential members of the religious hierarchy opposed modernizing and secularizing efforts in Portugal and rejected scientific explanations of geophysical phenomenon. This is not to reinforce a traditional/modern binary; not all religious orders were anti-science, and many scientists held strong religious beliefs. Indeed, at this time, the Jesuit order was a leading force in exploration and scientific scholarship (Harris, 2010), while many of the leading Enlightenment thinkers that promoted modernist views of disasters held deist theological beliefs (Barnett, 2003).

However, prominent religious authorities across Europe framed the Lisbon earthquake as punishment from God for the sins of the Portuguese people and for the religious apathy of those in positions of power. This was not a fringe position. Vattel's argument in favour of international disaster relief, which combined a humanitarian ethics and a rational, scientific understanding of natural disaster, was a radical challenge to the dominant interpretation of disasters at the time (Hutchinson, 2000:4). While the interpretation of "natural" disasters as punishment from God in the pre-modern period was not as hegemonic as is often suggested (Cutler, 2009:535; Udías, 2009:41) it remained a very popular and persuasive understanding into the late eighteenth century – traces of which linger today even in the most secularized societies. Lisbon played a key role in this transformation.

Proponents of the theological position in Portugal, such as the powerful Jesuit Gabriel Malagrida, offered a number of solutions in the aftermath: the first and foremost was for the people to repent and for those in power to adhere more strictly to religious teachings (Paice, 2008; Shrady, 2008). Malagrida singled out groups in the population that bore special responsibility for God's wrath, in this case Protestants,[4] and vociferously criticized those who busied themselves constructing shelters for the victims of the earthquake and reconstructing the city of Lisbon (Paice, 2008; Shrady, 2008). Religious authorities called for the relocation of the Portuguese capital and the abandonment of Lisbon, which they framed as a condemned city bound to suffer further calamities (Kendrick, 1955:88). The religious ideology of the time contained a social theory of the world complete with normative requirements of acting in that world that were at odds with the emergent modernist ideology.

In contrast, modernizing members of the Portuguese aristocracy, most notably, Jose de Carvalho e Melo, commonly referred to as Pombal, adamantly promoted modernism. Pombal endorsed a decidedly scientific view of the earthquake and disaster response, and he implemented a number of "modern" solutions: immediate relief (bury the dead and feed the hungry), rebuild the capital with earthquake-proof buildings, and systematic study of the earthquake through regional surveys. The materials and architecture used in reconstruction were tested for their ability to withstand ground shaking (Hamblyn, 2008), which was a first in Europe (Economist, 2008). Pombal's seismic surveys asked regional authorities to document when they felt the earthquake, what symptoms their region experienced and how long they lasted, the number of aftershocks, and the response of local authorities (Paice, 2008:174). These surveys essentially ushered in the modern field of seismology (Shrady, 2008:146).

The modernist ideology promoted by Pombal was deeply connected with the ideology of state sovereignty – which placed authority in the hands of the state above that of the church. In 1755, Lisbon was known for its extreme religious fervour, in which religious authorities associated with the Inquisition and Jesuits held great influence. The Lisbon earthquake was a seminal moment in the assertion of the sovereign state model because it was the first in which the state assumed primary responsibility for emergency rescue and reconstruction (Dynes, 1999:15). This involved a series of measures that centralized authority in the state and undermined competing sources of authority. These measures included ordering the clergy and all citizens to collect bodies; calling in troops from surrounding regions to assist with reconstruction; commandeering supplies, food, ships, and wagons; and preventing people from leaving the city so they could assist with the rescue and reconstruction. This last measure required the implementation of a national pass system that restricted mobility and required provincial governors to send people back to the capital (Kendrick, 1955:82). Thus, we can see in the earliest formations of disaster management the rise of modernist beliefs and authorization of the state, as well as the necessity of relief and a restriction on migration to achieve these outcomes.

In addition to these coercive measures, Portuguese authorities legitimized the power of the state by delegitimizing competing sources of authority. They accomplished this through the depiction of religious authorities as obstructionist to modernizing efforts and reconstruction. Although some prominent Jesuits opposed Pombal's measures, most religious authorities and private citizens contributed greatly to relief and reconstruction after the disaster, and most supported Pombal's reforms. Yet Pombal successfully framed the church, mostly the Jesuits, as obstructionist to diminish the contributions and the authority of the church and to reinforce Pombal's authority in particular (Kendrick, 1955:71). Continuing his assault against competing religious authorities, Pombal expelled the Jesuits from Portugal and its colonial holdings shortly after reconstruction began, further solidifying the authority of the state against its primary contender: the Catholic Church (Roehner, 1997). Identifying the state as the sole author of relief and reconstruction was a key act in legitimizing state authority over religious authority in Portugal.

Yet the relief and reconstruction measures also exposed tensions in Portugal's economic order. Wealth in Portugal was concentrated in the hands of the church and the Portuguese aristocracy, although there was a rising merchant class. Prior to the earthquake, Pombal had endorsed a number of economic reforms that would redistribute wealth toward Portuguese merchants and industries. The disaster and reconstruction offered an opportunity to achieve drastic economic reforms toward what could best be characterized as a nascent liberal economic order – a market-based system of distributing wealth rather than tradition and inheritance.

This vision of the Portuguese economy was threatened by the prospect of widespread looting. Immediately after the disaster, Portuguese authorities implemented two restrictions to prevent theft: the first was to prevent all migration out of Lisbon – everyone was expected to stay and assist with the rebuilding efforts. The second was to implement harsh measures for those caught looting or stealing; they were immediately executed following a summary trial (Kendrick, 1955:81). One English commentator, arriving a few weeks after the calamity, noted that Lisbon was "dotted with gallows every twenty-five yards" to punish anyone who attempted to flee or profit from the disaster (Paice, 2008:157).

Just as drastically, Pombal used the disaster to redesign Lisbon as a centre of commerce. Instead of rebuilding the city as it was, Portuguese authorities razed the remainder of the city and fundamentally altered its geography and architecture. The new city prefigured the importance of commerce and trade above that of the church, the crown, and the aristocracy. Pombal and his engineers created a city centred around an immense commercial area (Praca do Comerico) with widened streets running in a grid pattern. The streets descended from the centre based on "hierarchy of trades": gold street was closest to the centre, with leather and other goods further removed (Hamblyn, 2008:110). Most churches were not rebuilt, and those that were faced restrictions on the height of their towers and domes. The destroyed palace was not rebuilt, and old property lines, which reinforced the land holdings and wealth of the Portuguese aristocracy, were abolished in favour of new urban space that prioritized the interest of merchants and the growing middle class (Hamblyn, 2008:111).

Economic reforms targeted religious authorities in the colonies as well by creating trading monopolies modelled after the British system of trading companies (Maxwell, 1968). This new system was designed to increase the tax revenues of the Portuguese state from its colonial holdings for the purpose of rebuilding Lisbon, although it was also designed to undermine the Jesuit traders who were free from the state's tax regime. By 1759, Pombal moved further against the Jesuits, expelling them from Portugal and its colonies and confiscating their property on the grounds that they had supported the Tavora plot to assassinate the king. In one move, Pombal eliminated the Jesuits and limited the power of the aristocracy (Maxwell, 1968:630). Although Pombal's domestic economic reforms are not liberal by today's standards, they were clearly liberalizing in that they sought to implement a market-oriented economy in which merchants and the middle class, rather than the church and aristocracy, formed the backbone of the economy.

Pombal's efforts to reorient the economy toward Portuguese merchants exposed tensions not only with religious authorities and the aristocracy but also with Portugal's international trading partners. Lisbon was a centre of an international empire that relied on international trade, from its colonial holdings and from its ally and prominent trading partner Great Britain. Raw materials from the colonies, such as gold and sugar, passed through Lisbon to Great Britain in exchange for British food, cloth, and manufactured goods – resulting in a massive trade imbalance and the stunting of Portuguese domestic industry. Beyond that, British merchants in Lisbon controlled most of the trade into and out of Lisbon. Some commentators at the time referred to Portugal as a British colony (Kendrick, 1955). As former ambassador to London, Pombal was well aware of this trade imbalance and Portugal's reliance on Great Britain – a dependence aptly demonstrated by the earthquake.

In the aftermath of the tsunami following the earthquake, Britain was the primary provider of international assistance. In total, British aid was estimated to be worth 100,000 pounds (Paice, 2008:159), which would be equivalent to about 18 million pounds today. This unprecedented offer of assistance reflected the strong military alliance between England and Portugal and their intricate economic interdependence. While Great Britain benefitted from the unequal trade relationship, its economy was heavily dependent on gold from the Portuguese empire, and it sought to restore Portugal's economy as quickly as possible. The English couched their offer of aid in "humanitarian" terms, that is, the relief of human suffering, although their military alliance and trade relations with Portugal were clearly strong motivating factors (Hamblyn, 2008).

The Lisbon disaster provided an opportunity for Pombal to alter Portugal's trade imbalance with Great Britain. He implemented a number of mercantilist policies, including import taxes against British merchants to fund the reconstruction efforts, which violated the British-Portuguese trade agreements (Maxwell, 1968). He also set up Portuguese monopoly companies along the British model to give them an advantage, or equal footing, with their British counterparts (Maxwell, 1968). As noted earlier, this primarily affected Jesuit traders, but Pombal also used the reconstruction efforts to bolster Portugal's domestic industry by privileging Portuguese goods, merchants, and workers. These efforts did not have an immediate impact, but over time they reduced the influence of British merchants in Portugal. Thus, we see at this early stage the pressures that influenced a proto-embedded liberal compromise that characterized the later economic period – encouragement of international trade combined with the protection of key national industries and classes.

Concern over British trade relations affected the Portuguese response to Britain's offer of assistance. While Portuguese authorities accepted British aid, they engaged in what we might now term as delaying aid by increased inspection and confiscation. Upon arrival in Lisbon's badly battered port, ships carrying disaster relief had to wait to unload their cargo until Portuguese authorities had inspected and approved the cargo. The Portuguese authorities were also concerned with ensuring that distribution occurred through entities approved of by the Portuguese government and not by British merchants. Similarly, ships leaving the harbour after delivering aid were inspected to ensure that no people or goods were leaving that had been prohibited (Paice, 2008:190).

While Pombal sought to alter the balance of trade with Great Britain, it was clear that both states prioritized their special relationship, which extended beyond trade. The two states had been staunch allies against France and Spain for a considerable time. Consequently, Britain's decision to send aid, and Portugal's decision to accept aid, reflected realist considerations – namely, the belief that it was necessary to restore Portugal's power, and its role in balancing the power of

France and Spain in Europe, as quickly as possible (Paice, 2008). The importance of power calculations and military alliances extended to other offers of disaster relief as well. Portugal's other ally and key trading partner, the free Hanseatic city of Hamburg, also sent significant amounts of aid (Shrady, 2008:46).

Spain and France, Portugal's longstanding enemies, also offered disaster aid, although these offers were more contentious. At the time, the relationship between Portugal and Spain was an uneasy peace. The states had a rather lengthy history of animosity, and Portugal had achieved independence again from Spain in 1640. Since then, power disparity and international alliance patterns complicated their relationship – England was allied with Portugal, while Spain allied with France. By the mid-1750s, England and France were in conflict on the North American continent, and their two allies on the Iberian peninsula faced a difficult and precarious situation. Portugal and Spain managed to stay out of war with one another during the early portion of the Seven Years' War, although in 1761 (6 years after the earthquake), Spain, with its French allies, invaded Portugal, in part to damage the English economy. Despite familial connections between the crown heads of states, relations between Portugal and Spain were not friendly nor did they have extensive trade relations, despite their geographic proximity (Paice, 2008). The Portuguese king initially declined Spanish offers of assistance, although Portugal eventually accepted financial assistance worth around 34,000 pounds but refused any foods or goods (Paice, 2008:161). Paice speculates that it was familial connections that ultimately persuaded the Portuguese king to accept some aid from Spain (Paice, 2008).

France also offered assistance, although relations between France and Portugal were antagonistic due to the conflict between England and France. There were reports that the Versailles court received the news of the Lisbon earthquake with some merriment and that French ships prevented the arrival of English disaster aid (Shrady, 2008:46). Nonetheless, the French ambassador did offer assistance from France, which Pombal declined (Paice, 2008:161; Shrady, 2008:46). Both Paice and Shrady contend that the refusal of French aid was due to animosity of France toward England and Portugal.

Alliance patterns and considerations of military power very clearly influenced decisions on international disaster relief at the outset of the practice, although they do not explain the actions of all relevant actors in this case. Religious beliefs continued to influence some actors, most notably, the fiercely Protestant Netherlands. The Netherlands did not offer assistance to Portugal and publicly justified its position by labelling the earthquake as punishment from God (Hutchinson, 2000:5). At the time, there was little reason to think the Netherlands would not assist Portugal – they had a strong trade relationship, they fought on the same side in the War of Austrian succession in the 1740s, and both remained neutral early on in the Seven Years' War. According to Shrady, the Netherlands' decision was not a product of international alliance structures or trade relations but of its Dutch Calvinist heritage, which had condemned Portugal's "superstition, idol worship and popery" (Shrady, 2008:46). Religious ideology, rather than realist power considerations, remained influential in the international order of 1755.

The last ideological strand I explored in the previous chapter, and that will occupy much more space in the remainder of this chapter, is humanitarianism. Although beliefs about how to respond to the suffering of others had begun to change in the early 1700s (Halttunen, 1995), a clearly defined humanitarian ideology and movement did not emerge until nearly a century after the Lisbon earthquake. The Lisbon tragedy played a significant role toward spreading a basic belief central to humanitarianism: that alleviating suffering was a mark of civilization (Rozario, 2003). Concern with those suffering in other states emerged, in part, from a changing belief that international order rested on shared values of civilized states, at the time mostly limited to European states.

Writing of the 1755 Lisbon earthquake, Vattel argued that providing assistance was a "duty of humanity" among "civilized" nations (read "European") (De Vattel and Chitty, 1849:136). Vattel's appeal to "civilization" and "duties of humanity" was not novel; it was a common narrative that formed the basis for an international (European) order (see Gong, 1984; Bowden, 2004). Consequently, the same arguments did not apply to disasters around the globe, even in the vast colonial holdings of the European empires – it was a European humanitarianism. It was not until the following century that the concept of "humanity" expanded beyond the European continent (Finnemore, 2004). Although the decisions made in Great Britain and Hamburg to offer assistance to Portugal were rooted in trade relations and alliances in Europe, humanitarian sentiments played a role (Paice, 2008). This is not to suggest that humanitarian sentiment was dominant in Britain. Indeed, prominent clergy in Britain, such as John Wesley, depicted the disaster as punishment from God and an indictment on Catholicism (Shrady, 2008:133).

The combination of these ideological positions resulted in a system of disaster relief that persisted until the late 1800s. This system, if one could call it that, reflected an unstable balance between distinct ideological belief systems of modernity, state sovereignty, market economics, international power balances, and an emergent humanitarianism. The system entailed disaster relief as a state-led, bilateral, and ad hoc response to disasters that mirrored international alliance patterns and trade relations in Europe. It was not until the late nineteenth century that humanitarian entrepreneurs sought to construct a new international disaster relief system based on a rebalancing of the ideological foundations that had given rise to the existing system.

The Battle for the Red Cross

For nearly a century, the type of disaster relief seen in the aftermath of the Lisbon earthquake remained the norm; this changed in the mid-1800s with a growing humanitarian movement. Although humanitarian sentiment was evident in 1755, it was not until the 1860s that the humanitarian movement gained significant traction. At this time, a large number of humanitarian, philanthropic organizations were starting to emerge, in part to deal with the negative consequences of industrial capitalism. At this time, humanitarians began to "discover hunger" and other social ills (Croll, 2011).

Among the various charitable organizations that sprung up around this time, the Red Cross movement most profoundly contributed to the rise of humanitarianism in the form of international humanitarian law and in the realm of disaster relief. The International Committee of the Red Cross (ICRC) was established in 1863 to provide humanitarian assistance for war combatants and to promote legal restrictions on war (Meron, 1987). Henry Dunant, the ICRC's founder, actually envisioned an expansive humanitarian organization that also provided relief in the aftermath of disaster as well as war (IFRC, 2009:3), although most Red Cross societies focused exclusively on war relief efforts. In fact, the ICRC was hesitant to expand its mandate beyond wartime activities to include things such as public health or disaster relief. However, the Red Cross maintained a strong sovereigntist orientation, and national chapters were often closely aligned with their home state and operated with a relatively high degree of autonomy from one another and the ICRC. Consequently, some national chapters expanded their mandate to include disaster relief.

While the French Red Cross appears to have been the first to provide disaster relief (in response to the 1876 Paris floods), it was the American Red Cross (ARC, 1924), under the leadership of Clara Barton, that pushed the Red Cross fully into disaster relief, including international relief. Responding to domestic disasters beginning in 1881 (Gilbo, 1981:1), the American Red Cross (ARC) first contributed internationally in 1890, in response to famine in Russia. Due to the prevalence of disasters in the United States (compared to the relative paucity of disasters in Europe), the ARC quickly developed expertise in disaster relief and played a seminal role in international disasters, including the famous 1908 Messina earthquake (Moorehead, 1998:99).[5] Indeed, it was the effectiveness and expertise of the ARC's response in Messina that prompted Giovanni Ciarolo to propose an international disaster relief body to the League of Nations.

A key reason for the success of the humanitarian movement, and the Red Cross in particular, was its connection to the rise of a "sensationalist mass culture" (Rozario, 2003:418). As Halttunen demonstrates, humanitarianism relied on a spectacle of pain and sympathy (1995:307). The American Red Cross learned early the advantage of appealing to the American population with shocking images and stories of suffering (Rozario, 2003). As the ARC and the other national Red Cross societies shifted to wartime activities rather than disaster relief during the World War I, they capitalized on the brutality of the war to increase the humanitarian appeal and their profile. The Red Cross movement emerged from World War I "strong, prestigious and internationally respected" (Moorehead, 1998:257), making it the ideal platform from which disaster relief entrepreneurs could push for the institutionalization of disaster relief (on institutionalization, see Finnemore and Sikkink, 1998).

The ARC, in particular, was in a strong position to lead this movement. As the first Red Cross society to provide international disaster relief, the ARC could credibly project itself as the world leader in disaster relief expertise (Hutchinson, 1996), and in 1918 it even produced the first handbook on how to respond to natural disasters (Deacon, 1918). By the conclusion of World War I, the ARC was the largest Red Cross society in the world, with nearly a third of the US population listed as members. Its president, Henry Davison, and the ARC also enjoyed strong support from President Woodrow Wilson, who financially supported Davison's vision and actively promoted the Red Cross to other world leaders (Hutchinson, 2000).

While the ARC started as an independent, volunteer body under Clara Barton, like other Red Cross societies, it was co-opted by the state. In 1905, the ARC came under the control of the US Congress – although its funding came from volunteer contributions (Rozario, 2003:342). Consequently, the ARC was as an effective arms-length, quasi-governmental body tied to American foreign policy goals. To this end, the ARC, Davison, and President Wilson shared a vision of a "peacetime Red Cross". This Red Cross movement would include an expanded Geneva Convention; the establishment of national Red Cross societies in all states; and a new, broader mandate that would see the Red Cross engage in public health initiatives, eradication of disease, and disaster relief, in addition to its wartime responsibilities (Hutchinson, 1996:289). Davison and Wilson envisioned a voluntary philanthropic model of disaster relief that remained a bilateral activity and that benefitted states with the material capacity and expertise to participate.

The American proposal set it in conflict with the ICRC, who claimed control over the Red Cross movement and interpreted Davison's proposal as an attempt to usurp leadership over the Red Cross movement. The ICRC's humanitarian ideology was (and is) based on independence from states, impartiality, and neutrality, to be guaranteed by the neutral Swiss Committee. The ICRC insisted on a narrow mandate for the Red Cross: to alleviate suffering during war and to promote and expand international law on interstate war (Buckingham, 1964; Forsythe, 2005:36). In short, the ICRC sought to maintain a rigid barrier between humanitarianism and international politics – not to radically alter the international order but to limit the conduct of states during war through an independent and neutral International Non-Governmental Organization (INGO).

The United States, on the other hand, had a very particular view of humanitarianism, which was consistent with the distribution of power internationally. American officials saw their proposal for the Red Cross as the "natural outgrowth" of American leadership in the same way that the League of Nations was the "logical consequence" of America's victorious intervention in World War I (Moorehead, 1998:261). Davison suggested that "the people of America are in a position to do the most... are looked to for the most" in ending suffering (Davison, quoted in Moorehead, 1998:260). Upon realizing that the ICRC would not adopt the American proposal, Davison promoted the establishment of a second international Red Cross body – the League of Red Cross Societies (LRCS).

American diplomats framed the LRCS as an example of American leadership but also as a "truly international" and representative body distinct from an ICRC framed as "exclusive, elitist and Swiss" (Moorehead, 1998:259; Forsythe, 2005:36). The internationalist frame focused on empowering national Red Cross societies connected to national governments. Ideologically this simultaneously reasserted the sovereignty of states as well as the importance of American leadership in the emerging humanitarian order against the Swiss model, where decisions were made by a small committee in Geneva. The articles of association for the LRCS nicely demonstrate the values central to this vision of humanitarian order: "national Red Cross societies in all states"; a mandate to "improve health, prevent disease, and mitigate suffering"; a call to "bring the benefits of the fields of science and medicine to humanity"; and to "coordinate relief work in the event of national and international disasters" (Hutchinson, 1996:304).

The LRCS was a key institution in the civilizing mission that promoted Western societies as the model for development, the benefits of scientific knowledge and technological progress, and American forms of social and economic organization that emphasized a voluntary charitable model of disaster relief. Davison's proposal maintained and reinforced the existing practice of promoting the interests and power of donating states by emphasizing the importance of bilateral relationships between states through their respective national Red Cross societies. The LRCS enabled states to use disaster relief as a means to accrue and exert power internationally. The LRCS was to act only as an overarching body charged with promotion of the Red Cross and relief coordination functions. In this system, national societies could choose whether to participate in an overseas disaster relief program and how much assistance to provide, but they were not required to respond to any particular disaster nor were individual societies under the auspices or control of the larger body (Buckingham, 1964:65).

The LRCS system, which is now the International Federation of Red Cross and Red Crescent Societies (IFRC), and its founding principles continue to form the basis of international disaster relief. The United Nations, which I discuss further in the next section, attaches special importance to the Red Cross; the seminal 1965 United Nations General Assembly (UNGA) Resolution 2034 called on all member states to establish "national Red Cross or Red Crescent societies" (UNGA Resolution 2034/1965). Katoch (2003:49) contends that working with the Red Cross or the Red Crescent is one of the "four foundations of UN General Assembly resolutions on disaster relief". Given its prominence in the international humanitarian order, the IFRC is the largest humanitarian aid organization in the world, consisting of 189 national societies; in 2010 the IFRC spent over $515 million (US dollars) in humanitarian aid (IFRC, 2009). Although international disaster relief now flows through a number of channels besides the Red Cross societies, the Red Cross occupies a special position in global civil society, and disaster relief remains primarily a bilateral activity (Fischer, 2003; Fisher, 2007; Katoch, 2003).

The International Relief Union

The LRCS was not the only model of international disaster relief proposed in the early twentieth century. As the ICRC opted out of disaster relief, the president of the Italian Red Cross, Giovanni Ciarolo, sought to institutionalize disaster relief through the League of Nations. Ciarolo envisioned a very different form of international disaster relief from the ARC model and pursued the creation of an International Relief Union (IRU). As it had with Davison's overtures, the ICRC rebuffed Ciarolo, who then opted to push for disaster relief through the League of Nations (Macalister-Smith, 1986). Ciarolo was thoroughly modernist and saw the expansion of humanitarianism and disaster relief as an important civilizing mission. In contrast with Davison, Ciarolo proposed a different resolution to the tensions with state sovereignty and liberal economic order, advocating for compulsory, universal, subsidized disaster insurance as the primary mechanism for disaster relief (Clary, 1985; Landis, 1998; Davies, 2013). He supported multilateral assistance and challenged voluntary, philanthropic humanitarianism by proposing compulsory annual assessments, to be distributed by the ICRC (Ciarolo, 1922a, 1922b).

As the ARC and ICRC contested the ideological foundation of humanitarianism, Ciarolo's proposal presented a much more profound challenge to realist and economic liberal ideology. Consequently, the US government and the ARC opposed Ciarolo's proposal. Acting as the American representative on this issue (since the United States was not a member of the League of Nations), the ARC strongly opposed compulsory financial obligations, which it referred to as a tax. Compulsory taxation, the ARC suggested, raised questions about how to collect this tax, control over redistributing the funds, and whether countries with higher risk of natural disaster would be taxed equally as those with lower risk. The ARC reasserted its vision of disaster relief as voluntary and charitable, asserting "the altruistic spirit [of the Red Cross] would be lost and it would become merely a dispenser of money collected by compulsory taxation" (ARC, 1924). In essence, the United States opposed Ciarolo's proposal because the dominant powers in the United States were sceptical of liberal internationalism and world government, preferred a liberal economic order based on "private capitalism and philanthropy", and saw bilateral disaster relief as a means to expand its power and sphere of influence (Hutchinson, 2000:30–31). Ciarolo's vision represented a more drastic rebalancing of the ideological foundations of international order and disaster relief.

Ultimately, the League of Nations endorsed an IRU that bore little resemblance to Ciarolo's initial proposal. The committee excluded human-made and other disasters that could have been prevented "by prudent administration" (Preparatory Committee, 1925:1259), and rejected creating a positive international obligation on states to participate in disaster relief (Preparatory Committee, 1925:1260). The committee dismissed the idea of compulsory insurance on the grounds that setting insurance premiums requires an estimation of risk that in the case of natural disaster are "unforeseen by definition" (Preparatory Committee, 1925:1261). On the matter of compulsory financial contributions, the committee concluded that enforced contributions would actually undermine humanitarian feelings of generosity on which it relied, and that "more will be gained from freedom than obligation" (Preparatory Committee, 1925:1264). The final draft rejected an expansive conceptualization of disaster, a commitment to a universal right of disaster relief, the idea of disaster insurance, and fixed financial commitments from individual member states. Instead, it prioritized state sovereignty and consent (IRU Convention, Articles 3 and 4; Macalister-Smith, 1986:365; Fidler, 2005:463) and established an initial compulsory fund (Article 9), maintained by voluntary funding (Article 11 and 12; Macalister-Smith, 1986:366–368; Hutchinson, 2001:259). The only feature of Ciarolo's proposal that remained was the distribution of state contributions through the Red Cross (McAlister-Smith, 1986:365).

The IRU had a short, largely unproductive life span. Created in 1932, it responded to two relief operations in 1934 and 1935, and began to compile preliminary data on disaster frequency (Macalister-Smith, 1986:370; Walker and Maxwell, 2009). By 1938, the LRCS was not participating in the IRU, and by 1943, the IRU stopped functioning. Although the IRU technically existed through the post–World War II period, it contributed to no further disaster relief operations and was dissolved in 1966 (Macalister-Smith, 1986:372).

The impact of the IRU, however, extends well beyond its actual performance. Not only did contestation over the IRU profoundly shape future disaster relief initiatives, it is also highly instructive. Ciarolo's initial proposal represented a significant challenge to key aspects of international political and economic order. It attempted to resolve the ambiguity over state sovereignty by removing significant aspects of disaster relief from state control through universal rights to relief and to disaster insurance, by empowering the League of Nations to collect compulsory fees from member states and to redistribute those funds through an independent supra-state organization – the ICRC. Ciarolo's emphasis on compulsory contributions and the distribution of aid through a multilateral institution challenged the prominence given to states and undermined the leadership and influence of individual states, most notably the United States.

Ciarolo's proposal altered the conceptualization of humanitarianism from a voluntary, charitable activity that supported the basic tenets of a liberal economic order, to a compulsory, rights-based activity. In essence, it undermined the American view of economic order in which the state played a limited role in the redistribution of wealth and that resolved the problem of moral hazard by limiting the amount of collective assumption of risk. The ideas that most drastically departed from the existing political and economic order, such as a universal right to relief and universal insurance schemes, were excluded early from serious consideration. Although some of these issues would arise again under the United Nations' efforts to create a disaster relief mechanism, the domain of contestation had been severely narrowed through the process of creating the IRU.

Post–World War II Developments in Disaster Relief

Until the 1990s, the most significant aspect of the UN's involvement in disaster relief was the incremental increase in its capacity to coordinate and facilitate relief operations. This was not the product of grand ideological battles that characterized the establishment of disaster relief in the early twentieth century but of incremental changes and mission expansion on the part of institutions created for other purposes. To be sure, there was contestation over the UN's involvement in disaster relief, but the most significant disputes were already settled in the debates over the LRCS and IRU. It was not until the 1990s, with the International Decade for Natural Disaster Reduction (IDNDR), which I deal with in Chapter 4, that prompted renewed contestation over the management of natural disaster. These negotiations and agreements post-IDNDR demonstrate renewed ideological contestation that has challenged the centrality of modernism, liberal economic order, realism, and humanitarianism.

Hannigan contends that there have been four significant areas of contestation, all of which correspond to the ideologies explored here. The first is how to reduce vulnerability: by addressing social systems or through technical innovation (modernism; economic liberalism). The second pertains to the relative weight given to life-saving assistance relative to investment in development and prevention (humanitarianism, economic liberalism). Hannigan's third area of contestation concerns movements to maintain a politically neutral view of relief versus a human-rights–centred approach that challenged the emphasis on impartiality and neutrality (humanitarianism). The fourth area revolves around the debate over pursuing disaster risk reduction in all development initiatives or focusing on development goals (economic liberalism) (Hannigan, 2012:36). While I do not have sufficient space here to trace each of these points of contestation, it is worthwhile to trace how states and the UN responded to ongoing disasters and these points of contestation.

As noted, the UN did not resuscitate the IRU after World War II. In the immediate aftermath of that war, the UN focused on rebuilding, solving the immediate refugee crisis, and handling the emerging confrontation between the United States and the Soviet Union. States addressed natural disasters through the two existing mechanisms: bilateral ad hoc state assistance and the national Red Cross societies. The first significant attempt to institutionalize disaster relief at the UN occurred in the early 1960s. The 1965 UNGA Resolution 2034 specified that all states should develop disaster response capability, establish a national Red Cross or Red Crescent society, inform the UN of disaster relief they offer other states, set up a provisional fund under the direction of the Secretary General, and increase the UN's coordination function (UNGA Resolution 2034, 4). The structure of international disaster relief endorsed by the UN mirrored that of the IRU and reinforced its basic principles, with the primary role for the UN that of coordination. Enhancing coordination became the departure point for future UN resolutions on disaster relief through the 1990s and even today.

The problem of coordination emerged from the explosive growth in the number of NGOs and states participating in disaster relief (see Barnett and Weiss, 2011), as well as the proliferation of UN agencies related to disaster relief. Numerous UN specialized agencies originally developed to address post-war problems in Europe incrementally expanded their mandates and developed areas of expertise that were relevant in the context of disaster relief, much as the Red Cross societies had done in the previous century. Consequently, these agencies worked alongside state aid organizations, the Red Cross, and a growing number of NGOs (Green, 1977:30–31). The growing number of actors involved in disaster relief led to intensified pressure for greater coordination.

In 1971, UNGA Resolution 2816 established a coordinating agency with modest funding: the United Nations Disaster Relief Office (UNDRO). The primary function of UNDRO was to "mobilize, direct and coordinate relief activities" from voluntary contributions, though the coordinator was allocated $200,000 per year for emergency assistance, with the amount received by a single country capped at $20,000 (UNGA Resolution 2816:XXVI:85-5). The UNDRO started with a vague mandate, small staff, and minimal budget, although it did establish a working relationship with the United Nations Development Program (UNDP) and member states, and worked alongside other UN agencies and major humanitarian NGOs (Green, 1977:34).

Despite the modest success of the UNDRO, disasters produced a soaring number of casualties, refugee flows, and economic costs through the 1980s and 1990s (Crisp, 2007). In addition to prompting calls for strengthening the UN's coordinating function, these increases gave rise to greater concern with the domestic affairs of states that experienced disasters, mainly developing states in the South, portrayed by many as a threat to international order (Barnett, 2000). Rather than addressing the deeper political and economic problems that produced disasters, the international community promoted humanitarian assistance as a substitute for resolving core problems and a mechanism for containing potential migration flows (Duffield, 2007; Keen, 2008:119; Donini, 2010). Aid donors sought increased access to victims of disaster in their home states, simultaneously placing greater restrictions on international mobility to ensure displaced persons did not seek protection in third countries.

Just as the end of World War I ushered in an era of profound change and optimism for international intervention that gave rise to the LRCS and IRU, the end of the Cold War signalled a shift that would empower the UN to enforce international peace and security and expand humanitarian assistance. As it happens, the first post–Cold War military intervention had a profound impact on disaster relief and humanitarian assistance. The Persian Gulf crisis demonstrated both the capacity of the UN to act collectively in the maintenance of international peace and security free of Cold War constraints and its ineffective coordination of disaster response (Crisp, 2007:485). The use of no-fly zones in Iraq to protect Iraqi citizens after the Gulf War represented a significant shift in violating sovereignty in the name of humanitarian assistance.

It was in this environment that the UN General Assembly attempted to strengthen the UN's disaster relief regime. In 1993, UN General Assembly adopted Resolution 46/182, the most significant effort to establish international disaster relief since the creation of the IRU. Negotiations over the resolution demonstrated both the international optimism following the end of the Cold War and the concerns exposed by the Persian Gulf War. The primary point of contestation surrounded the ideological tensions between humanitarianism, state sovereignty, and economic liberalism.

One of the primary functions of humanitarianism, since its inception, was to ameliorate some of the worst effects of capitalist economies. However, the need for aid continued, and increasing humanitarian assistance globally after disasters exposed tensions in the global liberal economic system, namely, its failure to alleviate the problem of poverty and underdevelopment, which contributed to disasters. As the number of less-developed states joining the UN increased, the problem of development was increasingly put to the fore, including in discussions on disaster relief. From this perspective, humanitarian relief was not a substitute for development. As Hannigan notes, this introduced several points of disagreement. In the negotiations over UNGA Resolution 46/182, developing states emphasized the importance of development funding alongside enhancing the protection of state sovereignty.

Northern states, led by the twelve European states sponsoring the resolution, understood the problems of disaster relief through the lens of state sovereignty and the inability to provide humanitarian assistance wherever the need arose and with minimal resistance from receiving states. These states proposed a new resolution designed to increase access to victims of disasters in their home states, essentially endorsing a limited form of sovereignty in which the UN would be free to intervene in cases where states were unable to protect people after a disaster. The Group of 77 (G77), which represented Southern developing states at the UN, understood disasters through the frame of economic underdevelopment and called for the preservation of non-interference as the foundational principle of state sovereignty – alongside increased development funding. These divergent views of disaster came to the fore during the negotiations of UNGA Resolution 46/182.

UNGA Resolution 46/182

The division between the Northern states and the G77 negotiating disaster relief reflected the growing North-South divide in the UN that emerged with the process of decolonization and ongoing Southern concerns about economic liberalism (Peterson, 2006). The two primary points of disagreement focused on the importance of non-interference as a constitutive component of state sovereignty (Smith, 1999:189) and the neglected relationship between disasters, underdevelopment, and liberal economic principles (Middleton and O'Keefe, 1997; Wisner, 2004; Strömberg, 2007). By all accounts, the negotiations over UNGA Resolution 46/182 were very difficult and intense and were understood largely in terms of a divide between North and South (Van Schalk, 1991; Awoonor, 1992).

The European states that sponsored the original resolution sought to ensure easier access to those in need of humanitarian assistance by depicting sovereignty as limited or tied to responsibility. The representative for Malta, speaking for the European sponsors of the resolution, criticized states that used sovereignty to prevent access to humanitarian assistance claiming this violated international norms "of which the concept of state sovereignty is an important but not exclusive element" (Camilleri, 1991:27). In contrast, for the G77, state sovereignty was the inviolable foundation of the international political order. The chair of the G77 claimed that discussion of the proposed reform would proceed only on thebasis of "respect for the fundamental principles of national sovereignty, on the basis of request and consent of the affected country and the allocation of new and additional financial resources" (Awoonor, 1991b). Throughout the negotiations, states from the G77 consistently rejected a rights-based approach, rooted either in a human right to assistance or the UN's right to intervene, and emphasized that they would support the resolution only if it fully protected an absolutist conception of state sovereignty and if consent from affected states was required prior to any assistance (Baba, 1991). This mirrors the reluctance of American and British negotiators at the League of Nations, who also resisted a rights-based approach to disaster relief, although on the grounds that it would create a positive obligation on states.

The G77 framed disaster assistance in terms of the international economy, stressing the need for increased financial resources for disaster prevention, mitigation, and relief, and a greater financial commitment to development (Awoonor, 1991a; Ehlers, 1991). Some members of the G77 called for reform of major economic institutions such as the General Agreement on Tariffs and Trade (GATT), the International Monetary Fund (IMF), and the World Bank (de Domingues, 1991). Numerous G77 members stressed that, while coordinating disaster relief was important, the UN should support speedier development, scientific research, and training of personnel (Elarby, 1991). In short, the G77 sought to construct an international political order founded on inviolable state sovereignty combined with a fundamental change to the international economic order that produced vulnerabilities and that restricted the transfer of resources and knowledge necessary to reduce vulnerabilities to natural and human-made hazards. These same tensions are now evident in negotiations on climate change and disaster risk reduction (DRR), which I address in the next chapter.

For their part, Northern states balked at increasing compulsory funding for development, technical assistance, or humanitarian assistance. Although the UN system of disaster relief had consisted of a small compulsory fund since 1971, Northern states maintained that humanitarian assistance was a voluntary, charitable activity. Mirroring its position on the IRU in the 1920s, the American representative insisted that voluntary contributions were a "more effective" way to finance humanitarian assistance (Moore, 1991). The Japanese delegation stressed that humanitarian assistance must remain voluntary as a "matter of principle" (Hatano, 1991).

The resolution that the UN General Assembly eventually adopted represented a compromise position that, for all intents and purposes, maintained the status quo: strong emphasis on state sovereignty, voluntary financial commitments, and the continued separation of disaster relief and development (Smith, 1999:191). At its core, UNGA Resolution 46/182 reasserted two foundational principles of disaster relief[6]: humanitarianism and state sovereignty. The first two clauses of the resolution stress the importance of humanitarian assistance for victims of disaster and that it "*must be* provided in accordance with the principles of humanity, neutrality and impartiality" (clause 2). The third and fourth clauses articulate the importance of state sovereignty, noting that affected states have the primary responsibility to assist those affected by natural disasters and that the "sovereignty, territorial integrity and national unity of States *must be* fully respected" by those providing assistance (clause 4). These are the only "musts" in the resolution; other clauses in the resolution "call on", "urge", or "request" states to contribute to economic development or technology transfers or note that certain principles such as sharing expertise "should be" respected. The resolution makes no mention of the right to intervene or a right to humanitarian assistance for victims, nor were states affected by a disaster required to accept assistance. Rather, the resolution "called on [states] to facilitate" the implementation of relief by local and international actors.

In addition to reaffirming the political and economic foundation of humanitarian disaster relief, UNGA Resolution 46/182 also prioritizes humanitarian assistance and disaster relief over prevention, mitigation, development, and technology transfers. This is evident by the institutions created to bring UNGA Resolution 46/182 into effect, all of which exclusively pertain to emergency humanitarian assistance. The resolution noted that states should assist other states with prevention, early warning, technical assistance, and increased development funds, but only by specifically created or modified disaster relief institutions. The resolution established a consolidated appeals process (CAP) to prevent multiple appeals for charitable funding for a single disaster, and a central emergency fund to support initial relief efforts prior to the influx of voluntary contributions. The resolution also created an inter-agency standing committee to increase cooperation between the key UN agencies, country-level coordinators to facilitate coordination between the UN and affected states, and an emergency relief coordinator. The funding mechanism of the new disaster relief regime was nearly identical to that of the IRU: a small, compulsory emergency fund supplemented by voluntary funds gathered through a public consolidated appeals process. Although the G77 sought to tie disaster relief to increased development funding, no further mechanisms were set up to facilitate technology transfers or development assistance.

As one might expect, the adoption of UNGA Resolution 46/182 did not permanently resolve the tension between Northern states advocating for a right of intervention and committed to the existing liberal international economic order, and Southern states protecting a strong version of state sovereignty and addressing the problem of development assistance and technology transfer. The issue came to the fore again in negotiations on DRR and in 2008 in response to Myanmar's refusal of aid after Cyclone Nargis. In response to this crisis, some Western countries advocated forced humanitarian intervention to assist victims of the disaster based on a right to relief (Hardcastle and Chua, 1998) or as part of the responsibility to protect (Saechao, 2007).

UNGA Resolution 46/182 also did not permanently resolve the ambiguities of disaster relief and economic liberalism. It did maintain the separation of economic development, technology transfers, and liberalized trade regimes from that of humanitarian assistance, which were to be negotiated in other forums. Issues surrounding prevention, mitigation, and early warning were also separated from disaster relief through the establishment of a separate mechanism for frameworks for disaster prevention and preparedness. Although scholars and practitioners argue in favour of moving from voluntary funding for UN agencies participating in disaster relief to compulsory assessments (Walker et al., 2005), thus far, this has not been taken up at the UN.

The adoption of UNGA Resolution 46/182 reinforced the status quo in one other way as well: both sides endorsed modernism, holding Western societies as the preferable model of social organization and endorsing scientific approaches to disaster management, including administrative organization, early warning systems, technological development, and technology transfers. During negotiations, the representative for France explicitly called for a scientific approach to disasters and a rejection of "metaphysical fatalism and media exhibitionism" (Kouchner, 1991). For their part, the G77 pushed for greater and accelerated investments in these initiatives, some calling for an accompanying resolution for comprehensive technological cooperation (Marker, 1991). As noted, the final resolution clearly reinforced a modernist understanding of disasters as preventable through technological solutions.

Summing up the UN's approach to disaster relief, which now spans close to 50 years, Katoch (2003:49) identifies four foundations: (1) the primary responsibility for assisting victims of natural disaster falls on the affected country, (2) the UN system's role is to "coordinate and facilitate" international offers of assistance, (3) the UN disaster response system requires continued efforts for improvement, and (4) the UN cooperates with the Red Cross movement.[7] Generalizing further, I suggest that the UN system retained the basic principles enshrined in the IRU and the LRCS model: a state-based order in which states hold the primary responsibility for responding to disaster, a modernist conception of disasters privileging scientific and technical knowledge, a voluntary conception of humanitarian assistance, an economically liberal approach that emphasizes the importance of property protection and international trade, and a realist international order in which states can use disaster assistance as a form of power. Despite the plethora of NGOs and UN entities involved in disaster relief, the vast majority of disaster relief funds, operations, and relationships are bilateral (Fischer, 2003; Fisher, 2007:12).

If the basic principles of disaster relief went largely unchanged as a result of decades of negotiations at the UN, this period of contestation did contribute to a greater understanding that disaster relief, on its own, was a wholly inadequate solution to the problem of disasters. Consequently, scholars, practitioners, and policy makers have drawn increasing attention to reducing disaster risk. The UN declared the 1990s as a Decade for Natural Disaster Risk Reduction (DNDRR), and following on that, states, in cooperation with NGOs, have negotiated three sequential frameworks for disaster risk reduction: the Yokohama Strategy (1995–2004), the Hyogo Framework for Action (2005–2015), and the 2015 Sendai Framework for Disaster Risk Reduction (2015–2030). In the next chapter, I examine how the international negotiations over DRR reflect the ongoing instability of the ideological foundations of international order. As a novel framework for dealing with the problem of disaster, DRR introduces a number of contingent resolutions to the problems that disasters pose for international order – made necessary by the repeated failure of disaster relief.

To be sure, there continue to be modest changes in international disaster relief since the 1990s. I argue that the most difficult and contentious issues were carved off and subject to distinct processes of negotiation, namely, DRR, climate change, and disaster migration. This allowed the existing disaster relief system to continue its order-sustaining function without significant alteration. The primary concern motivating recent innovations on disaster relief concern improving coordination and information on disasters. For instance, the 2005 Humanitarian Response Review resulted in the UN implementing a "cluster system" to make relief more consistent and coordinated (Weiss, 2016:99). According to Weiss, the push for this "cobbled together" and "gap-filling" reform was due to repeated demonstrated failures of the UN humanitarian assistance in places like Darfur (Weiss, 2016:100). Despite this reform, Weiss reports, the existing system is still hampered by the structure of the disaster relief system whose origins, I have argued, date to the pre–World War I period. Serious reform, which for Weiss would include "consolidation, command, control and compulsion" rather than "coordination, cohesion, and complementarity" (2016:102–103), requires a fundamental shift in a number of the ideological foundations of the international order. Reforms of the UN disaster relief system did not arise from a substantial challenge to the existing international order and the role that disaster relief plays in sustaining that system but rather from an attempt to hold the existing system together in the face of repeated failures.

Disaster Relief Beyond the UN

One of the most impressive and innovative recent developments in disaster relief has occurred outside the halls of the UN. Since the late 2000s and early 2010s, civil society actors and private corporations have begun to marshal big data and social media networks to enhance national and international disaster response. The Digital Humanitarian Network (DHN) that Patrick Meier depicts in his seminal book illustrates how social media networks can use data collected from social media to assist the UN, NGOs, and state actors in disaster rescue and recovery efforts (Meier, 2015). For Meier, who spearheaded digital mapping of Haiti after the 2010 earthquake, new technologies have greatly increased the capacity of humanitarian civil society actors to provide disaster relief.

As civil society actors demonstrated the potential of social media to assist in the aftermath of disaster, private corporations began developing their own disaster relief tools. Given Google's advanced mapping expertise, it is not surprising that it is at the centre of many recent innovations. Google has developed a "person finder" for post-disaster scenarios, as well as a Google Crisis Map that helps responders "navigate the post-disaster terrain". In addition, Google established a Google Crisis Response team that helps NGOs deal with a wide range of crisis scenarios. Another Google initiative, in collaboration with the International Rescue Committee (IRC), is the Crisis Info Hub, which provides information on rights and services available to refugees based on their current location. Google is working with UNICEF and the World Health Organization (WHO) to map disease outbreaks to assist states and NGOs in the prevention of contagion. While it is not actually delivering relief or providing legal rights and protection, Google is increasingly involved in producing knowledge that helps with the coordination of disaster response entities and increasing their ability to connect with disaster-affected populations.

Facebook is similarly launching initiatives that put service providers in contact with those in need. Its "disaster maps" feature, developed in cooperation with UNICEF, IFRC, and the World Food Programme (WFP), uses Facebook data to show these emergency responders and relief organizations where people have migrated after disaster, thus allowing them to deliver relief where people are

(Deahl, 2017). This initiative will work in conjunction with its Safety Check tool, launched in 2014, which allows users in certain designated areas to inform friends and family that they were safe and/or not in the location of the disaster (Metz, 2016). To date, the safety check feature has been used in many natural disasters, as well as in response to mass shootings and terrorist attacks.

While these technologies are new, they illustrate the ongoing influence of humanitarianism, modernism, and economic liberalism as key ideological foundations of international order. The humanitarian aspect of this endeavour is at the forefront of Meier's "digital humanitarians", as well as Google's and Facebook's initiatives. In addition to rhetorically drawing on "humanitarianism", they all emphasize the voluntary, charitable nature of disaster relief. For Meier, alleviating the suffering of others globally is a defining characteristic of being human. He suggests that "these technologies actually make us *more* human" by facilitating the ability to turn humanitarian sentiment into global action (18). The voluntary nature of the commitment is evident in his call to use these technologies to turn what is, in essence, leisure activity to humanitarian action. In one section, he offhandedly comments that, with the right technological shift, "gamers could really save the world" (191). Google similarly reinforces a charitable, voluntary understanding of humanitarianism by presenting its own initiatives as charitable and emphasizing its work with nonprofits and volunteers.

In addition to reinforcing humanitarianism as a key aspect of international order, these public-private partnerships are also thoroughly modernist. The "big data" movement, of which these initiatives are a part, sustains belief in the ability of technological development to reduce human vulnerability to natural hazards. As with previous modernist initiatives, there are strong "civilizing" tendencies. Consequently, the primary policy challenge is not at all different from those proposed by previous generations – the spread of these technologies globally in the service of humanity. As Meier notes, only a small fraction of the global population is on these social media platforms, and those who are, are not representative of the population as a whole. For the full potential of these technologies to be realized, it is necessary to ensure their global spread. The potential negative impacts associated with these technologies – big data, global internet connection, and social media – are largely ignored. While Meier acknowledges the problems of "too much data" and "false data" as significant obstacles to overcome, the benefits are viewed as outweighing the costs, which are largely presented as easily remedied by a combination of technological development (artificial intelligence [AI] to replace human categorization) and government regulation. In this context, the role of governments is to facilitate the spread of these technologies and to encourage their use in disaster situations through public information campaigns (Meier, 2016:32).

The significance of the big data movement is difficult to assess at such an early stage. As it pertains to this book, one of the most obvious implications is to reassert modernist beliefs about technology and human progress, and to empower humanitarian actors – yet it also clearly has important implications for other foundations of international order, namely, state sovereignty and economic liberalism. Governments and inter-governmental organizations (IGOs) are not at the forefront of this shift. As Meier's work demonstrates, civil society has played an important part in the push for a technological solution to the problem of coordination. Increasingly, however, multinational corporations are developing these technologies and collecting big data, and bringing both to bear on the problem of disaster relief. The growing relevance of private corporations to disaster relief suggests changing beliefs about the authoritative place of the state and a further entrenchment of economic (neo)liberalism.

In terms of economic liberalism, the acknowledgement of the importance of private corporations in the provision of disaster relief – and the subsequent policy recommendation to facilitate their global spread and decrease regulation to unlock their full potential – signals a further entrenchment of the near hegemonic status of this ideology. As both Weiss (2016) and Therien and Pouliot (2006) note, the UN's willingness to embrace public-private partnerships signalled a radical departure from the UN's previous ambivalent position on economic liberalism. While the Bretton Woods institutions clearly shaped many aspects of the global order prior to the Global Compact, the balance between state-led and market-led economic growth was still hotly debated in the halls of the United Nations.

The Global Compact, reached in 2000, sought to bring private corporations into the UN network of humanitarian assistance, alongside states and NGOs (Ruggie, 2001). Until recently, this had little direct impact on disaster relief, as the Global Compact focused mainly on establishing corporate social responsibility as it pertained to sustainable development. Yet the Global Compact illustrated the UN's embrace of private corporations, which was a significant departure from its earlier position and resulted from the failure of pre-existing state-led approaches to development (Therien and Pouliot, 2006; Weiss, 2016). Therien and Pouliot note that the shift to "complex multilateralism" and public-private partnerships was based on the belief that "only private corporations have the technology and expertise required to deal with today's economic challenges" and that "markets provide the only hope of pulling billions… out of abject poverty" (Therien and Pouliot, 2006:63).

The extent to which the expansion of public-private partnerships can spur development and reduce poverty remains an open question (Weiss, 2016:182). Berliner and Prakash (2015) suggest that the Global Compact has had very little impact on corporate adherence to human rights and environmental protection standards, although they do not assess its overall impact on poverty alleviation. Beyond their actual effectiveness, other scholars note that public-private partnerships raise questions of democratic accountability and produce unintended side effects (see Schaferhoff et al., 2009), and/or may operate principally as a way to legitimize and spread a hegemonic capitalist system (Levy and Newell, 2002). For Stephen Gill, the rise of private authority is leading to "new constitutionalism" in which key public policy questions are determined by corporations and directed first and foremost to the protection of corporate rights and to prevent public backlash (Gill, 1998:23).

These questions certainly arise in the case of disaster relief, and it is important to keep in mind how the involvement of private corporations in disaster relief may be geared toward increasing profit, growing market share, or even encouraging governments to view the global expansion of these corporations as public goods. The philanthropic roots of these ventures, or the "bluewashing" of corporations through affiliation with the UN, speak to the various processes that corporations use to legitimize the basic tenets of economic liberalism. Without casting aspersions on the multiple motivations of these corporations, it is clear that major corporations such as Google, Twitter, and Facebook are becoming significant players in disaster relief and that in doing so they contribute to the legitimization of these corporations in particular and to economic liberalism (or capitalism) in general.

The most profound impact of private actors assisting with disaster relief may be to challenge the foundational status of the ideology of state sovereignty. As a reminder, I would note that authority is not necessarily zero-sum; a rise in private authority does not necessarily indicate a decline in state authority and the rise in public and private authority is often complimentary and interdependent. Nonetheless, nation-statism promotes the belief that the state has ultimate authority within a given territory, suggesting that when two or more authorities are in tension with one another, it is ultimately the state that determines which one is authoritative.

The authorization of state authority rests partially on the production of knowledge to make society legible and through the provision of basic public goods, including emergency response. The production of big data by private corporations increases their power to make societies legible and thus steer public policy. While governments have their own technologies to produce knowledge about disasters and vulnerability, the data produced by Google, Twitter, and Facebook have the potential to be significantly larger and more accurate than that available through the usual state-based mechanisms. This explains why state-based actors and IGOs are so interested in partnering with these organizations.

Following Weiss, I would suggest that it is still too early to determine the ultimate effect on belief in nation-state sovereignty. With regard to knowledge production, the disaster relief tools discussed here do not yet displace state forms of knowledge production, and they have faced a number of challenges. As Meier noted, state officials did meet with digital humanitarians and draw on their data, but it was only one among many sources of information they used to shape their responses. Moreover, the social media–based approach to disaster response resulted in a number of errors, false reports, and an excess of data. Similarly, Facebook's early move into disaster response demonstrated a selection bias: one of the first uses of Safety Check occurred after terrorist attacks in Paris, but it was not activated in Baghdad or Beirut, which suffered terrorist attacks the same day (Metz, 2016). In addition, the effectiveness of these tools requires a level of internet connectivity that is not common in many disaster-affected regions. This is not to suggest these are insurmountable problems: in both cases, Facebook has responded by changing the algorithm that activates Safety Check to draw on global information to activate the feature and by investigating how to ensure internet connectivity in the event of disaster (Metz, 2016). Nonetheless, Facebook does show that there are limitations that may erode the authority of these sources of information or at the very least caution against significant displacement of the state as the source of information related to emergency response.

More important, states may be able to successfully incorporate these sources of knowledge within the framework of the state. As noted in the previous chapter, the state did not establish a monopoly on the production of knowledge, it attempted to establish a monopoly on authorizing and legitimizing sources of knowledge production. In many cases, this meant taking on the role of knowledge production itself (the census is a prime example), but in many other cases it meant authorizing and legitimizing non-state sources of knowledge (schools, universities, research centres, media corporations). The state has various mechanisms to achieve this authorization: regulation, funding, citing, or simply using them in their own public declarations. Thus, it is possible that states will simply legitimize these initiatives by using and publicizing them and, in so doing, maintain their authority to authorize knowledge production by certain private-sector actors. An interesting case may emerge if and when a state that has limited their citizens' use of these private social media services is confronted by IGOs or civil society actors for limiting the effectiveness of their disaster response. It will be in instances of failure, on the part of the state or public-private partnership (PPP) initiatives, that the impact of the growing importance of PPPs on state sovereignty will be decided.

As far as it pertains to the provision of public goods, it also still too early to assess. In all of these cases, the new initiatives aim to improve the emergency response, most of which are state-based – and all of whom operate with state consent. If these technologies do improve the ability of the state to provide public goods central to its self-authorization, then these initiatives may be viewed as enhancing rather than limiting the authority of the state. We cannot overlook, however, that these technologies also facilitate civil society groups and NGOs to bypass the state as the source of knowledge on disasters and, in some cases, in the provision of relief. As the cases in Chapters 6 through 8 of this book show, in instances where the public perceives non-state entities as more effective or even as the primary responders over state entities, it may undermine the legitimacy of the state and the ability to self-authorize its authority.

Conclusion

This chapter has developed the case that the basic foundations for international disaster relief emerged in the early modern period in response to the Lisbon earthquake, and tensions between these foundations were contingently resolved during contestation over the LRCS and the IRU in the interwar period. Recent innovations around disaster relief coordination have also addressed the failures of existing beliefs, although their full impact – particularly on state sovereignty – is yet to be determined.

Disaster relief is unambiguously modern in its approach to disasters and entails a civilizing mission that justifies intervention in areas of the world prone to disasters. This interventionism, combined with the potential for redistribution of wealth and the collectivization of risk, presented a number of ambiguities in need of resolution. Political and humanitarian elites succeeded in giving prominence of place to states and an "allied" humanitarian global civil society, meaning that the humanitarian order would reflect more closely the political and economic interests of powerful states, most notably the United States (Sending and Neumann, 2006). This resulted in the institutionalization of disaster relief as a voluntary, philanthropic, and bilateral activity loosely coordinated at the international level through a semi-autonomous space of global civil society. It is also reinforced an economic liberalism that minimized compulsory redistribution and moral hazard associated with the collective assumption of risk associated with property ownership.

New fissures appeared in the subsequent decades due to the expansion of the United Nations and the decolonization process, prompting a North-South divide that exposed the ambiguities and tensions between humanitarian assistance, state sovereignty, and economic liberalism that were never fully resolved. The resolution to these new tensions ultimately reinforced the status quo by reaffirming state sovereignty; separating economic questions from that of disasters; and privileging disaster relief over prevention, mitigation, and reconstruction. The formal structure of the UN system bears a strong similarity to the IRU, created in 1927, although it is characterized by a much greater degree of complexity and administrative capacity, recently expanded through PPPs.

How does this system influence how states manage disasters and international offers of assistance? First, the structure of disaster relief is constitutive of the relevant actors in the system: actors that can accept or reject aid, and the entities that may legitimately engage in disaster relief. Thus, disaster-affected states are forced to assess the identities, interests, values, and intentions of a wide range of actors offering disaster assistance. Because the particular form of institutionalization of international disaster relief empowers states and global civil society to participate in disaster relief, governments must consider multiple offers of aid from a wide variety of actors, requiring states to take into account bilateral relationships, to assess whether offers of aid represent a humanitarian or political intervention, and to determine whether disaster relief undermines the states' perceived ability to protect its citizens.

The disaster relief regime also constitutes subjects in the international order in terms of modernity. Consequently, governments employ a standard of modernity in their decisions to accept or reject offers of aid. As the cases in subsequent chapters will demonstrate, both developed and developing states accept the modernizing standard of civilization and respond to offers of aid through this standard.

This chapter demonstrated the intricately intertwined relationship between disaster relief and economic liberalism, and that the current resolution to the tension between humanitarian assistance and economic liberalism is contingent and contested. Consequently, the last section of this book will demonstrate that governments often tie acceptance of disaster relief to other economic outcomes: acquiring development loans, granting loan relief, modifying or enforcing trade restrictions, altering trade negotiations, easing economic sanctions, and protecting domestic markets and businesses.

As I have argued throughout this chapter, the current disaster relief regime represents a contingent resolution to the ambiguities exposed in creating a humanitarian order in a system composed of competitive, sovereign nation states and a capitalist economic system. Because the current resolutions are contingent and contested, states encounter these contingent resolutions when responding to offers of aid. The decision to accept or reject aid reflects an ongoing process of resolving the ambiguities explored in this chapter. These contingent resolutions were further entrenched during the UN's development of the disaster relief regime, a key aspect of which was the development of a parallel track of disaster risk reduction, a topic examined in the next chapter.

Notes

1 According to the IFRC, there are 189 national Red Cross/Crescent societies.
2 For good overviews of the UN's current approach to coordination of humanitarian assistance – the cluster system – see Hicks and Pappas, 2006, and Jahre and Jensen, 2010.
3 There is a growing literature on the problem of coordination in the provision of humanitarian assistance (Byman, 2000; Harmer and Cotterrel, 2005; Balcik et al., 2010).
4 The typical scapegoats in Europe at this time – Jewish and Muslim populations – had already been largely driven out of Portugal.
5 Based on Moorehead's (1998) account, these early ventures were mostly funded by co-nationals and co-religionists of the affected groups in the United States, funnelled through the ARC.
6 There are 12 guiding principles in this resolution, but the first four identify the primary foundations of disaster relief.
7 Though the relationship between the Red Cross and UN disaster relief agencies has not always been overly close or effective, the Red Cross continues to enjoy special status in the UN disaster relief system (Green, 1977:33–7).

References

American Red Cross, 1924. Reply from the Government of the United States of America. Resolution XVI, Articles V and VI. Insurance of peoples against Calamities, etc., Geneva: League of Nations – Official Journal.

Awoonor, K., 1991a. Address to UN General Assembly, GA Resolution 46/182. In *46th Session General Assembly, Provisional Verbatim Record of the 78th Meeting*. New York: United Nations Department of Public Information.

Awoonor, K. (Uganda), 1991b. Address to UN General Assembly, GA Resolution 46/182. In *46th Session General Assembly, Provisional Verbatim Record of the 41st Meeting*, New York: United Nations.

Awoonor, K. (Uganda), 1992. Declaration of the Minister for Foreign Affairs of the Group of 77. In *46th Session General Assembly*, New York: United Nations Department of Public Information.

Baba, J., 1991. Address to the UN General Assembly, GA Resolution 46/182. In *46th Session General Assembly, Provisional Verbatim Record of the 78th Meeting*, New York: United Nations Department of Public Information.

Balcik, B., Beamon, B., Krejci, C., Muramatsu, K., & Ramirez, M., 2010. Coordination in humanitarian relief chains: Practices, challenges and opportunities. *International Journal of Production Economics*, 126(1), pp. 22–34.

Barnett, J., 2000. Destabilizing the environment—Conflict thesis. *Review of International Studies*, 26, pp. 271–88.

Barnett, M.N., & Weiss, T.G., 2011. *Humanitarianism Contested: Where Angels Fear to Tread*, London: Routledge.

Barnett, S.J., 2003. *The Enlightenment and Religion: The Myths of Modernity*, Manchester: Manchester University Press.

Berliner, D., & Prakash, A., 2015. "Bluewashing" the firm? Voluntary regulations, program design, and member compliance with the United Nations global compact. *Policy Studies Journal*, 43(1), pp. 115–38.

Bowden, B., 2004. In the name of progress and peace: The "standard of civilization" and the Universalizing Project. *Alternatives: Global, Local, Political*, 29(1), pp. 43–68.

Buckingham, C.E., 1964. *For Humanity's Sake: The Story of the Early Development of the League of Red Cross Societies*, Washington, DC: Public Affairs Press.

Byman, D., 2000. *Strengthening the Partnership: Improving Military Coordination with Relief Agencies and Allies in Humanitarian Operations*, Santa Monica: Rand Corporation.

Camilleri, V., 1991. Address to UN General Assembly, GA Resolution 46/182. In *46th Session General Assembly, Provisional Verbatim Record of the 42nd Meeting*, New York: United Nations.

Ciarolo, G., 1922a. *Proposal by the President of the Italian Red Cross Society*, Geneva: League of Nations—Official Journal, July 1922.

Ciarolo, G., 1922b. *Scheme for an International Organization to Assist in the Relief of Peoples Overcome by Disaster*, Geneva: League of Nations—Official Journal, November 1922.

Clary, B.B., 1985. The evolution and structure of natural hazard policies. *Public Administration Review*, 45(special issue), pp. 20–8.

Crisp, J., 2007. Humanitarian action and coordination. In S. Daws & T. Weiss (eds.). *The Oxford Handbook on the United Nations*, Oxford: Oxford University Press.

Croll, A., 2011. Starving strikers and the limits of the "humanitarian discovery of hunger" in late Victorian Britain. *International Review of Social History*, 56(1), pp. 103–31.

Cutler, A., 2009. Acts of god? Review of geology and religion: A history of harmony and hostility. *Nature Geoscience*, 2, p. 535.

Davies, A.R., 2013. *Saving San Francisco: Relief and Recovery after the 1906 Disaster*, Philadelphia: Temple University Press.

Deacon, J.B., 1918. *Disasters and the American Red Cross in Disaster Relief*, New York: Russell Sage foundation.

Deahl, D., 2017. Facebook's new disaster maps aim to improve how organizations respond to crises—The Verge. *The Verge*. https://www.theverge.com/2017/6/8/15761720/facebooks-disaster-maps-relief.

de Domingues, F.C. (Cuba), 1991. Address to UN General Assembly, GA Resolution 46/182. In *46th Session General Assembly, Provisional Verbatim Record of the 42nd Meeting*, New York: United Nations Department of Public Information.

DeVattel, E., & Chitty, J., 1849. *The Law of Nations: Or, Principles of the Law of Nature, Applied to the Conduct and Affairs of Nations and Sovereigns*, Philadelphia, PA: T. & J.W. Johnson.

Donini, A., 2010. The far side: The meta functions of humanitarianism in a globalised world. *Disasters*, 34(s2), pp. S220–37.

Duffield, M., 2007. *Development, Security and Unending War: Governing the World of Peoples*, Cambridge: Polity Press.

Dynes, R.R., 1999. *The Dialogue between Voltaire and Rousseau on the Lisbon Earthquake: The Emergence of a Social Science View*, http://udspace.udel.edu/handle/19716/435.

Economist, 2008. Fire and brimstone: The Great Lisbon Earthquake of 1755 was the Hurricane Katrina of its time. April 3, 2008.

Ehlers (Uruguay), 1991. Address to UN General Assembly, GA Resolution 46/182. In *46th Session General Assembly, Provisional Verbatim Record of the 42nd Meeting*, New York: United Nations Department of Public Information.

Elarby (Egypt), 1991. Address to UN General Assembly, GA Resolution 46/182. In *46th Session General Assembly, Provisional Verbatim Record of the 39th Meeting*, New York: United Nations Department of Public Information.

Fidler, D.P., 2005. Disaster relief and governance after the Indian Ocean Tsunami: What role for international law. *Melbourne Journal of International Law*, 6, p. 458.

Finnemore, M., 2004. *The Purpose of Intervention: Changing Beliefs about the Use of Force*, Ithaca, NY: Cornell University Press.

Finnemore, M., & Sikkink, K., 1998. International norm dynamics and political change. *International Organization*, 52(4), pp. 887–917.

Fischer, H., 2003. International disaster law treaties: Trends patterns and lacunae. *International Disaster Response Laws, Principles and Practice: Reflections, Prospects and Challenges*, Vol. 25, V. Bannon (ed.). Geneva: International Federation of Red Cross and Red Crescent Societies.

Fisher, D., 2007. *Law and Legal Issues in International Disaster Response: A Desk Study*, Geneva: IFRC.

Forsythe, D.P., 2005. *The Humanitarians: The International Committee of the Red Cross*, Cambridge: Cambridge University Press.

Gilbo, P.F., 1981. *The American Red Cross*, New York: Harper and Row.

Gill, S., 1998. New constitutionalism, democratisation and global political economy? *Pacifica Review: Peace, Security & Global Change*, 10(1), pp. 23–38.

Gong, G.W., 1984. *The Standard of "Civilization" in International Society*, Oxford: Clarendon Press.

Green, S., 1977. *International Disaster Relief: Toward a Responsive System*, New York: McGraw-Hill.

Halttunen, K., 1995. Humanitarianism and the pornography of pain in Anglo-American culture. *The American Historical Review*, 100(2), 303–34.

Hamblyn, R., 2008. *Terra: Tales of the Earth—Four Events That Changed the World*, London: Picador.

Hannigan, J., 2012. *Disasters without Borders*, Cambridge: Polity Press.

Hardcastle, R.J., & Chua, A.T.L., 1998. Humanitarian assistance: Towards a right of access to victims of natural disasters. *International Review of the Red Cross*, 38, pp. 589–610.

Harmer, A., & Cotterrell, L., 2005. *Diversity in Donorship: The Changing Landscape of Official Humanitarian Aid, Humanitarian Policy Group*, London: Overseas Development Institute.

Harris, S.J., 2010. Jesuit scientific activity in the overseas missions, 1540–1773. *Isis*, 96(1), pp. 71–9.

Hatano, (Japan) 1991. Address to UN General Assembly, GA Resolution 46/182. In *46th Session General Assembly, Provisional Verbatim Record of the 39th Meeting*, New York: United Nations Department of Public Information.

Hicks, E.K., & Pappas, G., 2006. Coordinating disaster relief after the South Asia earthquake. *Society*, 43(5), pp. 42–50.

Hutchinson, J.F., 1996. *Champions of Charity: War and the Rise of the Red Cross*, Boulder, CO: Westview Press.

Hutchinson, J.F., 2000. Disasters and the international order: Earthquakes, humanitarians, and the Ciraolo project. *The International History Review*, 22(1), pp. 1–36.

Hutchinson, J.F., 2001. Disasters and the international order-II: The International Relief Union. *The International History Review*, 23(2), pp. 253–98.

IFRC, 2009. *90 Years of Improving the Lives of the Most Vulnerable*, Geneva: International Federation of Red Cross and Red Crescent Societies.

IRU League of Nations, 1927. Convention establishing an International Relief Union.

Jahre, M., & Jensen, L.-M., 2010. Coordination in humanitarian logistics through clusters. *International Journal of Physical Distribution & Logistics Management*, 40(8/9), pp. 657–74.

Katoch, A., 2003. International natural disaster response and the United Nations. In International Federation of the Red Cross and Red Crescent Societies. *International Disaster Response Laws, Principles and Practice*, Geneva: IFRC, pp. 47–56.

Keen, D., 2008. *Complex Emergencies*, Cambridge: Polity Press.

Kendrick, T.D., 1955. *The Lisbon Earthquake*, Philadelphia, PA: J.B. Lippincott Company.

Kent, R.C., 1987. *Anatomy of Disaster Relief: The International Network in Action*, Sussex: Pinter Publishers.

Kouchner, B. (France), 1991. Address to UN General Assembly, GA Resolution 46/182. In *46th Session General Assembly, Provisional Verbatim Record of the 39th Meeting*, New York: United Nations Department of Public Information.

Landis, M.L., 1997. Let me next time be tried by fire: Disaster relief and the origins of the American Welfare State 1789–1874. *Northwestern University Law Review*, 92, pp. 967–1034.

Levy, D.L., & Newell, P.J., 2002. Business strategy and international environmental governance: Toward a neo-Gramscian synthesis. *Global Environmental Politics*, 2(4), pp. 84–101.

Macalister-Smith, P., 1986. Reflections on the convention establishing an International Relief Union of July 12, 1927. *Legal History Review*, 54, pp. 363–74.

Marker, J., 1991. Address to UN General Assembly, GA Resolution 46/182. In *46th Session General Assembly, Provisional Verbatim Record of the 41st Meeting*, New York: United Nations Department of Public Information.

Maxwell, K.R., 1968. Pombal and the nationalization of the Luso-Brazilian economy. *The Hispanic American Historical Review*, 48(4), pp. 608–31.

Meier, P., 2015. *Digital Humanitarians: How Big Data Is Changing the Face of Humanitarian Response*, New York: Routledge.

Meron, T., 1987. Geneva Conventions as customary law. *The American Journal of International Law*, 81(2), pp. 348–70.

Metz, C., 2016. How Facebook Is transforming Disaster Response | *WIRED*. Wired. https://www.wired.com/2016/11/facebook-disaster-response/.

Middleton, N., & O'Keefe, P., 1997. *Disaster and Development: The Politics of Humanitarian Aid*, London: Pluto Press.

Moore, (United States), 1991. Address to UN General Assembly, GA Resolution 46/182. In *46th Session General Assembly, Provisional Verbatim Record of the 41st Meeting*, New York: United Nations Department of Public Information.

Moorehead, C., 1998. *Dunant's Dream: War, Switzerland and the History of the Red Cross*, Cambridge: Cambridge University Press.

Neiman, S., 2004. *Evil in Modern Thought: An Alternative History of Philosophy*, Princeton, NJ: Princeton University Press.

Paice, E., 2008. *Wrath of God: The Great Lisbon Earthquake of 1755*, London: Quercus Publishers.

Peterson, M.J., 2006. *The UN General Assembly*, London: Routledge.

Preparatory Committee, 1925. *Report of the Preparatory Committee for the Ciarolo Scheme*, Geneva: League of Nations—Official Journal.

Roehner, B.M., 1997. Jesuits and the state: A comparative study of their expulsions (1590–1990). *Religion*, 27(2), pp. 165–82.

Rozario, K., 2003. "Delicious horrors": Mass culture, The Red Cross, and the appeal of modern American humanitarianism. *American Quarterly*, 55(3), pp. 417–55.

Ruggie, J.G., 2001. global_governance.net: The Global Compact as learning network. *Global Governance*, 7(4), pp. 371–8.

Saechao, T.R., 2007. Natural disasters and the responsibility to protect: From chaos to clarity. *Brookings Journal of International Law*, 32(2), pp. 663–707.

Schaferhoff, M., Campe, S., & Kaan, C., 2009. Transnational public-private partnerships in international relations: Making sense of concepts, research frameworks, and results. *International Studies Review*, 11, 451–74.

Sending, O.J., & Neumann, I.B., 2006. Governance to governmentality: Analyzing NGOs, states, and power. *International Studies Quarterly*, 50(3), pp. 651–72.

Shrady, N., 2008. *The Last Day: Wrath, Ruin, and Reason in the Great Lisbon Earthquake of 1755*, New York: Viking.

Smith, C.B., 1999. The politics of global consensus building: A comparative analysis. *Global Governance*, 5(2), pp. 173–201.

Strömberg, D., 2007. Natural disasters, economic development, and humanitarian aid. *The Journal of Economic Perspectives*, 21(3), pp. 199–222.

Therien, J.-P., & Pouliot, V., 2006. The Global Compact: Shifting the politics of international development. *Global Governance: A Review of Multilateralism and International Organizations*, 12(1), p. 55.

Udías, A., 2009. Earthquakes as God's punishment in 17th- and 18th-century Spain. In M. Kolbl-Ebert (ed.). *Geology and Religion: A History of Harmony and Hostility*, London: Geological Society of London, pp. 41–8.

United Nations Gender Assembly, 1965. Resolution 2034 (XX) Assistance in cases of natural disaster. *1390th Plenary Meeting*, December 7, 1965.

United Nations General Assembly, 1971. GA Resolution 2816 (XXVI) Assistance in cases of natural disaster and other disaster situations. *2018th Plenary Meeting*, December 14, 1971.

Van Schalk, 1991. Address to UN General Assembly, GA Resolution 46/182. In *46th Session General Assembly, Provisional Verbatim Record of the 78th Meeting*, New York: United Nations Department of Public Information, pp. 42–5.

Walker, P., Wisner, B., Leaning, J., & Minear, L., 2005. Smoke and mirrors: Deficiencies in disaster funding. *The BMJ*, 330, pp. 247–50.

Walker, P., & Maxwell, D.G., 2009. *Shaping the Humanitarian World*, London: Routledge.

Weiss, T., 2016. *What's Wrong with the United Nations and How to Fix It*, Cambridge: Polity Press.

Wisner, B., 2004. *At Risk: Natural Hazards, People's Vulnerability, and Disasters*, London: Routledge.

4

DISASTER RISK REDUCTION

As disasters worsened through the late twentieth century and development efforts failed to reduce vulnerability for many communities, negotiations to strengthen the international disaster relief regime in the late 1980s and early 1990s began to emphasize the need to move beyond relief and focus on development. Shifts in the dominant understanding of disasters as the product of socioeconomic systems rather than natural, exogenous events slowly influenced policy. In 1987, the UN General Assembly declared the 1990s as an International Decade for Natural Disaster Reduction (IDNDR), which led to the establishment of the UN Office for Disaster Risk Reduction in 1999 and the negotiation of disaster risk reduction (DRR) frameworks, of which there have been three: the Yokohama Strategy and Plan of Action for a Safer World (1995–2005), the Hyogo Framework for Action (2005–2015), and the Sendai Framework for Disaster Risk Reduction (2015–2030). While I refer to these initiatives as the DRR regime in this chapter, it is worth mentioning that I employ the concept quite loosely. As Hannigan (2012) has noted, no single concept (network, regime, system, landscape of governance) accurately captures the shift toward DRR, as it relies on voluntary commitments, vague targets and unspecified timelines, and imposes no penalties on non-compliant states (18). Rather than see this as an absence of regime or governance, however, I view this as a defining feature of the DRR regime.

The United Nations International Strategy for Disaster Risk Reduction (UNISDR) defines DRR as "the concept and practice of reducing disaster risks through systematic efforts to analyze and reduce the causal factors of disaster" (UNISDR, 2004). Unlike relief, which deals with the effects rather than the causes of disaster, risk reduction necessitates dealing with the structural causes of human vulnerability, and thus it is a serious effort to address the tensions and inconsistencies of the foundations of international order, in particular economic liberalism. The reason that risk reduction casts economic liberalism into sharper relief than the other ideologies explored thus far is because it is most directly concerned with the issues of poverty and humanity's exploitation of the natural environment. Consequently, most of the ideological work performed by the DRR framework relates to sustaining this particular belief system.

This chapter suggests that the DRR framework attempts to sustain economic liberal beliefs by employing a number of problematic claims about the relationship between development and environmental degradation. One of the more prominent is the "poverty-environment trap" narrative that asserts a causal link between poverty and environmental degradation, which in turn increases disaster risk (Barbier, 2010). In this frame, poverty leads to environmental degradation due to a combination of higher birth rates associated with poverty and the need to engage in environmentally degrading practices to provide short-term sustenance (Gray and Moseley, 2005). As Gray and Moseley note, responsibility for environmental degradation is attributed to the actions of the poor and to poverty more generally. The obvious policy conclusion is that to reduce poverty, environmental degradation, and disaster risk, societies need to "develop", which necessitates engaging with the global liberal economy. Gray and Moseley (2005) show that placing blame on the poor is misguided: empirical studies suggest that the poor implement a variety of measures to prevent environmental degradation and that in cases where such a pattern is evident, the poor and marginalized have often been forced to live in marginal areas and to engage in such practices due to displacement caused by large-scale corporate activity that has forced people from their land and caused massive environmental damage. Similarly, Barbier (2010) shows that the poverty-environment trap narrative places disproportionate blame on the poor for environmental degradation and that the most effective solution is to increase the assets of the poor – an approach at odds with many of the policies associated with global economic liberalism.

The types of economic activity and land usage encouraged under the global liberal economic system contributed extensively to environmental degradation. Gray and Moseley argue that the emphasis on the poor obscures that the wealthy use a far greater proportion of the world's resources than do the poor, who bear most of the cost of this usage (2005:16). In short, the scale of environmental degradation caused by the poor and marginalized pales in comparison with, and is a product of, large-scale environmental destruction caused by the wealthy and by corporations who are seen as essential for the process of development.

Consequently, the DRR frameworks had to deal with the ambiguous status of development in producing or reducing disaster risk. In this chapter, I suggest that "sustainable development" and "resilience" have played a key role in DRR to resolve the tensions exposed by the failure of development. While the development narrative outlined above highlights the problem of underdevelopment for disaster risk (thus necessitating development), sustainable development and resilience addresses the problems of development itself. "Sustainable development" identifies how to eliminate the environmental "bads" associated with development, while "resilience" identifies how to deal with vulnerability and recurrent disasters that are inevitable within the current international order.

In this chapter, I argue that the development of the DRR frameworks initiated a novel trajectory in managing disasters internationally and came about from the recognition that the number and impact of disasters had steadily increased over time, despite (or because of) international efforts to provide disaster relief and the spread of economic liberal policies. The growing impact of disasters demonstrated the insufficiency of the existing model of disaster management, and of social order in general, which emphasized the spread of scientific knowledge and placed responsibility in the hands of states to reduce disaster risk, with international cooperation remaining a voluntary commitment.

Empirical developments and changing understandings of disaster disrupted the existing dominant beliefs about how the world works and how states, non-governmental organizations (NGOs), and international non-governmental organizations (INGOs) ought to act in it. Acknowledgement that disasters were getting worse, which was stated at the outset of the IDNDR and the 1994 Yokohama Strategy and Plan of Action for a Safer World (YS), suggested that the existing order, based on a complex mix of modernism, sovereignty, economic liberalism, realism, and humanitarianism, had failed to solve the problem of disasters and may have contributed to a worsening situation.

The Yokohama Strategy (YS)

Leading up to the IDNDR and the YS, new scholarly approaches to disaster management had begun to challenge the dominant frameworks of disaster as "natural" and as events that had an indiscriminate impact on all members of a disaster-affected society. Vulnerability studies, associated with the works of Phil O'Keefe, Ben Wisner, Kenneth Hewitt, Mark Pelling, and others, drew attention to the causes of vulnerability to disaster and the differential impact of disasters on certain populations, notably those excluded or marginalized within societies. This body of work was strongly rooted in a critique of capitalist economic systems that produced poverty and vulnerability to natural hazards (Varley, 1994:6). This shift in understanding disasters mirrored, in some respects, the shift that had occurred around the time of the Lisbon earthquake outlined in the previous chapter. This new understanding of disaster was incompatible with existing beliefs and practices and required a new set of ideas, concepts, and policies. Initiating such a change had the potential to modify the existing order and was subject to deep contestation over who would shape this order. Taken up by the United Nations, international efforts to address vulnerability entailed a shift from humanitarian disaster relief to DRR.

Given the strong critique of economic liberalism within the vulnerability approach and the investment of the most powerful states in maintaining the existing international order, incorporation of vulnerability into the UN system resulted in significant alteration. YS, as the first international framework to address vulnerability, did not fully shift to a sociopolitical understanding of disaster, instead maintaining a naturalized understanding. For example, the YS document includes the phrase "natural disasters" 61 times (privileging natural and geologic explanations of disaster over human causes of disaster) and a preference for technical solutions rather than social and economic changes (Cannon, 1994:17).

Beyond maintaining a naturalist understanding of disaster, YS also separated vulnerability from the critique of liberal economic policies and the problem of poverty. YS mentions poverty only twice – both times through the assertion that economic growth and development "eradicates" or "alleviates" poverty. This assertion is a clear restatement of economic liberalism and one that does not fully incorporate the insights of the vulnerability approach. Shorn of its roots in Marxist critique, the concept of vulnerability in YS functions to legitimize economic liberalism and to call for a renewed commitment to economic growth and development.

In place of fully incorporating the vulnerability framework, YS relied on the concept of sustainable development. In the economic liberal belief system outlined in Chapter 2, disaster causes disorder – understood as a disruption of economic production, the system of private property, and the functioning of the market. In the vulnerability approach, disaster is not a cause but a symptom of social systems that produce poverty and environmental degradation. By the 1990s, there was growing acceptance that environmental degradation associated with economic development contributes to human vulnerability to disasters (Wisner et al., 2004; O'Brien et al., 2006). The move to DRR necessarily involved greater attention to human use and exploitation of the natural environment – a particular challenge to those committed to preserving the liberal economic order. The key mechanism employed by the authors of YS, as well as the Hyogo and Sendai Frameworks, to resolve this tension was the frame of sustainable development.

The concept of sustainable development represented a particular resolution to the problem of economic growth, development, and vulnerability to natural hazards. A product of ongoing contestation over the Western model of economic development within the environmental movement, "sustainable development" at once encouraged economic growth and the exploitation of the natural world while limiting permanent damage to the environment, which was seen as a contributor to vulnerability (Blewitt, 2014). The concept promoted the belief that continued exploitation of the environment into the distant future was not only possible, provided it was done properly, but was necessary to lift people out of poverty and reduce their vulnerability to natural hazards. Against a more radical environmental challenge that privileged the preservation of the natural environment, sustainable development offered a moderate solution by "ameliorating, without challenging" the impetus to maintain economic growth (Robinson, 2004:370).

The sustainable development frame operates by distinguishing between good and bad types of development: those that are capable of producing wealth and well-being for present and future generations from those that do not (Kemp and

Martens, 2007). By incorporating sustainable development, YS and subsequent DRR strategies drew on an existing framework that provided a ready-made solution to the tensions exposed by the connection between liberal beliefs about economic growth and the associated increase of human vulnerability to natural disasters. In this framing, sustainable development would reduce disaster risk while simultaneously alleviating poverty through economic growth and by designing development projects that would not expose the human population to "natural" hazards. The good/bad development binary is clearly articulated in YS, which observes "some patterns of consumption, production and development have the potential for increasing vulnerability to natural disasters . . sustainable development can contribute to reduction of this vulnerability" (United Nations, 1994:7). Sustainable development was the framework employed to reduce the tensions exposed by the recognition that economic growth and resource exploitation, two key aspects of economic liberalism, contributed to human vulnerability to disaster.

The shift to DRR also disrupted key tenets of humanitarianism, which had served as the resolution to the tensions exposed by disasters in the modern state system beginning in the late eighteenth century. Under this resolution, humanitarian assistance was the most appropriate international response to human suffering and remained distinct from poverty alleviation. With its focus on reducing vulnerability, the DRR strategies rarely mention humanitarianism; in fact, it is not mentioned at all in YS. YS calls for a move away from disaster response/relief in favour of "prevention, mitigation and preparedness", a sentiment repeated in both Hyogo and Sendai. The Yokohama document explicitly states that "disaster response alone is not sufficient, it yields only temporary results at a very high cost. We have followed this approach for too long" (United Nations, 1994:2). The approach adopted in YS and beyond endorses a shift in funding priorities from humanitarian relief to development, which represents a significant departure from the traditional humanitarian focus on alleviating suffering rather than preventing suffering – a distinction that remains divisive in the humanitarian community (Rieff, 2003; Calhoun, 2008:73).

While DRR signals a shift away from a humanitarian approach to disaster management, key aspects of humanitarianism remain firmly in place. All of the recommendations for international action in Yokohama are "humanitarian" in the sense that they are meant to alleviate human suffering and entail only voluntary and charitable action on the part of donor states. For instance, to the extent that YS calls for developed states to commit resources to DRR, it only recommends "extrabudgetary resources" based on "voluntary contributions" (United Nations 1994:13). This ultimately resulted in the establishment of the Trust Fund for the International Strategy for Disaster Reduction, which was financed through voluntary contributions (UNISDR, 2006). Similar to the foundation of the International Relief Union (IRU), the creators of the Yokohama framework did not intend to create positive international obligations or to establish responsibility for reducing disaster risk outside the state.

The implications for humanitarianism emerging from the IDNDR and YS are still unclear. For some, the shift from *alleviating* suffering to *preventing* suffering further reinforced a divide between humanitarianism and development, while for others YS and subsequent frameworks further blurred this already fuzzy line (Slim, 2000). Many NGOs, INGOs, and state actors do both humanitarian relief and DRR. In many respects, DRR is being incorporated into the humanitarian enterprise. The Sphere project, which was initiated by some of the world's largest humanitarian NGOs (including International Federation of Red Cross and Red Crescent Societies [IFRC], Oxfam, World Vision, CARE), aims to establish the basic principles and minimum standards for humanitarian assistance. This project affirms the humanitarian imperative to "prevent or alleviate human suffering arising out of disaster or conflict" (Sphere, 2011). The Sphere project now includes DRR as one of its areas of emphasis and calls for humanitarian actors to assist in recovery and prevention as a basic principle of humanitarian action.

Even as the line between DRR and humanitarian relief has blurred, one trend has emerged: 25 years of promoting DRR has not yet resulted in changing funding priorities. Since the 1990s and the emphasis on DRR, funding for emergency disaster relief has continued to rise relative to development assistance (Fearon, 2008:71). While risk reduction is now often included in development projects more than it was prior to IDNDR, states continue to fund disaster relief over DRR. The Organisation for Economic Co-operation and Development (OECD) reports that DRR funding spiked in 2004 and 2005; but on average, it makes up less than 1% of development assistance and has not changed significantly since 2006. Meanwhile, funding for emergency relief, relief co-ordination, and emergency food aid has increased steadily over the same period (Kellet and Sparks, 2012:13).

The DRR frameworks all emphasize that states and the international community should invest more in risk reduction than in relief, but there are compelling reasons why actors prefer to fund relief. Governments gain from being seen as responsive and generous in the aftermath of disasters to a much greater extent than contributing to mitigation and preventive measures (Schwarze and Wagner, 2007; Vogel, 2012; Kunreuther and Michel-Kerjan, 2013). These political calculations arise from the affective dimension that favours relief over prevention. Media and civil society actors mobilize to a much greater extent in response to appeals to help those affected by disasters than to appeals to help those vulnerable to disaster or to reduce risk. Much of this stems from the humanitarian narrative that characterizes post-disaster scenarios in which those affected by disaster are represented as passive victims of natural forces (Malkki, 1996; Hewitt, 1997; Watson, 2011). In contrast, discussion of preventive and mitigation measures tend toward technical language with less of an affective component. Consequently, even as policy makers shift focus to risk reduction rather than relief, they face strong political incentives and affective narratives drawing on humanitarian beliefs to the contrary.

As it pertains to state sovereignty, the DRR framework retains the strongly statist framework characteristic of the relief regime and reinforces a symbiotic relationship between sovereignty and humanitarianism by emphasizing that states have the primary responsibility for reducing disaster risk within their territories. Most of the recommendations in YS target national governments, reinforcing nation-statist beliefs. It does fundamentally challenge, however, a strong sovereigntist position and an inherently competitive international system characteristic of realist beliefs. State responsibility features five times in the document; however, the document stresses the need for improved international and regional cooperation 18 times. These passages indicate that even as YS recognizes the responsibility of states, it sees cooperation as essential and recommends that states adopt similar institutional mechanisms for reducing disasters, for greater cooperation within and between states, and to strengthen the UN's capability to assist in these processes. Thus, the document simultaneously calls into question the capability of states to preserve the existing international order through self-help alone while recognizing that responsibility for DRR ultimately resides in the state and limiting the responsibility of the international community for DRR initiatives.

As for its modernist aspect, YS actually represented a significant shift toward accepting "traditional" forms of knowledge. Of course, the document is replete with references to research, science, technology, and early warning systems as the solutions to DRR. The conference that led to the adoption of YS featured scientific and technical presentations about reducing disaster risk. However, YS calls for "recognition and propagation of traditional knowledge, practices and values" as well as "coping mechanisms" (8,11). In total, YS endorses traditional responses to disaster eight times, with nine references to scientific knowledge and another eight to technological solutions. The numbers belie the relative importance placed on these forms of knowledge; the scientific and technological solutions are foregrounded at the opening of many sections, while the references to traditional knowledge tend to be placed toward the end of sections, as a last contribution. Nonetheless, the inclusion of traditional forms of knowledge at least implies a slight modification of modernist ideology with room for indigenous forms of thought and solutions to reduce disaster risk. This is evident in other areas as well – on matters of biodiversity preservation, a number of international agreements have called for the "integration of indigenous knowledge into scientific policies and programs", although actual integration has been slow to materialize (Mauro and Hardison, 2000). As we will see in later sections of this chapter, this "post-modern" approach to disaster management was short-lived, with the Hyogo and Sendai documents increasingly emphasizing technology and science and downplaying traditional knowledge.

Hyogo and Resilience

As the ten-year time frame of YS came to a conclusion, the international community began to negotiate the follow-up framework – the Hyogo Framework. It did not represent a significant departure from YS and sought to build on its insights and successes, and similarly emphasized DRR and reducing vulnerability. Hyogo, however, was not simply a continuation of YS; it signalled an important shift in DRR – away from the naturalist understanding of disaster that dominated YS and toward an understanding of the social roots of disaster. Notably, climate change had become a much more significant concern in international relations – Hyogo was adopted the same year as the Kyoto Protocol (UNISDR, 2005), and this awareness was reflected in the framework. While YS made no reference to climate change, the Hyogo Framework featured repeated references to the relation between climate change and disasters. For the first time, international agreements on disaster privileged human causes of disaster over or on par with "natural" causes. Consistent with this transformation, Hyogo also rarely referred to disasters as natural disasters.

An important corollary of this shift was the assertion that disasters are, at this point, unpreventable due in part to the damage already done by greenhouse gas emissions. Consequently, Hyogo places far less emphasis on prevention and mitigation of disaster as the appropriate means to reduce disaster risk. Compared with YS, references to these as appropriate strategies to disaster declined by more than 50%. While prevention and mitigation are still included in Hyogo, in a world marked by climate change, preventing and mitigating disasters requires an international strategy on climate change, which is not what Hyogo was negotiated to do. International climate change negotiations take place under the UN Framework Convention on Climate Change (UNFCCC), which is meant to lead to a reduction in greenhouse gas emissions, and create an agreement in principle on how to assign responsibility for climate change and how to bear the costs of mitigation and adaptation. Despite the significant amount of overlap between climate change adaptation and DRR, the two processes through which these are addressed internationally remain problematically isolated from one another (Schipper, 2009).

To be sure, there may be advantages in separating DRR from discussion on climate change adaptation (CCA). Mercer stresses that incorporating the discussion of DRR within the broader discussion of CCA risks neglecting non–climate change causes of disaster, of which there are many (Mercer, 2010:252). DRR also places greater emphasis on local solutions and development than does CCA, which tends to focus on global solutions to reduce greenhouse gas emissions (Mercer, 2010). For Mercer, CCA should be "embedded within DRR" as "one factor amongst many contributing to community vulnerability" (Mercer, 2010:252); yet until there is a clear and effective global solution to the problem of climate change, it is unlikely to be featured prominently in DRR.

I suggest that the central reason that the two processes remain separate reflects the attribution of responsibility. The UNFCCC ostensibly allocates the responsibility and cost for climate change on developed and developing states, while the DRR framework is designed to reduce the need for and cost of humanitarian assistance after disasters (Mace and Schaeffer, 2013:6). According to Mace and Schaeffer, Hyogo does this by placing responsibility on disaster-prone countries to reduce their own vulnerability to disasters and their need for external post-disaster assistance (Mace and Schaeffer, 2013:6). In contrast to the UNFCCC, Hyogo (and the DRR framework generally) relies on the ideology of nation-state sovereignty to cement responsibility for reducing disaster risk on affected states. As with Yokohama (and Sendai), Hyogo stresses that the primary responsibility for managing disasters falls to individual states, and not on the international community as a whole, or on the developed countries most responsible for greenhouse gas emissions.

Here again we see the interplay between ideologies of nation-statism and humanitarianism. The Yokohama and Hyogo frameworks were meant to reduce reliance on humanitarian assistance after disasters, although they both maintained strong ties to the basic humanitarian principles of voluntary, charitable contributions. The voluntary nature of international efforts to reduce disasters operates through and reinforces the obligatory responsibility of states to reduce the risk of disaster within their territories – even if they are caused in part by global processes. One of the key tensions then that informed the negotiation of the Hyogo Framework (HF) concerned how to allocate responsibility for reducing disaster risks in a global system in which responsibility for producing vulnerability no longer falls on nature, God, or failed development within states but on the economic and environmental practices of developed states in a globalized economy. The first mechanism was to separate negotiations on allocating responsibility for climate change, as was done in the UNFCCC, from allocating responsibility for DRR. The second was to incorporate a new framework that altered existing beliefs about the responsibility for managing disasters. Just as YS relied on sustainable development to resolve some of the tensions over economic liberalism, modernism, and disasters, HF employed the framework of resilience.

Scholars examining resilience suggest that its rise as a mode of governance must be understood in terms of the ideological contestation over modernism, economic liberalism, and state sovereignty. The embedded liberal compromise outlined in Chapter 2 faced critique from a variety of positions: neo-liberalism, pragmatism, and various neo-Marxist-inspired positions. It is far beyond the scope of this book to examine the distinct challenges each of these posed; however, it is worth noting that these challenges questioned whether the interventions by states to reduce disasters could have the predicted effects on reducing disasters (among other harms) that they claimed.

Developments in disaster studies demonstrate these shifts. In the 1950s, scholars in the field of disaster studies began to examine social aspects of disaster, even though they maintained a very modern understanding of disasters as "distinct events…that exist out there" (Tierney, 2007:506). Subsequent research in this field focused on how social problems such as disasters are constructed, which in turn garnered additional attention to how social groups are differentially affected (Spector and Kitsuse, 1987; Dynes and Rodriguez, 2010). Further research into social sources of disaster challenged the classic view of disaster and began to examine "human causes of vulnerability" (Wisner, 2004). Hewitt notes that this was a substantial shift from viewing damage from disaster as "accident, negligence or deliberate sabotage" to an intrinsic feature of modern human activity (Hewitt, 1983:17). This new understanding challenged the modernist assertion that disasters are "sudden ruptures in the social order" due to natural phenomena to which central authorities can respond with predictable outcomes; rather, disasters were understood as a product of "larger political-economic" forces associated with modernity itself and of historical processes of marginalization (Hewitt, 1997; Pelling, 2003; Tierney, 2007:510). In short, trends in the study of disaster shifted from viewing disasters as problems to be solved through modern forms of thinking to viewing disasters as symptomatic of modern social order.

In response to these challenges to the prevailing order, new approaches to disaster management were developed, such as risk analysis, sustainable development, disaster resistance, or "invulnerable development" (McEntire et al., 2002). These novel approaches to managing disaster are, broadly speaking, modern in orientation, if not classically so, in that they still espouse a commitment to the utility and predictability of human interventions. These new approaches sought to expand "scientific" inquiry to explain how "human misunderstanding of nature" produces vulnerability (Pelling, 2003), to identify the various social and contextual factors that cause vulnerability (Wisner et al., 2004:19), to support forms of development that do not degrade the environment (Mileti, 1999), or to apply probability and risk analysis to employ insurance as a technology for managing and mitigating natural disaster (Ewald, 1991; Beck, 2000). All of these approaches base their claim to legitimacy on scientific forms of knowledge production and on the assertion that this knowledge will produce relatively predictable and certain outcomes.

Resilience, on the other hand, rests on an "ontology of uncertainty" that requires an ability to adapt and respond to change (O'Malley, 2000:462). Uncertainty arises because disasters are understood as neither predictable nor preventable but as a "permanent part of human experience that can never be fully anticipated or prevented" (McEntire et al., 2002:269; see also Welsh, 2013). Because disasters are understood as "intrinsic to the internal functioning of society", they are an inherent part of the human condition that can never be prevented (Duffield, 2012:481). While some forms of disaster may be prevented through human intervention, this will simply open up new possibilities of disaster and "normal accidents" that arise periodically because of the nature of a system itself (Perrow, 1981, 2011).

This is a direct challenge to the certainty and predictability offered in modernism, nation-statism, and economic liberalism. Reid suggests that the resilience frame captures this post-modern view of disasters and creates subjects that cannot "conceive of changing the world, its structure and the condition of possibility… that accept the disastrousness of the world … and accepts the necessity… to change itself in correspondence with the threats now presupposed as endemic" (Reid, 2012:74). This reorientation of our understanding of disasters requires governance through local "preparedness and pre-emption" (Anderson and Adey, 2011:1098) rather than centralized prediction, prevention, and mitigation. Seen from this perspective, the resilience framework addresses a key contradiction exposed in the ideological compromise that has characterized disaster management: that scientific knowledge and technological innovations implemented by centralized states would have a certain and predictable effect on reducing vulnerability.

The post-modern challenge to the ability of centralized state planning to prevent disasters fits ideologically with neo-liberal thought that similarly challenged the wisdom of state-led interventions, instead favouring market rationality (Collier, 2014). The neo-liberal move has been characterized by the reduction of the public sphere, wherein a number of these functions were "returned" to civil society and other private actors through processes of privatization and individual responsibility (Garland, 1996:453).

Once put into practice, however, the tensions and ambiguities in neo-liberalism have come to the fore. Critical of the claim that government intervention could reduce uncertainty, the neo-liberal project still required governance and "goal-directed" activity to bring about neo-liberal reforms (Schmidt, 2014:420). Neo-liberalism thus had an important role for the state to play, to "create and preserve" a system understood as maximizing freedom of action (Harvey, 2007:22). This required an expansion of the state, in terms of increased expenditures, regulatory frameworks, and oversight on the part of state authorities. This process was summed up by Vogel's pithy observation that "freer markets" means "more rules". According to Vogel, the implementation of neo-liberal policies has produced "stronger markets but not weaker governments" (Vogel, 2006:2). With neo-liberal reforms, states expanded in a number of areas, in regulating more and more areas of economic activity and in "infrastructural control" of financial institutions (Konings, 2012), and especially in expanding the security function of the state (policing, law and order, immigration restriction) (Laffey and Weldes, 2004; Gordon, 2005).

In this context, resilience emerged not as a logical extension of neo-liberal governance but as a response to the inconsistencies within neo-liberalism (Schmidt, 2014). Rather than emphasizing the security functions of the state or the need to regulate to maintain an operating market mechanism, resilience focuses on local communities and individuals who accept the necessity and value of preparing themselves (Reid, 2012:69). At the heart of the resilience framework is the ability of individuals and communities to "recover with little or no external assistance" – meaning state assistance (Manyena, 2006:433). Consequently, neo-liberal ideology attempts to govern communities through resilience by emphasizing the necessity of the market to building resilience. As Pingali et al. show, neo-liberalism promotes the market as the "quintessential resilient institution" that "recovers quickly and functions in the absence of government" (2005:518).

The resilience framework may be used to address ideological inconsistencies in modernism, economic liberalism, and nation-statism, although it contains tensions and ambiguities of its own (Dunn Cavelty et al., 2015:12). Claims that resilience reproduces an ontology of radical uncertainty and entails a rejection of modernist certainty obscure how the concept relies a great deal on scientific claims to knowledge. As numerous commentators have observed, the term itself is drawn from the natural sciences and refers to the ability of materials to regain their original shape after being subject to a variety of forces (Manyena, 2006:437–438; Brasset et al., 2013:222). For Coaffee (2013) the term operates as a "translation term", "bridging concept", or "floating metaphor" that attempts to bridge the "science-policy divide" by drawing attention to a different temporal understanding of responding to disasters, emphasizing "foresight and preparedness" (Coaffee, 2013:243). While resilience discourse may challenge some of the core assumptions undergirding modernist understandings of disaster, it draws its strength as a metaphor from the "scientific" credentials of the concept. In practice, the application of the concept to disasters rests on the assumption that societies, like natural materials, have a knowable breaking point and recovery period, and that human intervention can have a predictable impact by extending their breaking points and reducing their recovery period. Nowhere in the resilience literature is there an acknowledgement that building resilience itself could produce social systems subject to normal accidents and to uncertain outcomes that increase vulnerability. In this regard, it is an entirely modernist concept.

The debates over resilience in DRR may have important implications. If, as suggested by many critics, promoting resilience implies a passive acceptance of existing socioeconomic, political, and environmental conditions that produce recurrent disasters, combined with a downloading of responsibility onto the vulnerable, then it would seem a regressive framework unlikely to reduce disaster risk. On the other hand, if building resilience reduces the impact of disasters while empowering (and potentially emancipating) local communities to apply local knowledge and beliefs, free from the bureaucratic logic of centralized planning often associated with disaster management, then resilience appears far more progressive and capable of reducing vulnerability to disasters.

Resilience may also operate as a framework designed to resolve ideological inconsistencies in the existing foundations of international order. In this respect, it is more of a "shallow concept" without substantive meaning, (Joseph, 2013:51) whose impact is far less consequential in influencing policy change but is instrumental in sustaining a particular international order. Its effect then is primarily politico-ideological. The concept of resilience is useful politically precisely because it is ultimately ambiguous about whether it implies "bouncing back" to the same social order that existed prior to the disaster or if it implies a transformation to a new order (Alexander, 2013:2711). A program of "building resilience" implies initiating change to the existing order within a deep commitment to maintaining continuity of some central aspect of the social order. The use of an ambiguous concept with multiple understandings as the centrepiece of the international strategy for DRR implies a commitment to basic aspects of the existing social order and a low prioritization for reducing disaster risk.

Against this set of controversies, the Hyogo and Sendai frameworks present a useful point of analysis to gauge the extent to which the resilience frame has altered DRR strategies. Compared with YS, HF places heavy emphasis on resilience, and it is very revealing that nearly all references to the term in the HF focus on "building resilience". In HF, resilience is not understood as a pre-existing feature of local communities; rather, it is a characteristic that can be built by centralized planning and state intervention, albeit through engagement with local actors.

The HF also shifts further from a naturalist depiction of disaster and recognizes social causes of disaster, including climate change. Given the aforementioned critiques of resilience, this connection should not be lightly dismissed. But it is also worthwhile to note the increased emphasis on scientific knowledge and technological solutions in Hyogo, and the reduced recognition of traditional forms of knowledge. As I explore further in the next section on Sendai, the emphasis on building resilience does not produce a simultaneous shift away from a commitment to centralized planning and investment in science and technological solutions toward local knowledge.

Beyond the modernist narrative that permeates Hyogo alongside the resilience frame, patterns of funding do not seem to support the strongest fears about resilience: that it would download responsibility onto local communities and reduce commitments to assist those affected by disaster. While the emphasis on local solutions increased dramatically in HF relative to YS, so too did the emphasis on national plans and strategies. The decreased focus on prevention and mitigation does seem to suggest a general acceptance that disasters are no longer preventable, but this has not resulted in a decline in funding for disaster relief reduction – if only because the international community never invested much to begin with. In the period covering the Hyogo Framework (2005–2015), levels of official development assistance are largely unchanged, funding for DRR as part of Official Development Aid (ODA) continues at the same rate as before (less than 1%), and disaster relief has marginally increased since the adoption of Hyogo.

I would suggest that the recently negotiated Sendai Framework further supports the contention that resilience is a useful term whose primary purpose is to perform "remedial ideological work" rather than produce significant policy change that reduces the risk of disaster. The negotiations over Sendai reflected many areas of disagreement, and the importance of resilience was not one. The contestation over Sendai focused on the division of responsibility between the international and national levels of government, particularly on the issues of climate change, funding, and technology transfer. If resilience was being used to promote a post-modern understanding of disasters or neo-liberal socioeconomic policies, this is not immediately clear in the finalized agreement. Indeed, the agreement seems to indicate that building resilience requires greater direct investment by the state, often in scientific and technological solutions.

Sendai

The Sendai Framework continues much of the same focus as the Hyogo Framework, with its emphasis on sustainable development and resilience as key components of reducing disaster risk. Indeed, resilience is mentioned more often in Sendai than in Hyogo. Yet the broader shift to a post-modern/neo-liberal worldview that the resilience framework is claimed to signify is not readily apparent in the Sendai agreement. The increasing emphasis on resilience has not dramatically altered the ideological commitment to modernism, economic liberalism, nation-state sovereignty, and voluntary international commitments in DRR. In fact, the primary points of contestation over Sendai reflect longstanding sources of tension within these ideological belief systems, such as the location of responsibility for managing disasters, the voluntary nature of international cooperation, and the protection of domestic industries and intellectual property.

The most celebrated (and critiqued) feature of the Sendai agreement is the inclusion of seven global targets for DRR along the lines of the Millennium Development Goals (MDGs). These targets include (1) reduce disaster mortality, (2) reduce the number of people affected by disaster, (3) reduce direct disaster economic loss, (4) reduce substantially disaster damage to critical infrastructure, (5) increase substantially the number of countries with national and local DRR strategies, (6) enhance substantially international cooperation with developing countries through adequate and sustainable support to complement national actions, and (7) increase availability of multi-hazard early warning systems and disaster risk assessments (UNISDR, 2015). The inclusion of these targets is very telling.

First and foremost, the setting of targets demonstrates a relatively strong modernist belief system. As with the MDGs, the setting of targets asserts a concreteness that belies the complexity of these processes and asserts certainty in the ability to collect accurate data (Attaran, 2005). It also indicates a belief that it is possible to implement policies that will have a measurable impact on these indicators. The use of measurable targets reflects a modernist orientation committed to measurement and objectivity, and also implies a faith in humanity's ability to reduce disaster risk through purposeful intervention.

The establishment of common indicators and requirements for data collection on mortality rates, affected people, economic loss, disaster damage, and DRR strategies is in fact an exercise in state building and a concentration of authority in the hands of the state. The creation of certain forms of knowledge and measurements to make sense of a complicated reality is a very modern and statist approach (Scott, 1995:191). Just as national forms of measurement transformed local practices, common international indicators of disaster risk are meant to change how local actors understand the impact of disasters. A key goal of Sendai is to have local actors produce knowledge about local hazards in the standardized, international language of disaster risk.

Historically, the creation of official measurements was intended to replace local practices, often through imposition with no consultation. In that regard, the Sendai agreement constitutes a form of consultation, at the national level. And the Sendai agreement at least recognizes that disaster risk has "local and specific characteristics that must be understood ... to reduce disaster risk" (13). In this formulation, local knowledge is complementary to national, indeed international, knowledge. While this does not mark a full-scale retreat from modernism and centralized state authority, it does represent a compromise position in which local practices are not simply seen as obstacles to progress but as potentially contributing to DRR by providing insight that cannot be captured by central actors and standardized forms of knowledge. Notwithstanding this minor caveat, the establishment of targets as the centrepiece of the new DRR agreement does not fit with concerns raised about the resilience framework.

It is worth noting, however, that the Sendai targets are non-specific and vague. Compared with the MDGs on which they were modelled, Sendai's targets are extremely vague. The MDGs listed relatively firm commitments, such as to "halve . . the proportion of people whose income is less than $1 a day" or "reduce by two thirds ... the under-five mortality rate" (United Nations General Assembly, 2000). In one sense, the absence of firm targets may reflect the lack of political commitment at the international level to reduce disaster risk, which is evident by the complete absence of any financial commitments by developed countries to increase development assistance or technological transfers. The Sendai framework reinforces Hyogo's commitment to placing responsibility for dealing with the complex issue of disaster risk, as well as the effects of global climate change, in the hands of affected states.

However, the vagueness of the targets may also reflect the abiding uncertainty behind the DRR program, of which resilience is a key component. As noted earlier, uncertainty over human intervention is a key feature of a postmodern understanding of the social world and of the resilience framework. This uncertainty may be reflected in ambiguous targets such as "reducing", "substantially reducing" or "substantially increasing". Proclaiming a goal of halving disaster mortality rates would reflect a level of certainty in the policies created to reduce mortality that is simply absent in the case of DRR.

Of course, resilience itself is a vague and uncertain term. The Sendai Framework defines resilience as "the ability of a system, community, or society exposed to hazards to resist, absorb, accommodate to and recover from the effects of a hazard in a timely and efficient manner, including through the preservation and restoration of its essential basic structures and functions" (UNISDR, 2015:9). What is regarded as "essential basic functions" of society is left undefined, although it is often made apparent in the context of its usage. In the targets established in the Sendai agreement, resilience is mentioned only once, in the fourth target to "substantially reduce disaster damage to critical infrastructure and disruption of basic services". The target specifically mentions health and education facilities, and recommends "developing their resilience". In this context, resilience applies principally to physical infrastructure, mainly hospitals and schools. This is interesting both in terms of what it identifies as the primary institutions essential to basic societal functioning (health and education), and what is less essential; note that the market is not invoked. In essence, the Sendai Framework calls on states and local authorities to (1) designate some infrastructure as critical, (2) further regulate the building and maintenance of this infrastructure, and (3) prioritize its repair and accessibility in the aftermath of disaster. Rather than indicating a retreat of the state or surrender to the logic of the market, resilience calls for greater state involvement in regulating the construction of infrastructure and in prioritizing certain public goods in post-disaster emergency response.

In many ways, the Sendai agreement represents a significant return to a modernist foundation. While it does not endorse a naturalist understanding of disaster, the framework contains a heavy emphasis on science and technology as the appropriate means to reduce disaster risk (Carabine, 2015). The agreement contains many references to "scientific and technical work", empowering "research institutions", technology transfers, and the application of science and technology to decision making. The emphasis on technology transfers as the primary means to reduce disaster risk has increased dramatically over the course of the transition from YS to Hyogo and to Sendai (de la Poterie and Baudoin, 2015:128). According to de la Poterie and Baudoin, references to research, science, technology, technology transfer, and dissemination of data more than doubled from Yokohama and Hyogo, while references to traditional knowledge halved and community participation remained stagnant, indicating a shift toward a top-down, centralized approach to DRR that sees local communities as passive recipients of information and knowledge rather than key partners in reducing disaster risk (de la Poterie and Baudoin, 2015:136). In addition to reinforcing a hierarchy of responsibility within the state, this pattern indicates a reassertion of modernist belief in the importance of science and technology in solving the world's problems. Indeed, as I will show later in this section, technology transfer was a key point of contestation in the negotiation over Sendai – and not due to disagreements about the centrality of science and technology. Rather, disagreement over technology transfer reflected ideological contestation over the relative weight given to values associated with economic liberalism (protecting intellectual property) and humanitarianism (reducing suffering).

Part and parcel with its modernism, the agreement also displays a strong commitment to nation-state sovereignty, in reinforcing centralized implementation of modernist solutions and in how it allocates responsibility for DRR. One of the key functions of the Sendai agreement is to reconcile the inconsistencies of a statist world order, with growing recognition that disaster risk is produced in part through global forces beyond the control of individual states. As with the previous DRR frameworks, the first guiding principle of the document is that "each state has the primary responsibility to prevent and reduce disaster risk" (UNISDR, 2015:13). This principle is reinforced throughout the document in terms of how the recommendations are aimed at national actors, even as it calls on them to work with local, regional, and international actors. Yet cooperation is a dominant theme in the Sendai Framework that repeatedly calls into question the strong statist position that a centralized state authority is best positioned to protect citizens from disasters, as well as the realist belief that multilateral cooperation is imprudent.

The emphasis on international and local cooperation is not a challenge to state authority; indeed it empowers the state to identify or create local entities with whom it can work to collect data and to implement risk reduction strategies. It does, however, challenge realist assertions regarding the manner in which states secure themselves in an anarchic international society. Sovereignty in this case, as it was in Yokohama and Hyogo, is employed to reinforce a hierarchy of responsibility and simultaneously to limit international obligations. Reasserting that the primary responsibility to reduce disaster risk falls to individual states and removes responsibility from the actors most responsible for global climate change.

This is most clearly enunciated in the official US statement on the agreement. Explaining the United States' refusal to firmly commit to technology transfers, the US spokesperson asserted that the "core principle of Disaster Risk Reduction, [is] that each state has the primary responsibility for taking effective measures to reduce disaster risk" (Sumner, 2015). For the United States, and other developed states, international cooperation is fundamentally humanitarian – although the term does not appear in the document. Obligations on the parts of developed states to assist developing states reduce disaster risk are voluntary rather than obligatory or based on an assessment of responsibility for producing risk. The Sendai document calls for the "strengthening of… international *voluntary* mechanisms for monitoring and assessment of disaster risks" (UNISDR, 2015:18). The developed states were keen to prevent any discussion of responsibility for disaster risk and asserted this through the longstanding combination of nation-state sovereignty and voluntary international obligations.

Sendai also contains a strong commitment to economic liberalism, most notably captured in the prominence given to sustainable development. It is even more explicitly stated in the guiding principles of the agreement. After national responsibility, the agreement aims to protect people "and their property, health, livelihoods and productive assets, as well as cultural and environmental assets, while promoting all human rights" (UNISDR, 2015:13). Culture, environment, and human rights are included, but they are ultimately secondary to the primary purpose of ensuring the ability of people to continue to function as productive workers and consumers. Furthermore, the agreement provides an economic rationale for investing in DRR: it leads to "innovation, growth and job creation" (UNISDR, 2015:18).

This is not to suggest that the Sendai agreement reinforced the existing ideological commitments noted in this section without contestation. Negotiations over the Sendai agreement were intense and required last-minute compromises to produce a finalized agreement. Sendai was heavily contested by virtue of its temporal proximity to major negotiations set to take place later in 2015 on climate change (in Paris) and sustainable development (in New York). Of the three major international negotiations to take place in 2015, the meetings in Sendai to discuss DRR were expected to be the least contentious; however, given the widely acknowledged interconnection between climate change, development, and disaster risk, the meeting had the potential to bring in areas of contestation characteristic of negotiations on these other issues. The decision to negotiate on DRR separately from and prior to climate change and development was an intentional strategy to separate these issues as distinct and to reduce animosity in the hopes of fostering an agreement. Despite these intentions and expectations, the Sendai agreement was fraught with difficulties and disagreements over financing, technology transfers, climate change, and the role of conflict (Darby, 2015; Wilkinson, 2015).

Disaster-prone developing states sought to move the DRR framework further from the voluntary/humanitarian framework characteristic of negotiations on disaster relief toward compulsory and firm commitments, such as those sought in climate negotiations. One of their first priorities was to achieve increased funding for DRR, a proposal strongly resisted by the wealthy developed states (Rowling, 2015). Although one of the most celebrated aspects of the Sendai agreement included establishing seven targets to be achieved over the 15-year period of the agreement, not one of the targets calls for a specific increase in funding. The most relevant is the sixth target, which calls for "substantially enhancing international cooperation … through adequate and sustainable support" (UNISDR, 2015:10). An earlier draft of the agreement called for "additional and predictable support", but even that signalled a level of commitment that developed states were unwilling to make (Rowling, 2015). The sixth target, ostensibly the one related to international funding for DRR, falls far short of an actual, measurable funding target and does not include an actual reference to additional funding. The closest the Sendai Framework for Disaster Risk Reduction (SFDRR) comes to a funding commitment is the eleventh guiding principle calling on developed states to fulfil their "respective commitments of official development assistance" (UNISDR, 2015:14), a voluntary and aspirational commitment routinely unmet. Organizations hoping for clear and concrete funding commitments were disappointed. The spokesperson for Oxfam noted that the framework is built around "flimsy unambitious targets that will not galvanize bold action or create meaningful accountability" (Paul, 2015). World Vision and Action Aid also harshly criticized the failure of developed nations to provide a firm commitment to fund DRR in developing states (Darby, 2015).

A related set of concerns arose over technology transfers. The developing countries placed high priority on the inclusion of technology transfers as a way for developed states to assist with reducing disaster risk. In many ways they succeeded – the number of references to technology transfers in the Sendai agreement is far higher than in either Yokohama or Hyogo. Yet many developed states rejected a strong and firm commitment to this measure. The United States went so far as to propose a "mutual agreement" clause that would have allowed them to opt out in specific circumstances (Darby, 2015; Wilkinson, 2015). While developing states managed to have this clause removed in the final agreement, the United States' delegation released a statement the same day that the agreement was signed clarifying that "the transfer of technology must not be coerced and that the rights of private holders of intellectual property must not be abrogated" (Sumner, 2015).

American opposition to technology transfer in Sendai is broadly consistent with its position on intellectual property, such as the Agreement on Trade-Related Aspects of Intellectual Property Rights (TRIPS) agreement, which reflects a commitment to a liberal international economic order (May, 2013) and a view of international cooperation as a fundamentally humanitarian (voluntary) endeavour. The US State Department emphasized that they did not take the Sendai agreement to imply any change to the nature of its international obligations on technology transfer or that the framework provides a precedent for other negotiated frameworks, namely, UNFCCC (Sumner, 2015). The US position on funding and technical assistance demonstrates the mutually reinforcing ideologies of state sovereignty, economic liberalism, and humanitarianism. Rather than fundamentally challenging this set of ideologies, the Sendai agreement reinforces it and attempts to resolve the ambiguities and tensions within and between these belief systems by reorienting the framework toward establishing common measures and strengthened data collection.

A third area of contestation concerned the relationship between climate change and disasters. Climate change talks were scheduled for the fall of 2015, and the Sendai agreement was never intended to replace that set of negotiations; nonetheless, developing states attempted to include language around the responsibility for climate change and tying DRR funding to that responsibility (Rowling, 2015). An earlier proposal from developing countries incorporated a clause establishing "differentiated responsibility for causing disasters" (Mitchell, 2015). Tying responsibility for causing climate change to an assessment of costs for mitigating and adapting to the effects of climate change has long been a sticking point in climate change negotiations (Christoff, 2010), as it was with the Sendai agreement. In the final Sendai agreement, there is no mention of a responsibility to contribute to DRR based on a state's contribution to climate change. Kelman suggests that climate change is no more prominent in the Sendai agreement than in Hyogo, and perhaps even less so (Kelman, 2015:121). In the Sendai agreement, climate change is incorporated simply as one of a number of causes of disaster (Rowling, 2015).

The choice to negotiate DRR separately from climate change and development and to position it within the UN Office for Disaster Risk Reduction signals that DRR is still considered a humanitarian activity, despite the absence of references to humanitarianism in the document. The UNISDR coordinates mostly with humanitarian agencies and is funded through voluntary contributions (Mitchell, 2014). The voluntary nature of the recommendations in Sendai further supports this perspective. As I have argued throughout this book, the humanitarian narrative serves to allocate responsibility in particular ways, mainly by concentrating it in the hands of states, thus leaving international cooperation voluntary and charitable. While climate change talks entail negotiating and apportioning responsibility for causing and addressing climate change, the DRR framework positions responsibility in the hands of affected states with support from a neutral and charitable international community who holds no actual responsibility.

Non-State Actors and DRR

One of the most problematic aspects of the creation of international DRR frameworks is the gap between the global and local initiatives to reduce disaster risk. Gaillard and Mercer argue that these international frameworks tend to emphasize top-down solutions shorn of local knowledge and context (2013:94) and that this approach to DRR has failed to reduce disaster risk. Arguing for greater collaboration between disaster scientists/experts and local knowledge carriers in the form of community-based disaster risk reduction (CBDRR), they suggest this requires a shift from seeing scientific knowledge as intrinsically superior to local knowledge (Gaillard and Mercer, 2013:96). Such a move would signal a significant challenge to the modernist foundations of disaster management and in turn the international order. As I outlined in the previous sections, while YS seemed to embrace such as a move, the more recent international DRR frameworks of Hyogo and Sendai seem to be reinforcing the value of scientific knowledge over local knowledge.

Perhaps as a consequence of this shift, there have been very few efforts to implement CBDRR to date. Effective CBDRR entails partnerships between national and local governments, international organizations, community groups and NGOs, and corporations (Gaillard and Mercer, 2013:98). As was argued in the last chapter, strengthening public/private partnerships has potential implications for state sovereignty and economic liberalism. The greater involvement of corporations and civil society in disaster relief suggested (1) further entrenchment of economic liberalism and (2) a potential challenge to conventional understandings of state sovereignty. We see similar potential in the realm of DRR.

One of the most prominent global movements to initiate risk reduction at the local level is the Global Resilience Movement and its 100 Resilient Cities (100RC) initiative. Founded by the Rockefeller Foundation, 100RC facilitates public/private partnerships to help cities become more resilient to both "chronic stresses" and "acute shocks". Now funding initiatives in 98 cities from 46 states, the 100RC promotes an understanding of resilience as the ability to "withstand, respond to, and adapt more readily to shocks and stresses to emerge stronger after tough times, and live better in good times". While resilience is related to sustainable development and DRR, the 100RC project suggests that resilience is distinct from both concepts. Of sustainable development, 100RC states that sustainable development (SD) attempts to rectify the imbalance between development and "depletion of natural resources", while resilience works to ensure that "systems endure and even thrive in an imbalanced world". The tension should be immediately obvious, as is the ideological role that resilience plays along with SD. For 100RC, SD entails a promise that humans can strike the necessary balance between economic growth and resource depletion, while resilience entails a promise that communities can survive and prosper even if (when) SD fails to achieve this balance.

On DRR, 100RC suggests the concept narrowly focuses on "reducing damage caused by natural hazards"; in contrast, resilience has a broader focus and involves planning for "natural disasters and adverse socio-economic trends", such as "high unemployment, overtaxed or inefficient infrastructure", and even "endemic violence" and "chronic food and water shortages". The resilience framework reinforces a distinction between natural disasters and social vulnerability while simultaneously suggesting that resilience is sufficiently broad to deal with all of these factors. In addition to demonstrating the fuzziness of the resilience concept, 100RC also demonstrates key aspects of resilience identified in the academic literature – stresses and shocks are endemic in the social order, the state is insufficient on its own to manage disasters but essential to building resilience, and scientific knowledge is the foundation for resilient societies.

In practice, 100RC provides four specific forms of assistance. The first is to provide two to three years of funding for a chief resilience officer (CRO) in the city government. The CRO's job is primarily coordination: to connect various government departments, non-governmental organizations and local business in the city to raise awareness of and to prioritize resilience (Berkowitz, 2015). The CRO also acts as mechanism of global information exchange, as the CROs from each city work with one another on innovative ideas and best practices. The second practical impact of 100RC is to provide support for cities to produce a city resilience strategy (CSR). The CRS is a document produced by the CRO and partner organizations of 100RC to identify a city's existing capacities as well as the potential shocks and stresses facing each city, to engage with relevant stakeholders, and to show how existing initiatives may better incorporate resilience (Lipper, 2016). In management-speak typical of the 100RC project as a whole, the goal of the CRS is to produce "actionable initiatives and projects" (Lipper, 2016).

The third product associated with 100RC is a platform to put municipal governments in contact with various actors and services to help them "develop and implement their Resilience Strategies" (100 Resilient Cities, 2013). Last, 100RC acts as a "global network of member cities who can learn from and help each other" (100 Resilient Cities, 2013). This is achieved primarily through the interactions of CROs, but it also acts as a form of exclusive membership that is attractive for other cities to join. In this respect, the network signals a global movement in which certain cities operate as the civilizational standard for other cities to follow. The 100RC project is designed as spearheading a "global practice of resilience" (100 Resilient Cities, 2013).

The 100RC project has divided resilience into four distinct dimensions: "Health and Wellbeing, Economy and Society, Infrastructure and Development, and Leadership and Strategy". The constitutive components of these four dimensions provide a clear statement of the ideological roots of resilience thinking. At its core, resilience (like sustainable development and humanitarianism) performs ideological work in rectifying some of the inconsistencies and ambiguities in the existing foundations of international order. With regard to some of these foundations, resilience offers slight modification, whereas in others, resilience offers a fairly substantial revision of these existing ideologies.

Although it is often depicted as a post-modern paradigm, as implemented through the 100RC project, resilience is a thoroughly modern project. Because it focuses on cities, it takes urban living and the basic technological features of modern cities as basic standards globally. Under "Health and Wellbeing", the basic standards for all resilient cities include access to water and sanitation, durable sources of energy, medical clinics, and ambulances. In "Infrastructure and Environment", sea walls and levees, flood-resistant roads and bridges, airports, communication networks, and multimodal transport systems are all necessary requirements. In the "Leadership and Strategy" component, 100RC calls for evidence-based decision making, knowledge transfer, and timely and appropriate monitoring. While 100RC also emphasizes the importance of maintaining local cultural heritage and natural ecosystems that "reduce physical vulnerability of city systems", the resilient city is clearly a thoroughly modern city in which scientific knowledge and technological solutions are paramount. Although the absolute certainty of human progress and triumph over nature that Scott and others associate with strong modernist beliefs is absent in the 100RC project, it retains core aspects of modernist beliefs, including faith in science and technology, central planning, and the possibility of progress and reduced vulnerability to natural and human-made hazards.

The liberal economic foundations of the initiative are also on full display. Under the "Economy and Society" dimension, the 100RC project calls for a "vibrant economy with diverse revenue streams ... ability to attract business investment ... integration with the regional and global economy and measures to attract investment". Under "Health and Wellbeing", the plan calls for "diverse employment opportunities, including access to business investment and social welfare", as well as investment in "skills and training, fair labor policy, and development and innovation". In short, the resilient city framework envisions a city that is well integrated into the global liberal economic system, with the state functioning to facilitate this integration and to alleviate the potential negative impacts of doing so. To ensure smooth integration with the global economy, 100RC calls for "inclusive local cultural heritage; and encouraging cultural diversity while promoting tolerance and a willingness to accept other cultures". Here we see the liberal solution to the problem of promoting local solutions, local knowledge, and distinct identity at the core of resilience thinking. Local culture is to be promoted to the extent that it creates the desire and ability to respond to local challenges without external assistance but is sufficiently tolerant and cosmopolitan to facilitate engagement with a globalized liberal economy. The involvement of close to 75 private corporations as partners in 100RC further illustrates its commitment to economic liberalism.

On the matter of nation-state sovereignty, again resilience appears more as a very minor tweak than a fundamental challenge. Despite concerns raised by academics of the strong connection between resilience and the hollowing out of the state often associated with neo-liberalism, the 100RC project advocates for an expansive role for government. While much of the plan to build resilience entails making cities receptive to the corporate sector and involving the private sector in building resilience, it also includes a strong endorsement of effective municipal government, as well as effective law enforcement, "public healthcare and emergency management", "social welfare", and "fair labor policy" (100 Resilient Cities, 2013). In this respect, 100RC's understanding of resilient communities bears a strong resemblance to classic or Keynesian liberalism, with the government acting to ameliorate the negative aspects of a global liberal economic order. Rather than calling for the hollowing out of the state and austerity measures as a way to make cities more competitive in the global economy, 100RC depicts resilience as emerging from strong public and private actors working in partnership. In many ways, this simply reasserts the basic approach to nation-statism depicted in the previous chapter, where the state claims authority to facilitate and authorize the provision of public goods.

It should be noted, however, that the 100RC project does not explicitly critique austerity measures, and accepts such measures as a given. In its online frequently asked questions (FAQ), it suggests that "in a time of fiscal austerity and budget cuts" attentiveness to resilience will "maximize benefit for the cost, allowing governments to spend their funds in the best way possible" (100 Resilient Cities, 2013). In response to how resilience measures will be funded, 100RC notes part of what a resilience officer does is "making government more efficient, freeing up resources for other projects" (FAQ number 9). The website for 100RC sums up the relationship between resilience and neo-liberalism as such: "a chief resilience officer is a perfect tool for an age of austerity" (FAQ number 7).

In addition, a significant part of 100RC is connecting governments to private and non-profit actors who can already offer services that will build resilience at a lower cost than governments developing these capacities in-house. This aspect of the project promotes the neo-liberal belief that the private sector is more efficient at producing public goods than are governments. The creation of a government position (the CRO) whose job is to identify inefficiencies in other government departments and to connect governments to private actors to build resilience, would seem to meet many of the hallmarks of neo-liberal economic policy that expands government in order to facilitate the growth and success of private actors. As critics of resilience have suggested, this approach takes the world (and government policy) as it is and suggests it cannot be changed – as true of natural hazards as of austerity measures and declining budgets. Consequently, resilience performs an ideological function similar to that of humanitarianism, which is to focus on alleviating the hardships associated with the existing international order rather than working to change the system.

In one respect, the focus on cities does challenge a strong statist narrative. As noted in the previous chapter, many of the emergency services we associate today with the state emerged as municipal developments. It was only later that national states took on a complementary role to local services. While disaster relief highlighted the role of national governments in managing disaster, 100RC is once again emphasizing the role of cities as leading the way in disaster management. The objective of this initiative is not to influence national governmental policy or to establish international standards. Rather, its primary focus is urban, and then global. To the extent that national governments feature in 100RC, it is simply as another potential source of funding for resilience projects, alongside philanthropic organizations and private corporations.

Projects such as 100RC most directly challenge realist beliefs about the international system. Primarily funded by an American philanthropic organization, any sense of a "national interest" is clearly subsumed under a global humanitarian outlook to reduce human vulnerability. While 100RC features far more American cities than any other country, it includes cities from countries with whom the United States has strained relations, such as China, Myanmar, and Palestine; and others with whom the United States has mixed relations, such as Uruguay, Argentina, Brazil, India, the Dominican Republic, and Liberia. In short, there does not appear to be any significant national interest influencing the cities participating in 100RC. From a purely realist perspective, building the resilience of foreign cities to all forms of disruption, including those that would conceivably follow an attack, runs counter to the national interest. In addition, the 100RC initiative emphasizes a wide variety of hazards/security threats that are not military in nature; of the 66 challenges identified, only two are featured in traditional realist literature (and then only quite recently): cyberattack and terrorism. In terms of prioritizing the temporal point of action and the allocation of public spending, the resilience project shares much in common with a human security approach than either liberal interventionism (Chandler, 2012) or realist power politics. However, by focusing on cities and restructuring how cities allocate resources, 100RC also leaves the national context out of the conversation, including as a target of critique. Unlike the human security paradigm, which directly engages with and challenges the policies and spending priorities emphasized by a realist framework (Newman, 2001; Paris, 2001), resilience research largely eschews such direct engagement, or confrontation. Because resilience encourages acting in a world as it is rather than attempting to change that world, it adopts a similar ontological position as realists. Given that the world is a dangerous place, and that there is little that can be done to change that, both resilience and realism offer prescriptions for how to act in such a world.

From this perspective, resilience and realism may be quite commensurate and essentially reduced to a division of labour; local communities must build resilience to deal with a wide range of non-military challenges, while the national government prioritizes a narrow set of military security threats. Indeed, building resilience might be quite sensible to most realists, as it could free national governments to focus on accumulating power and military security while local communities secure themselves against the vast range of other challenges associated with modern life. From this perspective, it is the coupling of resilience with global humanitarianism that challenges realism, not resilience per se.

Consequently, resilience appears as an especially fluid concept that may be used to perform ideological work across a wide range of belief systems. Thus, it would be premature to draw firm conclusions about resilience outside its specific ideological uses. Unlike humanitarianism and sustainable development, which were developed and contested over time to produce resolutions to a specific set of tensions, resilience has not undergone a similar process. As this chapter demonstrates, usage of the concept by state actors and INGOs differs significantly from philanthropic organizations, and thus it is used to contingently resolve an ever-expanding set of ideological tensions and inconsistencies. The key role it has played in the context of this book's focus on disaster management has been to resolve some of the failures of DRR. It does so by suggesting that disaster risk can never be reduced to zero, and consequently individuals and communities must also foster resilience. In this context, building resilience has become another project for authorizing authority, from states to NGOs.

Conclusion

The DRR frameworks that have been in place since the Yokohama agreement all stress that DRR is more effective and should be prioritized over disaster relief. Yet these international agreements have failed to achieve even the modest institutional and funding success of international disaster relief, for which there is a central fund and numerous intergovernmental bodies to facilitate and coordinate action. This is not to say there has been no international funding for DRR programs. A number of states have contributed to DRR programs as part of their ODA. Notably, Japan committed 4 billion US dollars at the conclusion of the Sendai negotiations. To date, however, funding for DRR is less than

0.4% of all aid (Kellet and Caravani, 2013; Tanner, 2015), and disaster funding at the international level is still concentrated in disaster relief.

There are a number of reasons for the disparity in international funding directed toward relief rather than risk reduction. First, risk reduction is a very recent development in international disaster management, whereas disaster relief has a long history in international politics, as outlined in the previous chapter. Beyond this, political elites benefit politically from "generous" relief measures in the aftermath of prominent disasters much more than they do from contributions to risk reduction. This reflects the dominant narratives attached to disasters – emergency response motivated by humanitarian narratives and human interest stories garner significantly more attention than do technical stories outlining preventive measures and technical assistance – to the extent these are covered at all.

I would also suggest that the half-hearted transition away from relief to risk reduction reflects the way that disaster relief capacity has been developed compared to DRR capacity. As the previous chapters have shown, disaster relief was developed as a public good provided primarily by the state or by civil society actors funded by the state. As a result, developed states have a built-in capacity to provide international disaster relief, often in their military forces, but also through the extensive global humanitarian network. This relief capacity is easily exported to other states without major implications for private corporations. DRR, on the other hand, entails a number of measures that are not easily exportable – either due to their geographic specificity or to the political economy of reducing risk. Building codes, for instance, are unique to particular cultural and geographic situations: climate, building materials, and so on, while other risk reduction measures, such as early warning systems or technical innovations, may involve patents or other protections against widespread dissemination for the purposes of profit. As demonstrated clearly by the disagreements in Sendai, technology transfers are highly contentious because they entail giving away knowledge or technical systems that are profitable and protected by intellectual property rights legislation. Unlike relief, risk reduction has not been developed as a public good, but as a private good or a public/private partnership. Consequently, giving DRR in the form of technology transfers to other states is qualitatively different than giving relief: it has implications for private actors and their global competitiveness.

Understanding the difference between relief and risk reduction, in particular the unique ideological commitments behind both, is essential to understanding the ongoing prioritization of relief in the face of widespread acknowledgement that risk reduction is more cost effective, in human and financial terms. DRR itself emerged as a response to ideological tensions exposed by the increasing number of disasters and the inadequacy of disaster relief. Yet developments in the fields of DRR, namely, the increasing prominence given to climate change and the reliance on "sustainable development" to salvage belief in economic liberalism – mean that a meaningful and effective DRR agreement cannot be reached outside agreements on climate change and sustainable development (Mitchell, 2015).

Resilience, which has contributed to much hand-wringing over its potential to justify the withdrawal of the state from preventing disasters and protecting vulnerable people, seems more and more a convenient term for reducing ideological tensions rather than a radical ideological challenge to modernism, economic liberalism, or nation-statism. To be sure, resilience as a frame performs "remedial ideological work" by promoting the acceptance of disasters as an unpreventable aspect of life in a world marked by climate change. In such a world, the best we can hope for is to reduce risk and to be resilient, both of which entail significant effort by states to regulate building of infrastructure and to prioritize certain public goods in the disaster recovery period. Thus, rather than expanding market logic at the expense of the state, this framework actually further empowers states.

The negotiations on the three international DRR frameworks attempt to address areas of ongoing ideological disagreements exposed by disaster events. Many of these same areas of disagreement have surfaced in negotiations on climate change and development. In that respect, the Sendai Framework, as with Yokohama and Hyogo, represents a contingent and relatively minor attempt to resolve the problem of disasters – namely, how disasters reveal the inadequacy of existing ideological foundations of the international social order. With the Sendai Framework extending until 2030, international contestation around these issues is likely to emerge during climate change negotiations but also in response to specific disasters. However, the contested response to specific disasters it is unlikely to centre around DRR, given the weakness of the regime and the lack of concrete initiatives. Rather, disagreements over disaster management are likely to focus on areas that remain outside the DRR framework, such as migration, or in the context of international disaster relief.

Migration remains one of the primary human responses to disaster, and both the relief and risk reduction frameworks operate to reduce the necessity of migration in the aftermath of disaster. Discussion of migration is almost entirely absent from the negotiations over relief and risk reduction, but the next chapter will demonstrate that containing migration has been one of the central features of the modern international order. As order-sustaining regimes, relief and risk reduction have as one of their primary objectives the containment of human populations in the existing state-based international order. Consequently, it is essential to examine the international migration regime as a third DMR. This issue is once again coming to the fore as the world is in the midst of a perceived "migration crisis" and, if the predictions related to climate change come to fruition, will be facing large-scale migration for the foreseeable future. International actors have been promoting a disaster migration regime to respond to these challenges.

Following my discussion of disaster migration, I return to the issue of disaster relief in the final chapters of the book. As I have noted throughout this chapter, the discursive shift to DRR from disaster relief has not been accompanied by a significant shift in actual practice or funding priorities. Disaster relief remains the dominant mechanism that the international community uses to manage disasters, and the Sendai Framework suggests this is unlikely to change in the coming decade. To that end, I focus on contestation over disaster relief in the final section of the book. I examine how disasters and offers of disaster assistance have exposed ideological inconsistencies that then shape the response to international assistance.

References

100 Resilient Cities, 2013. 100 Resilient Cities: About us. https://www.100resilientcities.org/about-us/.

Alexander, D.E., 2013. Resilience and disaster risk reduction: An etymological journey. *Natural Hazards and Earth System Science*, 13, pp. 2707–716.

Anderson, B., & Adey, P., 2011. Affect and security: Exercising emergency in UK civil contingencies'. *Environment and Planning D: Society and Space*, 29, pp. 1092–1109.

Attaran, A., 2005. An immeasurable crisis? A criticism of the millennium development goals and why they cannot be measured. *PLoS Medicine*, 2(10), e318.

Barbier, E.B., 2010. Poverty, development, and environment. *Environment and Development Economics*, 15, pp. 635–60.

Beck, U., 2000. Risk society revisited: Theory, politics and research programmes. In J. Cosgrave (ed.). *The Risk Society and Beyond: Critical Issues for Social Theory*, New York: Routledge, pp. 211–29.

Berkowitz, M. 2015. What a chief resilience officer does, 100 Resilient Cities. https://100resilientcities.org/what-a-chief-resilience-officer-does/.

Blewitt, J., 2014. *Understanding Sustainable Development*, London: Routledge.

Brasset, J., Vaughn-Williams, N., & Croft, S., 2013. Introduction: An agenda for resilience research in politics and international relations. *Politics*, 33(4), pp. 221–28.

Calhoun, C., 2008. The imperative to reduce suffering: Charity, progress, and emergencies in the field of humanitarian action. In M.N. Barnett & T. Weiss (eds.). *Humanitarianism in Question: Politics, Power and Ethics*, Ithaca, NY: Cornell University Press, pp. 73–97.

Cannon, T., 1994. Vulnerability analysis and the explanation of natural disaster. In A. Varley (ed.). *Disasters, Development and Environment*, New Haven, CT: John Wiley & Sons, pp. 13–30.

Carabine, E., 2015. Revitalizing evidence-based policy for the Sendai Framework for Disaster Risk Reduction 2015–2030: Lessons from existing international science partnerships. *PLOS Current Disasters*, April 23, 2015.

Chandler, D., 2012. Resilience and human security: The post-interventionist paradigm. *Security Dialogue*, 43(3), pp. 213–29.

Christoff, P., 2010. Cold climate in Copenhagen: China and the United States at COP15. *Environmental Politics*, 19(4), pp. 637–56.

Coaffee, J., 2013. Rescaling and responsibilising the politics of urban resilience: From national security to local peace-making. *Politics*, 33(4), pp. 240–52.

Collier, S., 2014. Neoliberalism and natural disaster. *Journal of Cultural Economy*, 7(3), pp. 273–90.

Darby, M., 2015. Sendai as it happened: Final day of UN disaster risk reduction talks. http://www.rtcc.org/2015/03/18/sendai-live-final-day-of-un-disaster-risk-reduction-talks/.

de la Poterie, A.T., & Baudoin, M.-A., 2015. From Yokohama to Sendai: Approaches to participation in international disaster risk reduction frameworks. *International Journal of Disaster Risk Science*, 6(2), pp. 128–139.

Duffield, M., 2012. Challenging environments: Danger, resilience and the aid industry. *Security Dialogue*, 43(5), pp. 475–92.

Dunn Cavelty, M., Kaufmann, M., & Soby Kristensen, K., 2015. Resilience and (in)security: Practices, subjects, temporalities. *Security Dialogue*, 46(1), pp. 3–14.

Dynes, R., & Rodriguez, H., 2010. Finding and framing Katrina. In D. Brunsma, D. Overfelt, & J. S. Picou (eds.). *The Sociology of Katrina: Perspectives on a Modern Catastrophe*, Plymouth: Rowman & Littlefield.

Ewald, F., 1991. Insurance and risk. In C.G.G. Burchell & P. Miller (eds.). *The Foucault Effect: Studies in Governmentality*. Chicago, IL: University of Chicago Press, pp. 197–210.

Fearon, J., 2008. The rise of emergency relief aid. In M.T.W. Barnett (ed.). *Humanitarianism in Question: Politics, Power and Ethics*, Ithaca, NY: Cornell University Press, pp. 49–72.

Gaillard, J.C., & Mercer, J., 2013. From knowledge to action: Bridging gaps in disaster risk reduction. *Progress in Human Geography*, 37(1), pp. 93–114.

Garland, D., 1996. The limits of the sovereign state strategies of crime control in contemporary society. *British Journal of Criminology*, 36(4), pp. 445–71.

Gordon, T., 2005. The political economy of law-and-order policies: Policing, class struggle, and neoliberal restructuring. *Studies in Political Economy*, 75(1), pp. 53–77.

Gray, L.C., & Moseley, W.G., 2005. A geographical perspective on poverty-environment interactions. *The Geographical Journal*, 171(1), pp. 9–23.

Hannigan, J., 2012. *Disasters without Borders*, Cambridge: Polity Press.

Harvey, D., 2007. Neoliberalism as creative destruction. *The Annals of the American Academy of Political and Social Science*, 610(1), pp. 21–44.

Hewitt, K., 1983. *Interpretation of Calamities: From the Viewpoint of Human Ecology*, Boston: Allen & Unwin.

Hewitt, K., 1997. *Regions of Risk: A Geographical Introduction to Disasters*, Harlow: Longman.

Joseph, J., 2013. Resilience as embedded neoliberalism: A governmentality approach. *Resilience*, 1(1), pp. 38–52.

Kellet, J., & Caravani, A., 2013. *Financing Disaster Risk Reduction: A 20 Year Story of International Aid*, http://www.odi.org/sites/odi.org.uk/files/odi-assets/publications-opinion-files/8574.pdf.

Kellet, J., & Sparks, D., 2012. Disaster risk reduction: Spending where it should count. http://www.globalhumanitarianassistance.org/wp-content/uploads/2012/03/GHA-Disaster-Risk-Report.pdf.

Kelman, I., 2015. Climate change and the Sendai Framework for disaster risk reduction. *International Journal of Disaster Risk Science*, 6(6), pp. 117–27.

Kemp, P., & Martens, R., 2007. Sustainable development: How to manage something that is subjective and never can be achieved? *Practice and Policy*, 2(2), pp. 5–15.

Konings, M., 2012. Neoliberalism and the state. *Alternate Routes: A Journal of Critical Social Science Research*, 23, pp. 85–98.

Kunreuther, H., & Michel-Kerjan, E., 2013. Managing catastrophic risks through redesigned insurance: Challenges and opportunities. In G. Dionne (ed.). *Handbook of Insurance*, New York: Springer, pp. 517–46.

Laffey, J., & Weldes, M., 2004. Policing in global governance. In M. Barnett & R.D. Duvall (eds.). *Power in Global Governance*, Cambridge: Cambridge University Press, pp. 59–79.

Lipper, B., 2016. How to develop a resilience strategy, 100 Resilient Cities. https://www.100resilientcities.org/how-to-develop-a-resilience-strategy/.

Mace, M.J., & Schaeffer, M., 2013. Loss and damage under the UNFCCC: What relationship to the Hyogo Framework, Berlin. http://www.lossanddamage.net/download/7248.pdf

Malkki, L.H., 1996. Speechless emissaries: Refugees, humanitarianism, and dehistoricization. *Cultural Anthropology*, 11(3), pp. 377–404.

Manyena, S.B., 2006. The concept of resilience revisited. *Disasters*, 30(4), pp. 434–50.

Mauro, F., & Hardison, P., 2000. Traditional knowledge of indigenous and local communities: International debate and policy initiatives. *Ecological Applications*, 10(5): 1263–69.

May, C., 2013. *The Global Political Economy of Intellectual Property Rights: The New Enclosures?* London: Routledge.

McEntire, D., Fuller, C., Johnston, C.W., & Weber, R., 2002. A comparison of disaster paradigms: The search for a holistic policy guide. *Public Administration Review*, 62(3): 267–81.

Mercer, J., 2010. Disaster risk reduction or climate change adaptation: Are we reinventing the wheel? *Journal of International Development*, 22(2), pp. 247–64.

Mileti, D., 1999. *Disasters by Design: A Reassessment of Natural Hazards in the United States*, Washington, DC: National Academies Press.

Mitchell, T., 2014. Sticking points on the way to a new disaster reduction pact. http://www.trust.org/item/20140515090205-to3c1.

Mitchell, T., 2015. Will fault lines in Sendai signal a rocky 2015? *Overseas Development Institute (ODI)*, http://www.odi.org/comment/9145-will-fault-lines-sendai-signal-rocky-2015.

Newman, E., 2001. Human security and constructivism. *International Studies Perspectives*, 2(3), pp. 239–51.

O'Brien, G., O'Keefe, P., Rose, J., & Wisner, B., 2006. Climate change and disaster management. *Disasters*, 30(1), pp. 64–80.

O'Malley, P., 2000. Uncertain subjects: Risks, liberalism and contract. *Economy and Society*, 29(4), pp. 460–484.

Paris, R., 2001. Human security: Paradigm shift or hot air? *International Security*, 26(2), 87–102.

Paul, S., 2015. Governments fall short at World Conference on Disaster Risk Reduction, Oxfam International. Oxfam Press Release. https://www.oxfam.org/en/pressroom/reactions/governments-fall-short-world-conference-disaster-risk-reduction.

Pelling, M., 2003. *Paradigms of Risk, Chapter 1*, In M. Pelling (ed.). *Natural Disasters and Development in a Globalizing World*, London: Routledge.

Perrow, C., 1981. Normal accident at Three Mile Island. *Society*, 18(5), pp. 17–26.

Perrow, C., 2011. *Normal Accidents: Living with High Risk Technologies*, Princeton, NJ: Princeton University Press.

Pingali, P., Alinovi, L., & Sutton, J., 2005. Food security in complex emergencies: Enhancing food system resilience. *Disasters*, 29(s1), pp. S5–S24.

Reid, J., 2012. The disastrous and politically debased subject of resilience. *Development Dialogue*, 58, pp. 67–79.

Rieff, D., 2003. *A Bed for the Night: Humanitarianism in Crisis*, New York: Simon and Schuster.

Robinson, J., 2004. Squaring the circle? Some thoughts on the idea of sustainable development. *Ecological Economics*, 48(4), pp. 369–84.

Rowling, M., 2015. Ten things to know about the Sendai disaster risk reduction deal – News – Professional Resources – PreventionWeb.net. Thomson Reuters. http://www.preventionweb.net/english/professional/news/v.php?id=43375.

Schipper, E.L.F., 2009. Meeting at the crossroads?: Exploring the linkages between climate change adaptation and disaster risk reduction. *Climate and Development*, 1(1), pp. 16–30.

Schmidt, J., 2014. Intuitively neoliberal?: Towards a critical understanding of resilience governance. *European Journal of International Relations*, 21(2), pp. 402–26.

Schwarze, R., & Wagner, G.G., 2007. The political economy of natural disaster insurance: Lessons from the failure of a proposed compulsory insurance scheme in Germany. *European Environment*, 17(6), pp. 403–15.

Scott, J.C., 1995. State simplifications: Nature, space and people. *Journal of Political Philosophy*, 3(3), pp. 191–233.

Slim, H., 2000. Dissolving the difference between humanitarianism and development: The mixing of a rights-based solution. *Development in Practice*, 10(3–4), pp. 491–94.

Spector, M., & Kitsuse, J., 1987. *Constructing Social Problems*, Menlo Park, NJ: Transaction Publishers.

Sphere, 2011. *The Humanitarian Charter: The Sphere Project: Humanitarian Charter and Minimum Standards in Disaster Response*. 2011 edition. Geneva.

Sumner, W., 2015. Explanation of position: Sendai framework for disaster risk reduction 2015–2030. US Department of State. http://www.state.gov/p/io/rls/2015/239976.html.

Tanner, T., 2015. Can reducing disaster risk create a "triple dividend"? *Thomson Reuters*, http://www.trust.org/item/20150316103956-18oqn/?source=jtBlogs.

Tierney, K.J., 2007. From the margins to the mainstream? Disaster research at the crossroads. *Annual Review of Sociology*, 33, pp. 503–25.

UNISDR, 2004. What is Disaster Risk Reduction? Disaster Risk Reduction. https://www.unisdr.org/who-we-are/what-is-drr.

UNISDR, 2005. *Hyogo Framework for Action 2005–2015*, Geneva. https://www.unisdr.org/we/coordinate/hfa.

UNISDR, 2006. *GA Resolution 60/195: International Strategy for Disaster Reduction*. New York: United Nations.

UNISDR, 2015. *Sendai Framework for Disaster Risk Reduction 2015–2030*, Sendai. https://www.unisdr.org/we/inform/publications/43291.

United Nations, 1994. *Yokohama Strategy and Plan of Action for a Safer World*, Geneva: United Nations.

United Nations General Assembly, 2000. United Nations Millennium Declaration. 55th Session. September 18, 2000.

Varley, A., 1994. The exceptional and the everyday: Vulnerability analysis in the International Decade for Natural Disaster Reduction. In A. Varley (ed.). *Disasters, Development and Environment*. New Haven, CT: John Wiley & Sons.

Vogel, A., 2006. Who's making global civil society: Philanthropy and US empire in world society. *The British Journal of Sociology*, 57(4), pp. 635–55.

Vogel, R.M., 2012. Politics, diversity and the distribution of federal disaster assistance. *Economics and Business Letters*, 1(2), pp. 37–42.

Watson, S.D., 2011. The "human" as referent object?: Humanitarianism as securitization. *Security Dialogue*, 42(1), pp. 3–20.

Welsh, M., 2013. Resilience and responsibility: Governing uncertainty in a complex world. *The Geographical Journal*, 180(1), pp. 15–26.

Wilkinson, E., 2015. We leave Sendai with a new global deal on disaster risk, but does it go far enough? Climate and Development Knowledge Network. http://cdkn.org/2015/03/opinion-leave-sendai-new-global-deal-disaster-risk-go-far-enough/?loclang=en_gb.

Wisner, B., Blaikie, P., Cannon, T., & Davis, I., 2004. *At Risk: Natural Hazards, People's Vulnerability, and Disasters*, London: Routledge.

5

DISASTER DISPLACEMENT AND MIGRATION

In his seminal book on mobility in the modern world, Cresswell (2006) astutely observes that the meanings attached to human mobility are ideological, that is, while mobility is a recurring aspect of human existence, humans understand movement in terms of how it relates to social order. Drawing on the shift to modernity, Cresswell observes that order in the industrial period was markedly distinct from that of the feudal period and consequently entailed very different understandings of mobility. Economic and political changes in the industrialization process displaced people and rendered many landless; many moved to urban centres and engaged in wage labour, in the process challenging traditional understandings of mobility. New classifications of migrants were created, along with new institutions for both facilitating and restricting human movement. Today, the prospect of mass displacement due to climate change and other environmental stressors may challenge dominant beliefs about human mobility. These developments suggest that existing categories and institutions for managing migration will prove insufficient for understanding environmentally induced displacement. As Ransan-Cooper et al. (2015) observe, there has already been a proliferation of frames to capture the complex phenomenon of environmental migration.

Migration has long been one of the principal behavioural adaptations to disaster and environmental changes throughout human history. Prior to the relatively recent establishment of the restrictive international migration control regime, individuals and communities could move to escape immediate danger and to rebuild after disasters. In the current context, those dislocated due to disaster and environmental degradation are likely to face severe limitations on their mobility, particularly across international borders. In contrast, highly skilled workers and tourists traverse these borders with relative ease.

Thus, any discussion of disaster migration must start with an understanding of the current migration system that both facilitates and restricts human mobility. As Cresswell elegantly phrased it, "the production of some kinds of mobility often effectively immobilize others" (21). Similarly, Zolberg (2006) describes the current situation as a "complex structure ... [in which] certain populations ... can be assets from one perspective and liabilities from another" (70). Whether climate migrants will be viewed as assets rather than liabilities will depend on a great many factors related to the belief systems foundational to international order. As this chapter will show, the normative foundations of the current management of international migration do not bode well for the establishment of a disaster displacement regime, and those displaced by environmental factors will face severe restrictions on their mobility.

Studies on climate change suggest that degradation of natural environments is likely to have a significant impact on the levels and patterns of human migration, implying that the existing mechanisms for managing migration are inadequate (Black et al., 2011). Although the number of people expected to be displaced due to climate change is highly variable and contested, there is general consensus that displacement will increase markedly, placing greater strains on existing mechanisms for managing international migration. The current international migration regime already faces immense tensions and inconsistencies that erupt in periodic crises and require ongoing interventions to maintain. Mass migration movements in particular, such as those expected with climate change, will further expose the problematic foundations of the current order and lead to alterations of the existing order. Current migration crises suggest that existing practices for managing international migration are already being altered, with statist and realist beliefs gaining ascendency in the ideological milieu of the modern migrant management system.

While I refer to an international migration regime or system, migration is not governed like other aspects of the international system, such as trade or even disaster relief. International migration is not globally or even internationally managed; rather, it is "governed almost exclusively at the level of the nation-state" (Newland, 2010:331) and through a range of bilateral and multilateral mechanisms (see Aleinikoff, 2002; Robinson, 2018). There are a wide range of international institutions charged with overseeing aspects of international migration: the International Organization for Migration (IOM), International Labour Organization (ILO), and United Nations High Commissioner for Refugees (UNHCR) in particular are major managers of global migration, but the international system lacks a strong institution charged with overseeing and regulating state compliance with migration agreements in the manner that the World Trade Organization (WTO), for instance, governs trade. Nonetheless, global migration is among the most managed and controlled aspects of the modern international order.

The current migration regime is a mixed system that essentially categorizes migrants based on reason for migration and level of "desirability" for receiving states. Although migrants move for a variety of reasons and in response to a wide range of pressures, the international migration regime places migrants into broad categories such as economic migration, family reunification, humanitarian migration, and smuggled and trafficked persons – each with a distinct form and intensity of institutionalization at the national and international level. These streams, as presently instituted, have histories of norm entrepreneurship and contestation that attempt to balance the ideological foundations of international order in different measures, and for different purposes; for instance, there are separate international conventions on refugees, migrant workers, and smuggled persons. None of them are designed to respond effectively to disaster displacement or climate-induced migration.[1]

Recent attempts to accommodate environment-induced displacement either seek to incorporate it under existing provisions, such as those for refugees, or to create new principles and institutions, such as the Nansen Initiative or a special initiative under the United Nations Framework Convention on Climate Change (UNFCCC). To date, such proposals have met with limited success. Drawing on previous efforts to create migration-specific regimes, in particular the 1951 Refugee Convention and the 1990 International Convention on the Protection of Migrant Workers (ICRMW), I show that the dominant beliefs about international order severely limit the mobility of most migrants and offer little protection for environmentally induced migrants. While these belief systems exhibit ambiguities, tensions, and inconsistencies, the complexity of climate migration makes it difficult to distil into simple ideological terms, and does not lend itself easily to fully disrupt these belief systems.

Managing Migration

Torpey shows in his groundbreaking work on the passport how the process of creating states required increased control over the movement of people. Identity documents became the preeminent technology in this process, by fixing people's identity based on nationality and membership in a territorial state. The state came to establish "a monopoly of the legitimate means of movement", displacing the authority of religious organizations and private companies to sanction movement (Torpey, 2000:1). In essence, controlling migration was a necessary precondition for the establishment of a state-based international order.

State efforts to control migration in the maintenance of this order, however, led to what Guiraudon and Joppke (2001) refer to as control dilemmas, in which various pressures to solicit or stem migration flows produced inefficient policy outcomes or contrasting interests. In some contexts, this has led to policy gaps – the actual effect of migration policy does not match the intended effect due to "unintended consequences and inadequate implementation" (Cornelius and Tsuda, 2004:5). In addition, states faced a dilemma between facilitating the types of migration they wanted for purposes of economic benefit, nation building, and/or international influence and the types of migration they sought to prevent – usually based on beliefs about national identity, race, culture, and class. States, of course, were not the sole actors influencing migration policy even as they exerted authority over the process; corporations sought to influence migration patterns for purposes of access to labour, while civil society groups lobbied governments to accept or restrict migration in line with their views on the identity of the nation.

Further crises in the management of migration emerged in the context of repeated perceived failure of migration policies, most evident in the post–World War II period. Temporary migrants turned out to be more permanent than many thought, while diversity (national, cultural, religious) due to immigration proved to be more difficult to reconcile than many multiculturalists envisioned (Castles and Miller, 1998). Migrant labour policies had unexpected (and unwanted) outcomes, while control policies did not always produce their promised outcome and exposed the violence and dangers of the existing system. Advances in transportation technology combined with stronger push-and-pull forces gave rise to a dramatic increase in irregular and undocumented migration (Ghosh, 1998). Cornelius and Tsuda (2004) suggest that with the current levels of social and economic inequality globally, belief that control measures could reliably reduce irregular migration is misplaced. This has not been through lack of effort. In response to growing irregular migration, many destination countries began to implement harsher control mechanisms, such as remote-control policies that disrupt migration flows at the source through visa requirements, carrier sanctions, and territorial exclusions (Guiraudon and Joppke, 2003:13). Many states expanded the border internally through detention, deportation, and documentation requirements for access to employment, housing, and social services, as well as sanctions on employers for hiring undocumented migrants. Rather than halting irregular migration, these measures instead force migrants into more precarious routes of migration and forms of labour, further encouraging policy makers to expend greater resources on more stringent measures. These (ongoing) struggles to manage migration reveal the centrality of, and tensions between, the various belief systems that underpin current efforts to manage human mobility.

Migration Control and Economic Liberalism

Economic liberalism has profoundly shaped the current migration control regime – not only because the liberal economic system produces pressures that induce migration (push-and-pull factors); it also provides a simple and coherent account of why people migrate, as well as offering a solution to the management of migration – the market. There are various strands of market-based theories explaining international migration (see Massey et al., 2005), most of which assert that migration is the consequence of "uneven economic exchange between states of origin and destination" (Zolberg, 2006:64). Economically liberal perspectives view migrants in terms of their relative value in a global economy and migration as a positive mechanism that improves global equality by closing the gap between origin and destination states.

Indeed, free(r) human mobility is asserted as a basic precondition for an economic liberal system. In classic liberal economic ideology, few restrictions on migration are needed or desired. As long as there is an imbalance in labour or wages, migration is necessary. Restrictions on migration, if too severe or effective, lead to market inefficiencies, increases in the cost of labour and consumer goods, and escalating global inequality (Caplan and Naik, 2015). Conversely, economic liberalism suggests that the market will "naturally" control migration through a reduction in the disparity between the supply of labour and/or capital, either due to scarcity of economic opportunities in destination states or increased availability in home states. Cornelius and Tsuda suggest that "the most powerful determinants of international migration dynamics" are market forces and demography and not governmental preferences (2004:43). In short, in economic liberalism, the market is the most efficient and legitimate tool for regulating migration flows.

Liberalism does not simply provide an explanation for why migration occurs; it offers a normative assessment – migration produces positive outcomes. Migration is seen as essential to fulfilling one of economic liberalism's greatest promises – reducing economic inequality. In the various strands of economic liberal theory, migration is an essential tool to redistribute surplus labour (and capital), which eventually reduces differences in wages between areas (Arango, 2000:285). Caplan and Naik summarize the liberal position succinctly – a global liberal economy that embraces open borders would produce a "wealth explosion" by reducing the "capital/labour ratio in destination countries and raise it in origin countries" (2015:188). In line with economic liberalism, they suggest that allowing the market to regulate migration would drastically reduce global inequality over time. This is based on the claim that free movement within states produced significant spatial and class mobility within capitalist economies, and reduced, though not eliminated, regional inequality (Caplan and Naik, 2015). The European Union (EU) is a recent noteworthy example of this principle being put into practice at the supra-state level. Free movement within the EU was seen as a logical extension of its liberal economic foundation, although as Arango notes, the EU also demonstrates the limitations of the theory, amply demonstrating that individuals' decisions to migrate are not determined only by wage differentials between regions (2000:287).

Even as economic liberalism discourages restrictions on migration, it also discourages special protections for migrants. The value of migrants is based on supply and demand; migrants with unique skills and talents are in short supply and high demand, and thus can command higher salaries and access to sociopolitical rights in recruiting states. Conversely, less skilled labour is high in supply, with low to moderate demand. De Guchteneire and Pecoud (2009) suggest that it is this market-based valuation of migration that ultimately led to the failure of the ICRMW. Compared with other human rights agreements, the ICRMW has garnered very few signatories (51), none of which are migrant-receiving states, and is largely neglected – even by the UN (Grange and D'Auchamp, 2009). One reason for opposition to the ICRMW is that it would drive up the cost of low-skilled labour. In addition to increasing the cost of low skilled labour by granting such workers a wide range of socio-economic rights, the ICRMW undermined the "vertical hierarchization of migrants" based on their "rarity and market value" (Grange and D'Auchamp, 2009:30). According to Patrick Taran, the failure of the ICRMW stems from the desire for "cheap, docile, temporary, and easily removable" foreign workers (2009:151). This is consistent with economic liberalism, in which state intervention – either to restrict migration or to equalize the treatment of migrants – distorts the market for migrant workers.

The treatment of migrants as strictly economic actors is both normatively problematic and empirically inaccurate. An unregulated market has proven incapable of dealing with the abuse and mistreatment of migrant workers and indeed significantly contributes to it. The economically motivated account of migration central to classic liberal theory does not match reality – people's choices are severely constrained and thus do not approximate the basic unconstrained conditions of liberal economic theory; migrants do not leave their homes and familiar surroundings based on individual cost/benefit calculations of wage differentials (Castles and Miller, 1998:22; Zolberg, 2006:64). Historical connections between societies deeply influence migration patterns, and migrants rarely act individualistically; instead they often respond to family- and community-level decisions (Castles and Miller, 1998:24–26). Even in environments where mobility is largely unrestricted and there are significant push-and-pull factors, many people choose not to migrate. Thus, one of the core premises of economic liberalism, which is that individuals make choices based solely on economic rationale, is dubious. Once this is acknowledged, economically liberal objectives, such as reducing restrictions on migration (on mobility and treatment) and prioritizing the reduction of economic inequality above all else, are revealed as ideological positions, as is the belief that liberal economic policies alone can achieve that outcome, even if one were to accept its prioritization.

Economic liberalism problematically depicts the market responding to a pre-existing and naturally occurring global supply of, and demand for, labour. Critical migration perspectives show that, to a significant extent, the current supply of labour is the product of the expansion of economic liberalism (Arango, 2000:290; Massey et al., 2005:42). The liberal processes that produce the current supply (surplus) of labour include loss of land and traditional forms of subsistence; insufficient access to paid labour domestically; and, more recently, austerity measures that have reduced social support systems. Other aspects of the liberal economic system that push migrants out of their home states include environmental degradation associated with resource extraction and the global trade in arms and resources that fuel conflicts. The net product of this global economic system is vast disparity in economic opportunity and outcomes in which some parts of the world experience high levels of poverty, few economic opportunities, stalled development, environmental degradation, and violent conflicts – all of which are key drivers of migration. In short, economic liberalism offers a solution (migration) to a problem (surplus labour, economic inequality, conflict) that it is deeply implicated in producing.

For migration to work as portrayed in liberal economic theory, workers must be free to migrate to places where wages are higher. Yet this is not the case: most workers are highly constrained in their movement. Restrictions on workers' mobility are not solely the product of xenophobia and nationalist exclusions (which I explore in the next section), but of capitalist pressures. Historically, capitalists have used forced movement and fought against rights regimes for migrant workers to increase profits and to ensure that certain workers are unable to compete on an equal footing with others, while labour unions have also supported migration restrictions to reduce competition (Castles and Miller, 1998:48). The internal tensions within capitalism – particularly those between workers and capitalists – is fully on display in the case of labour migration. Efforts over the past 40 years to establish an international convention on the rights of workers to resolve some of the inconsistencies, and inequities, of the liberal economic system have failed to produce a strong migrant rights regime due, in part, to the global disparity in the production and demand for labour.

The demand for labour is also more complicated than depicted in economic liberal ideology. In the advanced capitalist economies, the economy is essentially divided between capital-intensive and labour-intensive sectors (Massey et al., 2005). This division produces a structural demand for foreign labour due to the unavailability of local labour, either due to lack of workers or potential workers resisting labour-intensive jobs regarded as low-status (Arango, 2000). This hierarchy of labour also reproduces a hierarchy of culture/race and gender, which in turn feeds back into the justification for the nation state and the division of the world into discrete political communities based on cultural similarities (Castles and Miller, 1998:35). This has given rise to a significant tension – industrialized societies need workers in certain sectors, but many of these societies do not want people with foreign cultural practices. Cornelius and Tsuda argue that global convergence in negative attitudes toward immigrants mirrors the global convergence in structured dependence on foreign labour (2004:4).

One policy area exhibiting international convergence is temporary migration. For much of the past 70 years, temporary labour migration has become the contingent ideological resolution to the tension between the need for labour mobility in a global economy and the need to restrict migration in a nation-state system (see Dauvergne and Marsden, 2014). Temporary migration functions as a resolution to this tension through the liberal rendering of the individual as an economic animal and a worker above all else. As long as the wages paid for temporary work are higher than those available in their home states, or families need to diversify their income in precarious economic circumstances, workers will migrate and support their families through remittances. This simultaneously produces increased demand for accumulation of skills. Skilled workers in greater demand can thus negotiate not only higher wages but also pathways to permanent settlement, family reunification, and other sociopolitical rights. In this economically liberal system, the market itself determines the types of rights and protections to which workers are entitled.

As a resolution to the problem of cultural heterogeneity implied by a global workforce, however, these temporary programs have given rise to new tensions and contradictions. In almost all states using temporary labour programs, temporary programs led to unintended and unwanted outcomes, such as permanent settlement, family reunification, and sizable minority populations (Joppke, 1999; Cornelius and Rosenblum, 2005). In many cases, the unintended outcomes are well known and expected. These programs are thus implemented not necessarily to achieve a specific outcome but, as Dauvergne and Marsden show, to serve an ideological purpose – to reduce the incongruity between two belief systems that offered very distinct views about the causes and purposes of immigration and the appropriate mechanism to regulate it. Temporary labour migration attempts to sustain belief that movement for purposes of labour, and directed by market forces, can and should be separated from the distribution of political membership, directed by the state.

Migration and Nation-Statism

In contrast to the liberal account of migration, nation-statist ideology asserts that the state has the final authority to decide who does and does not belong to the political community and thus, by extension, who may or may not enter its territory. Zolberg eloquently (over)states the relationship between migration control and state sovereignty as such: "there exists a universal and unambiguous consensus ... that every state has the right to restrict entry of foreigners ... as an essential attribute of political sovereignty, without which that concept would have little significance" (2006:66). While I would suggest this position is neither universal nor unambiguous, in its idealized representation, state sovereignty is essentially stripped down to a single function – control of the border. The use of nationalism to authorize the self-authorization of the state, which promoted the belief that the state should correspond to a distinct nation, gave further justification to restrict access and control membership in the national community. As Castles and Miller note, the historical development of the nation-state was "premised on the idea of cultural and political unity" (1998:15). In this formulation, the state is the legitimate entity to regulate migration.

As I argued in Chapter 2, the distribution of public goods has become the central component of the state's self-authorization, and in this respect, the distribution of membership is the foundational public good. According to Walzer, "the primary good we distribute to each other is membership in some human community" (1983:31). Belief in the nation state as the foundational human community means that the state's principal obligation is to manage the distribution of membership in the nation state. Currently there are two dominant models for distributing membership: *jus soli* and *jus sanguinis*, which allocate membership based on one of two criteria – soil or blood. Many states employ some combination of these two mechanisms in addition to alternative mechanisms for obtaining membership (see Weil, 2001). Regulating different routes to acquiring membership in the modern state system, such as reproducing in the territory, blood relations, marriage, and/or naturalization, is a basic function of the state as part of its self-authorization. Regardless of which mechanism of determining membership is dominant, controlling physical movement into the territory then is at the root of the ideology of state sovereignty and the reproduction of the nation in whose name the state purports to act.

Because few states attempt to directly control or regulate the reproductive practices of existing members of the political community,[2] the primary mechanism for controlling access is through restricting access to territory. This requires some control over who may enter the territory, who may reproduce in the territory, and who may qualify for naturalization once in the territory. The increase in migration has meant that "citizenship and naturalization have become prime areas of resolving the tensions of migration and the nation-state system" (Castles and Miller, 1998:16). To take one example, the problem of controlling births of non-nationals resident in the state's territory has played out in the politically charged atmosphere in the United States over concerns with "anchor babies". The term suggests that migrants are perceived as circumventing restrictions on access to membership in the United States by taking advantage of the *jus soli* route to American citizenship by non-nationals having babies in American territory (Huang, 2008). As Huang notes, this narrative has led to "immigrant women's bodies becoming the economic, demographic and political battleground for America's future" (406).

Yet there is an ongoing tension in nation-statism concerning its two rather distinct legitimizing narratives: the bureaucratic-administrative function and the ethno-national, or what in another context Anthony Smith has referred to as the Western and Eastern model of state building (see Smith, 1991; Schlesinger, 1994:319). The current nation-statist belief that the state controls its borders in service to the nation belies the early state formation process, which occurred in the absence of nationalist justification. Neuberger shows that in the post-colonial context, the push to increase the state's administrative efficiency is what led to nationalist policies, such as linguistic homogeneity – not the other way around (1997:234). According to Torpey, the earliest justifications for establishing identity and controlling movement was not to produce or enforce cultural homogeneity but for bureaucratic-administrative purposes of "extracting military service, taxes, and labor" (7). This bureaucratic rationale then contributed to the reification of national identity and to national identity becoming the primary identity associated with the state and the distribution of public goods. In many contexts, the state preceded nationalism, although nationalism emerged as one of the most powerful beliefs authorizing the authority of the state.

Even in a world in which there are various conceptions of the nation and the state and of various routes to membership in different political communities, the management of migration reflects the state's desire to maintain its authority by administering its borders and by preserving a national identity. The state does not need "the nation" to justify restriction of migration, but it helps; while "the nation" benefits from a bureaucratic-administrative state to manage migration and membership. Consequently, regardless of the actual content of migration policy and who the policy is designed to let in and keep out, the administrative-bureaucratic aspect of the ideology of state sovereignty requires demonstrated ability to implement the state's migration policy effectively. For example, while the content of Australian immigration policy changed from "white Australia" to a more multicultural immigration policy in line with a contested (and ongoing) shift in Australian national identity, Australia – the sovereign state – is expected to control its borders and enforce its migration policies (see Jupp, 2007).

Because of the varying and contrasting interests that influence migration policy, political leaders and migration managers may use migration policy to preserve some aspects of national identity and to challenge others (Dauvergne, 2005), while still maintaining state authority. This is a risky process, as the strategic use of migration policy may lead to organized resistance from groups opposed to changes in national identity. The rise of populist and anti-immigration parties in Western democracies suggests that less restrictive immigration policies and the resulting increase in cultural heterogeneity has triggered concerns over the use of immigration policies by certain elites to prioritize certain economic interests at the expense of preserving national identity and social order (see Thränhardt, 1995; Akkerman, 2005; Rydgren, 2008). As Akkerman notes, these parties are not necessarily anti-democratic or anti-liberal; in fact, many espouse a strong defence of political liberalism in their opposition to immigration. What this ongoing contestation reveals is that there are multiple and incommensurate ideological foundations of the nation-state order, with various actors attempting to reduce the discussion on immigration to a single ideological belief system.

The migration bureaucracies of most states attempt to manage migration in pursuit of several distinct goals that reflect beliefs about preserving social order. Preserving a relatively fixed national identity is but one consideration; states also seek to bring in various types of workers and entrepreneurs in order to provide public goods: to strengthen the economy and/or to sustain their population in the face of demographic decline and economic stagnation. They may also bring in family members of citizens to uphold commitments to basic political rights and to facilitate migrants' integration into the national community. Numerous states allow asylum seekers and refugees into the state out of humanitarian concern, to alleviate an international crisis, and/or to bolster their reputation and soft power.

On the opposite side of the ledger, states seek to exclude certain types of migrants: migrants of a particular race, culture, or religion that do not fit with the asserted national culture, but also those who do not fit the labour market needs of the national economy, those who might place excessive demands on the resources of the state, as well as those deemed to be a threat to national security. The general orientation of migration management at the national level is that, by keeping out the groups most likely to arouse organized opposition and thus call the authority of the state into question, the state may continue to facilitate wanted streams of migration that reinforce the authority of the state and its provisions of certain public goods.

The desire to restrict entry of unwanted migrants was the core issue in the debate over the ICRMW. Even though the convention explicitly reasserts the right of states to determine who may enter the state, many states opposed the agreement on the grounds that it undermined the right of each state to determine its admission policies (Batistella, 2009:62). The greatest problem from this perspective was that the convention granted rights to irregular migrants who are a direct challenge to the authority and ability of the state to control who enters the state and thus was interpreted as undermining state sovereignty (De Guchteneire and Pecoud, 2009:32). Others saw the convention as expanding the right to family reunification, which states interpreted as determining their immigration policies and limiting state control over admission (Grange and D'Auchamp, 2009:78). The convention also required that states submit to external scrutiny to ensure compliance with the convention's provisions, again interpreted as limiting state sovereignty (De Guchteneire and Pecoud, 2009:32).

Further demonstrating the centrality of nation-state sovereignty to the management of migration, De Guchteneire and Pecoud contrast the ratification processes for the ICRMW and the Palermo Protocols on Human Smuggling and Trafficking. After 19 years, the ICRMW, a convention expanding the rights of migrants and placing some limits on state sovereignty, had acquired only 41 ratifications (after 27 years, it is still at only 51). For comparison, it took less than two years to get 119 and 131 ratifications of the two protocols that expanded state control over migration "in the context of the fight against criminality" (De Guchteneire and Pecoud: 2009:24). Consequently, the ICRMW is somewhat unique amongst key UN human rights agreements in that it has not gained significant traction. The convention was promoted primarily by migrant-producing states, who are among the only signatories (Taran, 2009:18). No major migrant-receiving state has ratified the treaty, including those often associated with the promotion of human rights, such as the countries of North America and Europe; Australia; Brazil, Russia, India, China, and South Africa (BRICS), and most Gulf States (Wickramasekara, 2008). In short, countries that rely heavily on migrant labour oppose many of the provision contained in the convention. Taran notes that the agreement is often dismissed as "impractical and unrealizable ... irrelevant ... and essentially dead" (19). He goes on to suggest that the convention suffered from lack of awareness, promotion, and political will (22). Citing a report by Nune and Niessen, Taran concluded that many states opposed the convention on the grounds that it promoted more open immigration policies and that signing the convention would encourage more migration (Nune and Niessen, quoted in Taran, 2009). There is no appetite among migrant-receiving states to negotiate an international agreement on migration that reduces their ability to determine who may enter the state or that requires equal treatment for nationals and migrants (Wickramasekara, 2008).

Managing migration in an international order divided into nation states that participate in a global economy is a fine balancing act between facilitating desired forms of migration and restricting unwanted migration. These two pressures proceed hand in hand. The well-documented increasing securitization of migration, particularly in the West (Huysmans, 2000, 2006; Watson, 2009; Bourbeau, 2011), has not reduced international migration; rather, it increased the dominance of global economic liberalism, the structural demand for global migration, and high levels of migration. Once again, we see how the expansion of economic liberalism, including its neo-liberal strand, has not reduced the authority of the state but increased its regulatory oversight and authority. Of course, as I have argued throughout, we should not interpret areas of complimentarity to suggest a single, unified process. Liberal and nation-state ideologies offer contrasting understandings of the purposes of migration (who the community should want and not want) and the legitimate mechanism for managing migration (state or market), which produces tensions and inconsistencies that give rise to recurring crises. And yet the situation is more complex than the tensions that arise between economic liberalism and statist understandings of migration – our current understandings of migration are also profoundly shaped by realist, modernist, and humanitarian belief systems.

Migration and Realism

With few exceptions, realist academic literature has had little to say about migration, generally treating it as a matter of domestic politics virtually unconnected with the seminal concern in the field of International Relations (IR) – great power conflict (Guild, 2009; Hollifield and Wong, 2015). Despite this lack of attention from academic realists, realist ideology strongly influences dominant understandings of migration. While it shares many of the premises of nation-statism outlined in the previous section, it offers a subtle shift in thinking about migration. Myron Weiner, who is among the few realists to systematically address the issue of migration, articulates basic realist beliefs concerning migration, arguing that controlling borders is a basic condition of state sovereignty – a state that does not control entry to its territory is a colony and subject to "peaceful invasion" by another state (2006:91). In drawing attention to the necessity of managing borders to prevent invasion and to maintain sovereignty, realism frames migration as a matter of security between sovereign states.

This means that any claim that a migrant has to enter a territory is secondary to the state's right to control entry. As Elspeth Guild notes, the "security of the refugee against return to persecution is subordinated to the state's claim to collective security within its borders" (2009:81). Even the refugee, the entity who has the strongest claim to the protection of a foreign state in international law, does not override the state's claim to protect its own population. This is clearly reflected in international agreements regulating migration. In the 1951 Refugee Convention, the prohibition on refoulement does not apply to refugees who are "a danger to the security of the country in which he is" (Article 33). The ICRMW states that "nothing in the present Convention shall affect the right of each State Party to establish the criteria governing admission of migration workers and members of their family" (Article 79).

In the realist context, migrants are understood not necessarily in terms of their cultural background or as a source of labour; rather, they are seen as a contributing factor to conflict or cooperation between states. This shift means that "how countries treated their own populations – once regarded as primarily an internal affair – has become a matter of international relations" because mistreatment and incompetency create a burden for other states (Weiner, 1993:15). This is not to suggest that states that produce strong push factors will inevitably be in conflict with neighbouring states; indeed, migratory forces may produce compatible interests if one state is encouraging exit while another encourages entry (Weiner, 2006:95). On the other hand, when two or more states' perspective on exit and entry are not compatible, it "carries a high potential for conflict" (Weiner, 2006:96).

Zolberg suggests that states act politically when they manage migration, including consideration of where sending and receiving states fall on a "scale of strategic power" as much as on a "continuum of core and periphery" (2006:69). This has an impact on how states respond both to multilateral efforts to manage migration and to specific migration flows. To the extent that there are international rules pertaining to migrants, these rules reflect the international distribution of power (Hollifield and Wong, 2015:249). From this perspective, contestation over the Refugee Convention and the ICRMW was not a matter of states asserting control over entry to preserve national identity or their labour market; it was a struggle over which states would authorize the rules of international migration. For instance, the creation of the 1951 Refugee Convention reflected the direct interests of Western states to "avoid instability in the wake" of World War II and "later to give international imprimatur to the sheltering of enemies of these states' ideological adversaries" (Gammeltoft-Hansen and Hathaway, 2015:238). Once these specific interests gave way to new concerns, Western commitment to the refugee regime weakened.

In this context, a state's response to a particular migration stream or refugee flow requires an understanding of the bilateral or regional relations between the affected states. For example, when large numbers of Haitians or Cubans sought entrance to the United States in the 1990s, US political elites' primary concern was what the governments of Haiti and Cuba were doing to produce these migrant flows, and how they could work with – or against – these governments to stop or manage the flow (Weiner, 1993:16). In a number of cases, such as Haiti and the former Yugoslavia, migrant movements were understood as intentional flows meant to disrupt neighbouring states, and it resulted in armed intervention to stop the outflow of refugees (Greenhill, 2010).

These examples of coercive response should not overshadow the significant amount of cooperation between states in managing migration. Realist belief does not preclude the possibility of international cooperation on managing migration; however, it does suggest that international agreements on migration reflect a confluence of national interests and inter-state power relations. States implement policies to control migration in an environment where these policies necessarily affect citizens of other states and require the cooperation of others states to implement. Gammeltoft-Hansen and Hathaway depict the current refugee management system as "cooperative deterrence", in which states engage in cooperative measures to deter and prevent a shared threat – the entrance of unwanted migrants.

At a very basic level, international cooperation to manage migration requires a shared system of identification and documentation (Guild, 2009:11). This entails international agreement on basic standards of identity documentation, and there is strong international consensus on the passport's standard components: "one person per passport ... biometric information ... and similar global format", a set of components that remained relatively unchanged since the early twentieth century (Salter, 2003:85). These shared components, however, are not devoid of power considerations. The events of 9/11 demonstrated that powerful states could alter existing rules and agreements on identification standards when existing standards were seen as inadequate to meet the national interests of powerful states. The United States and EU pushed for biometric identity documents from certain states to strengthen the integrity of their passport document and to assist in the conduct of the war on terror (Amoore, 2006). These states used inducements, such as visa-free travel, to coerce states to adopt new, enhanced identification (Koslowski, 2008). The perceived targeting of their state, and their citizens, as a source of insecurity prompted retaliatory measures from some states, such as Brazil (Koslowski, 2008). The diplomatic conflict over required changes to identity documents and restricting migration demonstrates how the management of migration is interpreted in terms of inter-state relations.

Cooperation to deter unwanted migrants extends well beyond identification documents. Sending, transit, and destination states cooperate to contain asylum seekers and irregular migrants through "direct financial incentives ... provision of equipment ... deploy[ing] immigration officials in countries of origin and transit...and joint enforcement of anti-smuggling initiatives" (Gammeltoft-Hansen and Hathaway, 2015:252–4). Sending and transit states essentially participate in these arrangements due to power disparity; they gain foreign aid and assistance, as well as participation in regional trade agreements in exchange for preventing onward migration (Gammeltoft-Hansen and Hathaway, 2015:250). These cooperative arrangements bear a striking similarity to traditional alliance formations in realist thinking – temporary cooperative arrangements in the face of a shared threat or in response to material incentives. In the absence of these factors, cooperative arrangements would disappear.

Given the prevalence of realist beliefs and practices in the management of international migration, moving beyond temporary cooperative endeavours to a permanent international migration regime has proven elusive. Koslowski suggests that efforts to produce an international migration regime have failed because immigrant receiving states, who generally hold more power in the international system than do sending states, prefer to manage migration unilaterally or bilaterally and to build international institutions that facilitate this approach. Again, this is evident in the failure of the ICRMW. De Guchteneire and Pecoud (2009) argue that the agreement was pushed by the Group of 77 (G77), who stood to gain most from the convention's provisions, and was opposed by most of the powerful, migrant-receiving states, who stood to lose from the agreement. Even then, few members of the G77 ended up supporting the convention; for fear of disrupting relations with migrant-receiving states, many states who stood to benefit from the agreement did not ratify it (De Guchteneire and Pecoud, 2009:25). Competition between states also worked against ratification. Destination countries compete for the best and brightest immigrants but also to keep out unwanted immigrants. These states attempt to avoid becoming known as a "soft touch", and ratifying the convention was seen as opening their borders to further challenge and to undermining sovereignty (De Guchteneire and Pecoud, 2009). Countries of origin were also competing with each other for bilateral agreements to send their workers abroad and gain remittances; ratification of the convention was seen as jeopardizing these prospects in a competitive global market (De Guchteneire and Pecoud, 2009).

Consequently, migrant-sending states face a dilemma – in order to "reduce unemployment, gain remittances and transfer skills" (Castles, 2000:48), they need to export workers, which can lead to two basic, often oppositional goals: extend protections for their citizens abroad and secure bilateral agreements to facilitate these flows. The need to protect their migrant workers is well documented; as Taran notes, "violations of migrants' human rights are so generalized, widespread and commonplace that they are the defining feature of international migration today" (Taran, 2009:9). Most states are developing measures to maintain connections and offer protection to their citizens outside the state territory, in the process altering the state/territory nexus (Levitt and de la Dehesa, 2003). While it may be altering aspects of sovereignty, protecting one's citizens abroad is consistent with realist belief. Although he focused very little on emigration and protection of citizens overseas, Weiner notes that, while immigration "brings the outside in" – thus requiring state control – emigration "involves sending a piece of one nation into another society" (102). Consequently, nearly all states acknowledge their duty to offer some protection for their expatriates. The passport itself is a sort of basic agreement between states to treat each other's citizens justly as they travel abroad, as it includes a "request... to other governments that the bearer be afforded protection" (Salter, 2003:3). Reciprocal arrangements such as consulates and embassies; diplomatic agreements; and, to a lesser extent, extradition treaties facilitate the protective capacity of states to offer some protection to their citizens abroad. Many states "station labor attaches and welfare workers" at their embassies, or fund non-governmental organizations (NGOs) that provide support to expatriate communities (Castles, 2000:48).

On the whole, however, states are very limited in actually offering protection to their citizens abroad. This exposes several tensions in realist belief: the first is that cooperation, rather than military power, is essential to enhancing the security of overseas citizens. Second, insecurity is less the product of one's citizenship/nationality than is economic status. Low-skilled migrants, regardless of their citizenship and the military power of their home state, are more insecure than high-skilled migrants. Third, the state and migrants are heavily reliant on market-based mechanisms for their protection when outside their home state. There are several ways that "the market" operates in the protection of migrants. Migration itself is a tool for families and communities to ameliorate risk by diversifying economic income (see Massey et al., 2005:35–6). At the individual level, migrants may use travel insurance and health clinics to provide protection while outside their home state. At the macro level, the structural demand for certain types of migration may afford protection for some and increase insecurity for others. Emigration from wealthy industrial states, is predominantly for purposes of tourism, foreign investment, and international trade – all forms of migration in high demand with seemingly insufficient supply.

Modernism

Modernism also offers a framework for understanding migration; indeed, one might suggest that migration and mobility are defining features of modernist beliefs. Yet the place of migration in modernist belief is ambivalent. Again, returning to Cresswell, he poetically observes that "mobility is the lifeblood of modernity and the virus that threatens to hasten its downfall" (21). Consequently, modernist belief differentiates between types of migration – those to be encouraged and celebrated, and those to be prevented and condemned. At its core, modernism depicts certain forms of migration as a pre-modern form of adaptation and social organization and thus as a "problem" to be managed or solved through technological solutions and the modernization of non-modern societies. Modernists depict migration as "pre-modern" by emphasizing its prevalence as a form of social organization in earlier human history. To take one example, in their introduction to *The Migration Reader*, Lehav and Messina state, "[T]he phenomenon of migration is rooted in human prehistory, when people routinely travelled over great distances to hunt, fish, establish a livelihood, and secure a home" (2006:1).

This essentially begins the modernist story of human progress and civilization, in which humans become capable of providing their basic life necessities in sedentary, agricultural societies – and later in advanced industrial societies (see Christian, 2004:248–82). This civilizing process rendered migratory human communities, such as hunter-gatherers or nomadic peoples, as the prototypical pre-modern societies in need of civilization. Neumann and Wigen (2013) suggest that the dominant understanding in the field of IR is that sedentary populations within states symbolize civilization while nomadic populations are seen as backward and barbarian. Thus, mobile populations such as nomads, pastoralists, and gypsies fell under the violent gaze of the emerging state form (Scott, 1998). The first step toward being regarded as civilized was a sedentary population practicing agriculture, which could then lead to later developments such as trade, science, and art (Todorov, 2010:34). In this modernist narrative, migratory societies had not yet even taken the first step toward modernity and remained "barbarian" or "savage". Consequently, modernism encourages mobile populations to settle and modernize by conquering nature.

The negative perception of migration is evident even in the early stages of modernizing processes; mobile individuals and groups were stigmatized and subject to harsh control measures. In Europe, as larger numbers of people were displaced with the decline of feudalism and the advent of industrial capitalism, they became mobile and thus "illegible" to the managers of social order, which required "new forms of social surveillance and control" (Cresswell, 2006:12). As outlined in Chapter 2, this process led to the authorization of the national state to render the population legible and to control movement, displacing local authorities, corporations, and religious entities. By shifting the surveillance and control of the mobile poor from the local and private to the national state, it actually removed a number of barriers to migration within states. It is only with fully modern societies that the perceptions of migration began to shift. Gradually, in modern liberal societies, mobility came to hold positive connotations and to be understood as a right of citizenship (Cresswell, 2006). However, the tying of mobility to citizenship further legitimized restrictions on the mobility of non-citizens.

The concept of pre-modern or "primitive migration" (Piguet, 2013) operates through its opposite – "modern migration", whose purposes are civilized and modern. Modern migration seeks access to new markets and resources and to spread civilization; pursue art, leisure, and knowledge; and, more generally, conquer nature itself. To advance these purposes, modern societies pursued technological achievements to shorten distance and overcome natural barriers. In many societies, modernity itself is associated with "technological mobility – the car, the plane, the spaceship" (Cresswell, 2006:15). Indeed, MacCannell suggests that it is tourists that most fully epitomize modern migration, because they use technological mobility for the purpose of leisure. The tourist moves to interact with fixed attractions and immobile populations to experience the "simplicity, poverty, chastity or purity" of pre-modern "primitives" (MacCannell, 1999:41). Consequently, management of migration at the international level entails a double move: facilitating modern migration and restricting primitive migration.

Nowhere is this more evident than in the response to irregular migration flows, refugee flows, and even environment-induced migration. In all of these cases, migration is depicted as the strategy employed by less modern societies in pursuit of basic life necessities. Piguet notes that ecologically induced migration is often regarded as "primitive migration" as it resulted from the inability of early communities to "cope with natural forces" (2013:151). For modernists, the impact of environmental factors should reduce over time as human communities employ technology to subdue nature (Piguet, 2013). From this perspective, large-scale migration due to environmental factors represents a failure of these societies to modernize sufficiently. Indeed, such a perspective is common in the representation of climate migrants. To take one example, Biermann's and Boas's seminal report on climate refugees notes that "global warming may force millions of people – mainly in Asia and Africa – to leave their homes and migrate to other places" (2010:60). Their report identifies a number of less-developed areas in the Global South that are likely to be hit hardest by climate change, and thus migrate. Indeed, Bankoff argues that representations of certain populations as vulnerable reinforces "colonial imaginaries" of the South as a zone of danger, disease, and disaster, which contrasts with the North as modernized and capable of providing solutions for the South (Bankoff, 2001). These typical representations of refugees and climate migrants operate as a shorthand to identify "primitive migration" – the less-advanced regions of the world who are both more prone to disaster (due to their lack of modernity) and who migrate *en masse* in response. This is not to dispute the differential impact that climate change will have around the globe, and that poorer regions and people will bear the brunt of rising temperatures and sea levels. Rather, I raise this point to demonstrate how this narrative ties pre-modernity to vulnerability and migration, obfuscating how technological innovations and modern forms of migration have contributed to these processes.

One tension within modernist accounts of migration is that the technological solutions to advance modern migration, namely, how to move quickly and efficiently over great distances, also facilitate "primitive migration". The resolution to this tension is to enhance control over access to modern forms of transportation and a system for differentiating between the two types. Passports and visas serve both purposes – they can be used to restrict access to transportation, as in the case of airport travel and carrier sanctions, as well to differentiate the modern traveller from the primitive traveller. Due to the controls on access to transportation technology, the form of transportation technology itself becomes a marker of modernity. Current migrants who employ "early" modes of transportation and technology, such as walking, horse and carts, and (some) boats are symbolic of pre- or early modernity; airplanes and high-speed rail signal modernity. The figure of the migrant walking across borders or arriving on overcrowded boats is symbolic of "primitive migration". I suggest that it is the primitive/modern binary that explains why boat arrivals spark media coverage and securitizing narratives, while plane arrivals – of which there are a far greater number in most Western states – rarely warrant notice (see Watson, 2009). The form of transportation technology intersects with race and class in the construction of modern subjects, in which mode of transportation, phenotypic characteristics, and styles of dress are themselves markers of modernity and pre-modernity.

A second tension in modernism is that modernization itself produces significant levels of displacement. The construction of major modernizing projects, such as dams, have been among the leading direct causes of displacement – the Three Gorges dam in China alone is estimated to have displaced close to 1.3 million people (Bates, 2002:472). Other activities undertaken in the name of modernization – mass urbanization, removing natural barriers (trees, marshes, etc.) to facilitate the development of the land, and resource extraction – have all contributed to the worsening effects of natural hazards. Even as high modernism promoted mastery of nature, many of the projects associated with this belief system reduced nature's capacity to ameliorate inclement weather events and increased human vulnerability to natural phenomena. In short, modernism is a significant contributor to displacement; indeed, in many cases, it requires displacement (and thus mobility) in the pursuit of development.

In addition to differentiating positive and negative forms of migration, modernism offers a solution to the defining problem of managing migration movements – use of technology to establish identity. The cornerstone of modern migration management is establishing identity of individual migrants and consequently their purpose for migrating. Toward this end, the passport is the key piece of technology. The present form of the passport dates to the early twentieth century, but in response to concerns over the reliability of these documents, states have added a number of technological features to enhance its reliability as a marker of identity: "lamination, stamps, tamper-proof paper, or holograms" (Salter, 2003:3). More recently, new forms of paper and ink have been used, and the most recent versions are machine readable and may feature biometric information. Technological innovation has been promoted as the answer to the problem of establishing identity; however, in doing so, it has had a secondary effect of recreating the passport as a standard of civilization that identifies modern and non-modern states. States incapable of or unwilling to modernize their identity documents to be machine readable and to include biometric data are excluded from key migration agreements and subject to visa requirements and extra scrutiny (Koslowski, 2004).

Visas operate alongside the passport as key technologies in the management of migration because they (1) limit access to transportation technology, (2) can be applied to regions regarded as pre-modern or insecure, and (3) collect far more information and scrutiny than available on the passport. In practice, this is most evident in visa requirements – individuals from countries regarded as sufficiently modern (and thus unlikely to produce unwanted or "primitive migration flows") are often exempted from visa requirements, while those from insufficiently modern regions, and thus whose identity is less certain, are subject to greater scrutiny in the form of visas and security checks. Modernism thus offers a compelling account of migration in the current era, as well as two notable solutions to the problems associated with cross-border movement: development and technologically sophisticated identification.

Humanitarianism

The previous accounts paint a daunting picture of the management of international migration and the treatment of certain classes of migrants in the current international order. Whether migrants are seen as labour, threats to social cohesion, fifth columns, or primitive migrants, the movement of people across international borders is framed as a problem to be managed, often in highly coercive ways. There are obvious tensions between these accounts, and at various times, certain of these belief systems regarding migrants have been more dominant than others, resulting in variation in migration policies and patterns over time. One frame that seems to have had a significant impact on the understanding of migration, and thus ameliorating some of the highly coercive aspects of the previously discussed belief systems, is humanitarianism. Not surprisingly then, advocates for enhancing mobility for those displaced due to disaster or climate often rely on humanitarianism and the victim frame to sustain their claims.

Humanitarianism suggests that the preceding accounts of migration, and the tools for managing migration, are not appropriate in response to certain forms of migration. While few humanitarians would reject the modernist position that modernization and development ought to be the principal mechanism for alleviating poverty and vulnerability, they would suggest that, in certain instances, migration may be the only available option for some populations and that the existing international migration regime should facilitate such movements. Similarly, humanitarians emphasize that, while refugees often fulfil employment needs in host societies, the labour market is an inappropriate mechanism for determining entry for those fleeing persecution. The Canadian government's decision prior to World War II to deny access to Jewish refugees fleeing Nazi Germany on the basis that Jews did not fill a Canadian labour market need (Abella and Troper, 1983) would today be considered racist and a violation of humanitarian obligations.

Humanitarianism does not necessarily dispute the right of communities to restrict access to membership and territory on the grounds of maintaining a distinct community, but it does suggest that the refugee's right to protection supersedes the rights of communities to restrict entry (see Walzer, 1983). The humanitarian belief system suggests that normally operating constraints on the movement of people across international borders should not apply in instances when denying entry would result in serious human suffering. This is not an argument in favour of open borders; rather, humanitarianism accepts a general right of states to control their borders and that denying the right of the community to restrict access requires justification and subjects worthy of such consideration. Refugees are held up as the exemplar of the worthy migrant, which stems from their position as passive, depoliticized, innocent victims who share no responsibility for their own suffering (see Malkki, 1996; Scheel and Ratfisch, 2014). I argued in Chapter 2 that establishing the bona fides of the humanitarian subject necessitates an assignation of responsibility for poverty, vulnerability, and disaster. In the case of migration, it requires an account of responsibility for displacement. As Soguk relates (quoting Vernant), the refugee is "a helpless casualty, diminished in all circumstances, the victim of events for which, at least as an individual, she cannot be held responsible" (Soguk, 1999:11).

The idealized humanitarian subject, however, rarely matches lived reality, and refugees often behave in ways that violate the behavioural expectations created through the humanitarian representation. The violation of these expectations has led political actors to reconstruct the identity of asylum seekers from refugees (and thus worthy humanitarian subjects) to illegals, bogus refugees, smuggled humans, queue jumpers, or even terrorists – all of whom bear some, or all, of the responsibility for their movement across international borders. Each of these representations call into question the worthiness of asylum seekers when it comes to humanitarian protection.

For those who are deemed "worthy of exceptions to the normal immigration control rules" (Shacknove, 1993:516), humanitarianism demands of states special protections. To date, the humanitarian position on migration is most commonly associated (although not exclusively) with the 1951 Refugee Convention (and 1967 Protocol). According to Betts, the convention actually sets out "two sets of obligations: asylum and burden-sharing" (2015:363). Asylum obligations pertain to refugees who are in a state's territory, while burden-sharing obligations apply to refugees in other states (Betts, 2015) – or more accurately, to other states hosting refugees. By far, the stronger of the two obligations relates to asylum, and it is encapsulated in the principle of non-refoulement, which forbids states from forcibly returning a refugee to a territory where they would face persecution (Lauterpacht and Bethlehem, 2003). Indeed, one great accomplishment of the Refugee Convention was the establishment of a compulsory asylum obligation rather than voluntary obligations characteristic of most humanitarian endeavours. One of the most significant shifts from a humanitarian to a human rights approach is the assertion of compulsory obligations. As we saw in the case of the ICRMW, however, the human rights–based approach to managing migration has not gained significant traction. Similarly, the compulsory, "rights-like"[3] nature of the principle of non-refoulement has led states to try to circumvent these obligations, promote a strategy of non-entrée (Gammeltoft-Hansen and Hathaway, 2015), and confine refugees to areas where only voluntary obligations of burden sharing persist.

The principle of non-refoulement is broadly interpreted as prohibiting the return of refugees to their home state or to a non-signatory state that would return them to their home state (although this latter provision has been increasingly challenged) (Dummett, 2001). Helton suggests that, in addition to prohibiting rejection at the border, non-refoulement also implies "fair and non-discriminatory status determination" and provision of "tolerable living conditions" (1990:123–4). Subsidiary obligations that flow from the principle of non-refoulement include non-arbitrary detention, non-punishment based on mode of entry, and access to legal adjudication (Watson, 2009). In essence, the Refugee Convention obliges states not to return asylum seekers until their claims are adjudicated and their worthiness of the humanitarian exception is established.

This humanitarian arrangement stands in clear tension, however, with the statist basis of the existing international migration regime, and thus its crafters sought to maintain the ability of states to control their borders. In addition to reasserting the principle of state sovereignty, the convention has given states considerable latitude in their treatment of asylum seekers, including deciding whether non-refoulement actually entails a "right to remain" (see Goodwin-Gill and McAdam, 2007:267). Even if states interpret non-refoulement as an obligation to allow refugees to remain in their territory, states are the entities empowered to determine an asylum seeker's identity (although they may cede this responsibility to UNHCR); not surprisingly there is wide variation of recognition rates between, and within, states (Neumayer, 2005). Second, states have few obligations beyond non-refoulement; states may decide the duration of protection (permanent or temporary), as well as the sociopolitical rights to which refugees are entitled during protection. Third, because the obligations pertain once the asylum seeker has arrived in the state's territory, the state has few obligations prior to that. Consequently, states have implemented a number of measures to prevent asylum seekers from arriving in their territory; visa requirements, carrier sanctions, detention, principle of first asylum, safe third country provisions, and overseas interception operations are common and very effective in limiting the mobility of most of the world's refugees (Helton, 1990). The use of liminal and extraterritorial spaces, such as territorial waters, embassies, airports, and offshore territories, to reduce state obligations to asylum seekers demonstrate that the state's borders are not always where they were supposed to be and that states could and would manipulate borders to reduce their humanitarian obligations and enhance their ability to control entry.

The Refugee Convention defined refugees very narrowly, as those (1) outside their home state and (2) unable or unwilling to access protection from their home state based on a "well-founded fear of being persecuted for reasons of race, religion, nationality, membership of a particular social group, or political opinion" (Article 1(2)). The convention, by definition, excluded large swaths of humanity that may be forced to migrate for protection, such as those displaced due to poverty, generalized violence, complex humanitarian emergencies, and disasters. In doing so, the Refugee Convention violates basic humanitarian values, in essence by justifying restricting the mobility of people who are suffering. Yet these categories of displacement are far more ambiguous for establishing the worthiness of the displaced person and responsibility for displacement. Convention refugees more clearly meet the standard of worthiness for several reasons: they are not responsible for their suffering; they are individually targeted; their home state is complicit in their suffering; and they have, in essence, exhausted all potential remedies. In these other instances, one or more of these components is missing. In defining the refugee so narrowly, its crafters sought to reduce ambiguity and focus on instances where the worthiness of the humanitarian subject was relatively easy to establish. In practice, even this narrow definition is highly ambiguous and subject to differing interpretations. The standard for meeting "a well-founded fear of persecution" or establishing which "particular social group" is entitled to protection can vary quite significantly between states. Yet some of the proponents of the refugee regime attribute its modest success to its narrowness and relative lack of ambiguity over who does, and does not, qualify for protection – thus cautioning against expansion (Hathaway, 1991; Betts, 2015).

In addition to the compulsory obligation associated with asylum, the refugee regime also implies a second, voluntary set of obligations associated with burden sharing (Betts, 2015). In an effort to reduce the disproportionate cost borne by refugee-receiving states, most of whom are located in the developing world due in part to strict migration control measures of powerful states, the international community created several voluntary mechanisms for providing humanitarian support. These include funding the activities of UNHCR, participation in resettlement programs, and foreign aid to receiving states. While not codified in international law, this component of the refugee regime is far larger and is arguably more "humanitarian" in that it offers protection to a range of displaced persons who do not strictly qualify under the convention definition. Today, UNHCR provides assistance to over 43 million "people of concern", which include refugees, people in refugee-like situations, and internally displaced persons (UNHCR, 2016a). UNHCR's budget for providing life-saving assistance to those in its care is now close to $3.5 billion. UNHCR and the governments and private actors that support it routinely explain their support in terms of humanitarian beliefs.

The voluntary, burden-sharing aspect of the refugee regime exhibits similar weaknesses to other voluntary humanitarian obligations. Only a small portion of UNHCR's projected budgetary needs are met through its appeals for voluntary support. Resettlement is a voluntary program, and only three states contribute to refugee resettlement in significant numbers: the United States,

Australia, and Canada. Consequently, more than 86% of the world's 21 million refugees are located in the developing world (UNHCR, 2016b). The world's top refugee hosting states include Turkey, Pakistan, Lebanon, Iran, Uganda, and Ethiopia, most of which are straining to provide adequate support or protection and a number are threatening to refuse further entry. UNHCR is insufficiently funded. UNHCR refers to its current situation as "undertaking austerity measures... at high cost to other activities, particularly protracted situations" (UNHCR, 2015b). In 2015, voluntary contributions to UNHCR amounted to just over 40% of its projected budget, while additional appeals were chronically underfunded. UNHCR reported that over 30 appeals were funded at less than 45% of projected need (UNHCR, 2015b).

Meanwhile, developed countries are not increasing their financial contributions in line with increased need and were struggling to significantly alter their refugee resettlement policies to help alleviate the burden on developing states. Despite numerous international summits to increase resettlement and spread the burden more evenly, fewer than 1% of the world's refugees are resettled each year. With over 28,000 people displaced daily in 2016, developed states resettled less than one week's worth of newly displaced people (UNHCR, 2016a).

In 2016, there were renewed efforts to strengthen and further institutionalize the burden-sharing component of the refugee regime with the New York Declaration for Refugees and Migrants (NYD). This declaration proposed two distinct global compacts: one to establish a "comprehensive refugee response framework", the other on "safe, orderly and regular migration" (NDY, 2016). Early drafts showed that the declaration did not draw on or attempt to replicate the rights-based approach of the ICRMW. The refugee response framework was to address four basic objectives in dealing with refugee flows: "ease pressure on host countries, enhance refugee self-reliance, expand third country solutions, and support conditions in countries of origin for return in safety and dignity" (NYD, 2016). Three of these objectives mirror the longstanding preferred durable solutions of UNHCR: repatriation, local integration, and resettlement. While full impact of the NYD and the subsequent Global Compacts will require years to assess, early indications seem to suggest a slight reorientation of the relative preference of the three existing durable solutions, putting greater emphasis on resettlement and repatriation, and less on local integration, than is currently the case. The fourth, which emphasizes refugee self-reliance, appears to duplicate the move toward resilience addressed in Chapter 4, which is to say that it advocates for vulnerable populations to take greater responsibility for their own protection and recovery, and to justify and facilitate the removal of governmental support.

From a humanitarian perspective, the NYD is an improvement on the current system, as it calls for the international community to "strengthen and enhance mechanisms to protect people on the move" (NYD, 2016). For many, it is still insufficient, however, and does not adequately identify specific obligations on host countries or third states. Since the NYD, and the conclusion of the two Global Compacts, most indicators suggest little has changed. Crisp (2017) suspects that, rather than easing pressures on countries that host refugees through increased refugee assistance and development funding, refugees are being pressured to repatriate. Calls to build refugees' self-reliance face significant hurdles, primarily the unwillingness of states to get refugees out of camps and to give them "access to the means of production" (Crisp, 2017). On the matter of increasing resettlement, Crisp notes that the election of Donald Trump in the United States and his cuts to refugee resettlement have made this goal out of reach, even if other states stepped up to fill the gap, which they have not done. Crisp notes that, while the international community strongly supports fostering repatriation efforts, its members have done little to resolve ongoing conflicts.

Tinkering with the existing refugee regime is unlikely to resolve the primary tension exposed by refugee flows, which is that the existing refugee regime functions to contain refugee populations in the developing world. As Gammeltoft-Hansen and Hathaway observe, the developed countries of the world have essentially adopted a policy of non-entrée and pay lip service to the refugee regime as a symbolic commitment to ensure that poorer states continue to host refugees (2015: 236). By housing refugees in camps and participating in the prevention of "illegal migration", UNHCR acts as a police force for enforcing borders (Williams, 2002; Scheel and Ratfisch, 2014). Scheel and Ratfisch argue that UNHCR "provides a legitimization strategy by presenting restrictive controls as a necessary prerequisite for effective refugee protection" (2014:930). Similarly, Koch concludes that UNHCR's and IOM's facilitation of "state-induced returns" enables restrictive migration control policies (Koch, 2014). The existence of UNHCR camps creates the expectation that refugees should first flee to these camps to seek protection and await resettlement. This in turn facilitates the narrative that asylum seekers are "bogus refugees" or are "jumping the queue" despite the fact that there is no queue for resettlement positions and the number of refugees resettled each year is miniscule. Rather than presenting a challenge to the existing international migration management regime, the humanitarian foundations of the refugee regime naturalize the state as the normal form of political organization, including its restrictions on the movement of people (Soguk, 1999; Scheel and Ratfisch, 2014).

These points call into question the extent to which the current refugee regime fully embodies humanitarian beliefs. The history of refugee protection after World War II suggests humanitarianism was only one of a number of ideological frameworks that shaped the regime. The extensive labour shortages and the emergent Cold War were just as important in shaping the institutionalization of the post–World War II refugee regime as were humanitarian impulses. The willingness of states to resettle refugees often depended on their domestic labour needs and their foreign policy toward the refugee-producing state. Indeed, the narrow definition of the refugee contained in the 1951 Refugee Convention clearly reflects the concern with restoring order to Europe, delegitimizing communism, and ensuring states maintained as much control over international migration as possible. One might say that the refugee regime represents the bare minimum states could do to alleviate human suffering globally: only certain types of suffering were to be alleviated, and even then, those who were suffering had to arrive physically at sites of protection. These designated sites of protection were more and more often in refugee camps housed in neighbouring states that suffered similar economic and political problems as the refugee-producing state. Access to protection in developed states was legally enshrined in the Refugee Convention, but it was made increasingly difficult to obtain due to increased migration controls.

Disaster Displacement

In the context of the international management of migration that I have spelled out in this chapter, the question of whether and how to protect those displaced due to disaster and/or environmental factors looms large. This issue is particularly complicated by the complex and diffuse relationship between environmental degradation, climate change, natural disasters, and migration. The existing academic literature on environmental migration is broadly in agreement that there is no direct causal link between environmental stressors and international migration. In short, environmental factors operate alongside, and are mediated by, a range of political, social, and economic factors (Ambrosetti and Petrillo, 2016). There are a number of ways in which individuals, families, and communities respond to environmental stress, ranging from altering family structure to diversifying work and food supplies, to migration (Morrissey, 2013). In many contexts, environmental factors may actually reduce the mobility of vulnerable people rather than encourage migration (Morrissey, 2013). Even in contexts where migration becomes the dominant coping strategy, migration strategies may vary both spatially and temporally. How far people move and for how long is highly context-dependent. As a general trend, the "vast majority of those displaced by disaster are internally displaced" rather than internationally, although seldomly officially recognized as such (Esnard and Sapat, 2014:1). Ambrosetti and Petrillo (2016) show that "sudden-onset environmental disasters" are more likely to result in "short-term and short-distance" migration (Ambrosetti and Petrillo, 2016:83). Cai et al. suggest that permanent international outmigration is more likely in cases of temperature change and occasionally in cases when average rainfall changes than in cases of natural disasters, but then only in countries highly dependent on agricultural production (2016:146).

The complexity of migration flows, particularly as it pertains to the influence of environmental factors, leads to two problems for the management of environmental-, disaster-, and climate-induced migration: uncertainty around levels of international displacement and the absence of an easily applicable existing management model at the international level. The complex interaction of environmental and non-environmental factors leads to terminological confusion and difficulty estimating the number of environmental migrants, now and particularly in the future. As noted, there is no consensus on defining an environmental migrant or on the causal link between environmental degradation and migration, and the figures offered to date on the potential number of climate refugees is highly contested and largely unreliable (Gemenne, 2011). The intentional use of broad categories and impressively large numbers contributes to an alarmist perspective that has drawn significant criticism (Gemenne, 2011). Numerous commentators observe that the alarmist perspective on climate change–induced migration – intended as a humanitarian narrative to spur action on climate change – has had an entirely different effect, contributing to the securitization of migration and a push for increased control of state borders (McNamara, 2007; Piguet, 2013; Faber and Schlegel, 2017). Bettini argues that "doom and gloom narratives" are both "counter-productive and normatively problematic" because they foster hostility to migrants, disempower migrants, and de-politicize climate change by focusing on management/governance (2013:63). Seeking to counter the securitized narrative, Gibb and Ford, who propose expanding UNFCCC's mandate to take on the issue of climate migration, explicitly call for the framing of migration as "an adaptation to climate change and not a threat to national security" (2012:7). As I've attempted to show in this chapter, reframing migration in this way not only requires both definitional clarity and a better understanding of environment-induced displacement, it must also address strongly held existing beliefs about migration.

The second related problem is the lack of an existing model for managing environmental displacement. In some respects, environmental-induced migration most closely resembles forced migration, which is managed largely through the refugee regime outlined in the previous section. The Refugee Convention institutionalizes a humanitarian exception to existing control measures and is well established in international law and in practice. In many respects, the political refugee and climate refugee are both easily depicted as "worthy humanitarian subjects" who bear no responsibility for their displacement. Indeed, in 2006, there was a proposal to amend the 1951 Refugee Convention to incorporate climate refugees (Biermann and Boas, 2010). This would permit climate/disaster refugees to enjoy the same rights as convention refugees currently do, including asylum rights and non-punishment for mode of entry. And it would oblige receiving states to assess their claims and, if valid, admit the refugees for protection. It would also empower UNHCR to set up similar arrangements for disaster refugees as it does currently for refugees and internally displaced persons (IDPs).

And yet, for various reasons, the refugee regime is not particularly well suited to be applied to environmental displacement. One reason, as noted earlier, is that environmental displacement (to date) is predominantly internal to a state (Esnard and Sapat, 2014), and the refugee regime is not well designed for the protection of IDPs. It does not articulate a clear set of rights for IDPs or obligations for states, as it does in the case of refugees. While UNHCR provides protection for IDPs, it has only a weak mandate for this, operating under its "good offices" jurisdiction (Helton, 1990). As I stressed in the previous section, UNHCR's emphasis on offering protection in camps in neighbouring states, and more recently offering protection within conflict states, actually contributes to the global restriction of mobility by containing displacement and preventing further migration. Consequently, the vast majority of the world's displaced peoples are contained in their home state or in developing states least able to deal with the additional costs of providing protection. In the event of major climatic changes triggering massive displacement, affected and neighbouring states that traditionally house refugees are likely to be similarly affected and unlikely to offer protection. The existing climate change models suggest climate change will have a disproportionate effect on different regions of the world, with the Global South more severely affected than the Global North. The current model of containing and warehousing refugees in the Global South does not seem a suitable solution to the problem of environmental displacement.

Second, the UNHCR system for burden sharing and providing protection (which is distinct from the asylum obligations under the 1951 Refugee Convention) is already showing signs of strain. As Biermann and Boas note, it is highly doubtful that the UNCHR could credibly offer protection and support for a "stream of refugees that is possibly 20 times larger" (2010:11). While the 20-times-larger claim stems from questionable figures (see Gemenne, 2011), it is clear that disaster/environmental displacement is already as significant as conflict displacement. According to the Internal Displacement Monitoring Centre (IDMC), in 2016, over 19.2 million people were internally displaced due to disasters, over twice as many as the 8.6 million displaced due to violent conflict (IDMC, 2016). In the first eight months of 2017, another 9 million people were displaced due to disasters, double the 4.6 million displaced by conflict and violence (IDMC, 2017). Even if the number of environment-induced displacement remains equal to, or between two and three times greater than, conflict displacement, UNHCR is already facing significant challenges in providing protection for those under its current mandate. Charging UNHCR with offering protection to another 10 to 20 million (or more) annually would require a significant expansion of its mandate and budget, something that there appears to be very little political will to do. Biermann and Boas convincingly argue that amending the refugee convention to include environmental refugees would not resolve the problem of environmental displacement, is not politically feasible, and would raise dilemmas between the types of protection given to convention refugees and those covered under an amendment (2010:11).

Biermann and Boas suggest another alternative: that climate-induced migration should be managed under the rubric of climate change management, with UNFCC as the primary international institutional framework. Using a special protocol on climate refugees under UNFCC, they suggest that specified administrative areas, rather than individuals, would be nominated for resettlement, ideally within their home state, based on their vulnerability to environmental degradation associated with climate change. Their proposal would operate on the principle of common but differentiated responsibilities and would see the cost of these resettlement programs borne by those most responsible for climate change but administered through an independent fund supplied by signatories to UNFCC (2010:15). Acknowledging the complexity of establishing the link between climate change and migration, Biermann and Boas propose "incremental cost" coverage, in which costs would be distributed based on establishing the relative causal weight of climate change and need for resettlement. Biermann's and Boas's proposal is innovative and builds on existing beliefs about international order, but it significantly alters other aspects of that order. One might reasonably conclude that their proposal, like the proposal to expand the refugee convention that they critique, is similarly politically unfeasible. It seems highly unlikely that the United States would agree to pay for the relocation and resettlement of villages in other parts of the world, when the US federal government has refused to cover the cost of resettlement initiatives due to climate change within the United States itself (Roberts, 2017).

The UNFCC proposal represents a radical departure from the dominant understandings of responsibility that underpin the contemporary international order. Bermann and Boas propose differentiated responsibility for climate migrants; this stands in stark contrast to realist and humanitarian conceptions of responsibility. In realism, the state bears responsibility only for its own citizens and has a responsibility to promote the national interest. Such considerations are evident in refugee burden-sharing proposals. In international negotiations, states' positions on refugee burden sharing largely correspond to the relative burden they face from refugees – states who host large numbers of refugees have been the prime supporters of burden-sharing agreements, while states that do not have resisted (Dowd and McAdam, 2017:12). Dowd and McAdam note that scholars in the field acknowledge that state perspectives on responsibility sharing are shaped by the benefits that states will reap from such arrangements (2017:22). Even as many states rhetorically support the idea of sharing the responsibility for protecting refugees, the notion of differentiated and obligatory responsibility for migrants and refugees strongly violates realist conceptions of responsibility.

Humanitarianism offers a different understanding on responsibility sharing for refugees, albeit one that is not clearly articulated. The 1951 Refugee Convention suggests that responsibility for refugees should be shared but beyond that, it contains no clear expectations. Even the NYD, whose aim was to enhance international cooperation, did not result in a clear articulation of responsibility. As Dowd and McAdam note, the final declaration watered down its commitment from "an equitable sharing of responsibility" to "a more equitable sharing" (2017:2). The initial call to resettle 10 percent of the global refugee population annually was similarly omitted from the final declaration (Dowd and McAdam, 2017:30). Yet the vague treatment of responsibility in key humanitarian texts is not a retreat from humanitarianism but reflects the historically ambiguous treatment of responsibility in humanitarianism, which wavers between a refusal to assign responsibility entirely in order to remain independent and neutral, and a commitment to a charitable, voluntary conception of responsibility. To the extent that responsibility is assigned, it lies with the refugee-producing state. Refugees are held up as victims who bear no responsibility for their displacement, and receiving states as humanitarian actors who bear no responsibility for the refugees' initial displacement. The climate change model, on the other hand, identifies "differentiated responsibilities" and clearly ties the costs of dealing with climate change to responsibility for causing climate change (Biermann and Boas, 2010:12). This contrasts sharply with the conventional governance of refugee flows, in which "apportionment of blame for refugee flows is neither necessary nor appropriate in the refugee context" (Biermann and Boas, 2010:24).

In many other ways, Biermann and Boas maintain key features of the existing international order.They emphasize that a climate refugee protocol would reinforce the principle of state sovereignty, and thus resettlement of populations would occur only on the recommendation of the affected state. In protecting state sovereignty and by emphasizing internal resettlement, it reinforces the nation-statist perspective on international order, implying that nation states remain the natural ordering of the globe, whose citizens identify primarily with the nation state.

Perhaps the most "successful" initiative to deal with the issue of environmental displacement was the Nansen Initiative, launched in 2012 by Norway and Switzerland. The initiative sought to address "forced displacement related to disasters, including the adverse effects of climate change" (Nansen Initiative, 2015:6). Noting that many states are already offering protection to disaster displaced persons, the Nansen Initiative initiated a departure from the largely ad hoc nature of the current response to disaster migration to institutionalize and regularize "key principles" and "effective practices". Described as a "state-led, bottom up consultative process", the Nansen Initiative ultimately resulted in an "Agenda for the Protection of Cross-Border Displaced Persons in the Context of Disasters and Climate Change" (hereafter PA), produced in 2015 (Nansen Initiative, 2015). The PA did not advocate for a new convention or an expansion of the Refugee Convention, although it drew heavily on the Refugee Convention for identifying the appropriate response to cross-border disaster migration. Rather, it focused on identifying "three priority areas for enhanced action by States, sub-regional organizations, the international community, as well as civil society, local communities, and affected populations" (Nansen Initiative, 2015:7). These priorities are collecting data and enhancing knowledge, enhancing the use of humanitarian protection measures, and strengthening the management of disaster displacement risk in the country of origin (10).

These priority areas are reminiscent of several of the guiding principles of the Sendai Framework examined in the previous chapter, which is not surprising given the overlap between the two documents and that the Nansen Initiative was part of the negotiations on the Sendai Framework. The PA frequently refers to the implementation of the Sendai Framework as an important aspect of managing disaster displacement. Consequently, it is not necessary to restate in detail the analysis of the Sendai Framework provided previously; it should be sufficient to point out the obvious points of ideological continuity and challenge. The three priority areas demonstrate the persistence of the modernist, humanitarian, and nation-statist foundation of the international order.

The agenda's modernist foundation is evident first through its depiction of climate migration as a primitive form of migration. While the PA acknowledges the "potential" benefit of migration in response to environmental stressors, a key component of the PA is to create conditions in which such migration is not needed. In its call to "manage disaster displacement risk in country of origin", the preferred outcome is for "the country of origin to help people to stay" (Nansen Initiative, 2015:34). The PA further states that "when movement in unavoidable, policy options include …" (Nansen Initiative, 2015:34). In contrast with modern migration, which is needed, beneficial, and to be encouraged; primitive migration is a burden, a necessity, permitted only when unavoidable – and potentially a threat. The PA's depiction of mobility recalls Cresswell's depiction of modernism's understanding of mobility, in which certain forms of movement "threaten to undo many of the achievements of modern rationality and ordering" (2006:20).

Second, PA endorses modernism through the prioritization of data collection and knowledge enhancement as the necessary first step to make disaster-affected populations visible to states and the international community. In doing so, PA establishes as its first priority the modernization of existing states through further enhancing their capacity to categorize and count their population. Under the rubric of strengthening management of disaster displacement, the agenda prioritizes "technical and capacity building" in affected states to implement the Sendai Framework and to facilitate human mobility. To reduce vulnerability and build resilience, PA calls for the development of "mapping tools… early warning systems… infrastructure improvements such as sea-walls, dams, dykes and earthquake resistant buildings" (Nansen Initiative, 2015:34). All of these reflect a modernist belief in the necessity of technological solutions to the problem of primitive migration.

At the same time, the agenda reproduces Sendai's focus on building resilience, which implies a shift from high modernism and a recognition that not all problems are resolved through technological innovation. In addition, the Protection Agenda slightly modifies the modernist position on survival or "primitive migration". In Section 119, the document endorses the position that "if well supported and managed, migration has the potential to help people living in areas exposed to natural hazards …" (Nansen Initiative, 2015:47). This position reflects a shift in modernist thinking regarding migration: climate change is altering the belief that primitive or survival migration is exclusively a symptom of pre-modernity and that, in some contexts, staying put and conquering nature to facilitate continued permanent human habitation is no longer possible nor desirable. Of course, this shift is modest. As the quotation shows, survival migration is appropriate in so far as it is "well-supported and managed" (Nansen Initiative, 2015:47); and that states ought to support and manage migration so that people do not "resort to irregular migration with all its negative consequences, particularly for women and children" (Nansen Initiative, 2015:47). In this representation, unregulated survival migration is still viewed as primitive migration. In keeping with this shift, PA endorses effective and sustainable "planned relocation" (Nansen Initiative, 2015:48). While migration is viewed as a "potentially positive way to cope with" environmental stressors, the ultimate objective is to create communities that can "avoid situations that otherwise may result in a humanitarian crisis and displacement in the future" (Nansen Initiative, 2015:47) and, in the event it cannot be avoided, to facilitate orderly and planned mobility.

The Protection Agenda also explicitly reinforces a statist and humanitarian framework as its second priority area. As with other humanitarian approaches to migration, the PA reasserts that "the power to regulate the admission of foreigners on its territory is an inherent right of every state and an attribute of its sovereignty" (Nansen Initiative, 2015:24). However, the PA reasserts the humanitarian exception, in which the right to exclude "must be exercised with respect for…international human rights law" that prohibits return to a "situation deemed to be inhuman" (Nansen Initiative, 2015:24). This requires assessment of disaster-affected areas but also the worthiness of the migrant. The PA reinforces the distinction between voluntary and forced migration as the principle factor that warrants a humanitarian exception to existing restrictions on migration (Nansen Initiative, 2015:22). In its section on "identifying the displaced", the authors note that "although the difference between displacement and migration can be difficult to pinpoint, it is important to distinguish between voluntary and forced movement" (Nansen Initiative, 2015:22). In this binary, the forced migrant has special "protection and assistance needs" worthy of humanitarian consideration (Nansen Initiative, 2015:22).

Rather than expand the Refugee Convention or articulate a clear set of obligations for states, the PA draws loosely on the refugee framework, identifying "temporary admission, stay or non-return, as well as lasting solutions" as features of the existing humanitarian protection system that would be applicable to disaster migrants. Specific recommended practices include granting travel visas or suspending visa requirements, expediting migration assessments, permitting temporary stay under existing mechanisms, and possibly expanding refugee categories (Nansen Initiative, 2015:26). All of these reinforce the humanitarian approach to migration management, which is voluntary suspension of normally operating rules for those deemed worthy of an exception.

As it pertains to economic liberalism and realism, the Protection Agenda is more ambiguous. It embodies some of the key elements of economic liberalism, such as the general association of migration with unequal economic development, in this case, mediated through the phenomenon of disaster. According to the PA, the poor are among the most vulnerable to natural hazards, and levels of disaster displacement are "highly dependent upon a country's level of development" (Nansen Initiative, 2015:35). The PA calls for enhanced domestic mobility as well as for creating more job opportunities domestically to reduce vulnerability and thus susceptibility to disaster (Nansen Initiative, 2015:35). As a practical solution, the PA calls for the use of economic migration programs to help ease the pressure on humanitarian programs, such as through the use of existing migrant worker programs to assist those in disaster affected states. By calling for "seasonal worker programs … that prioritize people from countries … facing natural hazard" and "accreditation alignment to enable people … to compete for skilled employment opportunities" (Nansen Initiative, 2015:36), the PA seeks to align economically liberal accounts of migration with the humanitarian framework. By doing so, it legitimizes and reinforces the existing economically liberal management of migration.

Yet, the PA represents migration slightly differently than in conventional economic liberalism. The PA repeatedly presents migration as an adaption to or coping mechanism for environmental stress and failed development rather than as a means of redistributing labour and capital. In this frame, migration is a problem to be addressed, in part, by development rather than migration acting as a solution to the problem of development. A second alteration to liberal understanding of migration contained in the PA is the acknowledgement of market failures associated with global labour migration and the potential exploitation of climate migrants in ways similar to the ongoing exploitation of migrant workers. Consequently, the authors of the PA call for the enforcement of global labour standards, prevention of economic exploitation, and migration with dignity and respect for human rights (Nansen Initiative, 2015:37). Although it certainly does not go as far as the ICRMW, the PA does call for basic economic rights for migrants that would alter the economic valuation of migrant labour. To date, there have been few attempts to use migrant labour pathways to meet humanitarian demands, although a few countries have explored the possibility of such programs. Not all migrant advocates see this as a positive development, with some concerned that this may actually reduce overall opportunities for migration as humanitarian migrants replace "regular" economic migrants in the economic streams yet still count as fulfilling humanitarian targets. In addition, there is the possibility that overall protection for migrants will be reduced, as humanitarian migrants generally enjoy greater protection and rights than do economic migrants (Lehr, 2017).

Similarly, the PA exhibits a profound ambivalence toward realist beliefs. The agenda is founded on the necessity of greater international cooperation and advocates for states, NGOs, international organizations, and civil society to participate in protecting cross-border migrants. In this respect, it directly challenges realist beliefs about the provision of security. In fact, the PA states that "for the purposes of this agenda, 'security' refers to the physical security of disaster-affected people" (Nansen Initiative, 2015:53). Nowhere in the document does it reference the security of the state or identify migrants as a potential threat to receiving states. While the PA does reassert the authority of the state to determine who may enter the state and firmly roots the responsibility to protect citizens from disasters in the hands of the state, it grounds this responsibility in international human rights law and most directly associates protection with development rather than the accumulation of power. The PA explicitly states that "the discretionary power of the state... is not unlimited... and must be exercised with respect for applicable rules of international and national law" (24). By most measures, this directly challenges realist beliefs about international order.

On the other hand, PA refrains from tying responsibility for climate migrants to responsibility for climate change, as does Biermann's and Boas's UNFCC proposal. In essence, this absence signals an abiding presence and acceptance of realist beliefs. The PA consistently refers to "interested state parties", suggesting that those who have no interest in the issue of disaster displacement are largely excused from the exercise. At the outset, however, the PA does make an interest-based claim: the introduction, states that "potentially every state could be confronted with such displacement". Yet the overriding normative force behind the call for protecting cross-border displaced persons rests on humanitarian ideology and a commitment to "international solidarity" (Biermann and Boas, 2010).

The Protection Agenda represents a slight reconfiguration of the ideological compromise that sustains current international migration regimes. These regimes are, of course, experiencing significant tensions and crises of legitimation, which has given rise to new initiatives to deal with existing migration pressures and those expected in the future due to climate change. Some of the alterations offered in the PA, such as the strong challenge to realist beliefs that currently have a profound impact on the management of migration, are likely to face significant opposition – as has been evident in other proposed changes to migration management, such as the ICRMW. In other instances, the PA reinforces beliefs that have contributed to the current crises – namely, its endorsement of modernism, nation-statism, economic liberalism, and humanitarianism. While each offers a different framework for understanding and managing migration, together they legitimize the existing migration system that severely limits prospects for mobility and neglects the primary causes of displacement.

Conclusion

In this chapter, I have argued that disaster displacement and other forms of environmental displacement pose a significant challenge to existing international migration management regimes, which are already in crisis. The ambiguities, fissures, and inconsistencies within the dominant beliefs regarding human migration, and the practices that these beliefs sustain, are now proving to be too numerous to conceal without significant alterations. The number of international summits on migration in the past three years attests to the sense of crisis among the managers of migration. Despite a variety of innovative proposals, such as the expansion of the Refugee Convention, the New York Declaration on Refugees and Migrants, the additional protocol to the UNFCCC, the Nansen Initiative's Protection Agenda, and the Global Compacts, the most notable shift in the past two years has been the efforts of developed states to double-down on their policy of containment and non-entrée. Many of the proposals listed in this chapter represent rather modest alterations of existing beliefs and practices, but they have still proven difficult to implement due to the intransigence of the most powerful developed states.

Notes

1 I use variations on disaster displacement, climate migrants, and environmental refugees interchangeably to refer to populations who choose or who are forced to leave their homes due primarily to changes in the natural environment (earthquakes, flooding, desertification, fires, etc.).

2 China's one-child policy is an obvious exception. Beyond that case, it is clear that state policies regulate the reproductive practices of women, particularly in terms of access to birth control, abortion, and women's rights more generally. Indeed, in numerous cases, the state has sought increased control over women's bodies to facilitate the reproduction of the nation. My general point here, however, is that these mechanisms of control and regulation are indirect. In most states, the state does not have direct control over the reproduction of its members.

3 There is a significant body of literature on the "right" pertaining to asylum as contained in the Universal Declaration of Human Rights (UDHR). The negotiations over this clause, and its asserted right "to seek and enjoy asylum" rather than a "right to asylum" suggests a rights-like approach to asylum, in which everyone has the right to ask for asylum, and states have the right to decline.

References

Abella, I., & Troper, H., 1983. *None Is Too Many: Canada and the Jews of Europe*, Toronto: University of Toronto Press.

Akkerman, T., 2005. Anti-immigration parties and the defence of liberal values: The exceptional case of the List Pim Fortuyn. *Journal of Political Ideologies*, 10(3), pp. 337–54.

Aleinikoff, 2002. *International Legal Norms and Migration: An Analysis*. Geneva: International Organization for Migration.

Ambrosetti, E., & Petrillo, E.R., 2016. Environmental disasters, migration and displacement. Insights and developments from L'Aquila's case. *Environmental Science and Policy*, 56, pp. 80–88.

Amoore, L., 2006. Biometric borders: Governing mobilities in the war on terror. *Political Geography*, 25(3), pp. 336–51.

Arango, J., 2000. Explaining migration: A critical view. *International Social Science Journal*, 52(165), pp. 283–96.

Bankoff, G., 2001. Rendering the world unsafe: "Vulnerability" as Western discourse. *Disasters*, 25(1), pp. 19–35.

Bates, D.C., 2002. Environmental refugees? Classifying human migrations caused by environmental change. *Population and Environment*, 23(5), pp. 465–77.
Batistella, G., 2009. Migration and Human Rights. In R. Cholewinski, P. De Guchteneire, & A. Pecoud (eds.). *Migration and Human Rights: The United Nations Convention on Migrant Workers*, Cambridge: Cambridge University Press.
Bettini, G., 2013. Climate barbarians at the gate? A critique of apocalyptic narratives on "climate refugees." *Geoforum*, 45, pp. 63–72.
Betts, A., 2015. The normative terrain of the global refugee regime. *Ethics & International Affairs*, 29(4), pp. 363–75.
Biermann, F., & Boas, I., 2010. Preparing for a warmer world: Towards a global governance system to protect climate refugees. *Global Environmental Politics*, 10(1), pp. 60–88.
Black, R., Bennett, S., Thomas, S.M., & Beddington, J.R., 2011. Migration as adaptation. *Nature*, 478(7370), pp. 447–9.
Bourbeau, P., 2011. *The Securitization of Migration: A Study of Movement and Order*, London: Routledge.
Cai, R., Feng, S., Oppenheimer, M., & Pytlikova, M., 2016. Climate variability and international migration: The importance of the agricultural linkage. *Journal of Environmental Economics and Management*, 79, pp. 135–51.
Caplan, B., & Naik, V., 2015. A radical case for open borders. In B. Powell (ed.). *The Economics of Immigration: Market Based Approaches, Social Science and Public Policy*, Oxford: Oxford University Press.
Castles, S., 2000. The impacts of emigration on countries of origin. In S. Yusuf, W. Wu, & S. Evenett (eds.). *Local Dynamics in an Era of Globalization*, New York: Oxford University Press, pp. 45–56.
Castles, S., & Miller, M.J., 1998. *The Age of Migration*, 2nd ed., New York: Palgrave MacMillan.
Christian, D., 2004. *Maps of Time: An Introduction to Big History*, Berkeley: University of California Press.
Cornelius, W.A., & Rosenblum, M.R., 2005. Immigration and politics. *Annual Review of Political Science*, 8(1), pp. 99–119.
Cornelius, W.A., & Tsuda, T., 2004. Controlling immigration: The limits of government intervention. In W. Cornelius, P. Martin, & J. Hollifield (eds.). *Controlling Immigration: A Global Perspective*, Stanford, CA: Stanford University Press, pp. 3–48.
Cresswell, T., 2006. *On the Move: The Politics of Mobility in the West*, London: Routledge.
Crisp, J., 2017. New York Declaration on Refugees: A One-Year Report Card — Refugees Deeply. https://www.newsdeeply.com/refugees/community/2017/09/18/new-york-declaration-on-refugees-a-one-year-report-card.
Dauvergne, C., 2005. *Humanitarianism, Identity, and Nation: Migration Laws of Australia and Canada*, Vancouver: UBC Press.
Dauvergne, C., & Marsden, S., 2014. The ideology of temporary labour migration in the post-global era. *Citizenship Studies*, 18(2), pp. 224–42.
De Guchteneire, P., & Pecoud, A., 2009. Introduction: The UN Convention on Migrant Workers Rights. In R. Cholewinski, P. De Guchteneire, & A. Pecoud (eds.). *Migration and Human Rights: The United Nations Convention on Migrant Workers*, Cambridge: Cambridge University Press.
Dowd, R., & McAdam, J., 2017. International cooperation and responsibility-sharing to protect refugees: What, why and how? *International and Comparative Law Quarterly*, 66(4), pp. 863–92.

Dummett, M., 2001. *On Immigration and Refugees*, London: Routledge.

Esnard, A.-M., & Sapat, A., 2014. *Displaced by Disaster: Recovery and Resilience in a Globalizing World*, New York: Routledge.

Faber, D., & Schlegel, C., 2017. Give me shelter from the storm: Framing the climate refugee crisis in the context of neoliberal capitalism. *Capitalism Nature Socialism*, 28(3), pp. 1–17.

Gammeltoft-Hansen, T., & Hathaway, J., 2015. Non-refoulement in a world of cooperative deterrence. *Columbia Journal of Transnational Law*, 53, pp. 235–84.

Gemenne, F., 2011. Why the numbers don't add up: A review of estimates and predictions of people displaced by environmental changes. *Global Environmental Change*, 21(Suppl. 1), pp. S41–S49.

Ghosh, B., 1998. *Huddled Masses and Uncertain Shores*, The Hague: Martinus Nijhoff Publishers.

Gibb, C., & Ford, J., 2012. Should the United Nations Framework Convention on Climate Change recognize climate migrants? *Environmental Research Letters*, 7(4).

Goodwin-Gill, G.S., & McAdam, J., 2007. *The Refugee in International Law*, Oxford: Oxford University Press.

Grange, M., & D'Auchamp, M., 2009. Role of Civil Society in Campaigning for and Using the ICRMW. In R. Cholewinski, P. De Guchteneire, & A. Pécoud (eds.). *Migration and Human Rights: The United Nations Convention on Migrant Workers*, Cambridge: Cambridge University Press.

Greenhill, K.M., 2010. *Weapons of Mass Migration: Forced Displacement, Coercion, and Foreign Policy*, Ithaca, NY: Cornell University Press.

Guild, E., 2009. *Security and Migration in the 21st Century*, Cambridge: Cambridge University Press.

Guiraudon, V., & Joppke, C., 2001. *Controlling a New Migration World*, New York: Routledge.

Hathaway, J., 1991. Reconceiving refugee law as human rights protection. *Journal of Refugee Studies*, 4(2), pp. 113–31.

Helton, A., 1990. What is refugee protection? *International Journal of Refugee Law*, 2(1), pp. 119–129.

Hollifield, J.F., & Wong, T., 2015. The politics of international migration. In C.B. Brettell & J.F. Hollifield (eds.). *Migration Theory: Talking Across Disciplines*, New York: Routledge.

Huang, P., 2008. Anchor babies, over-breeders, and the population bomb. *Harvard Law and Policy Review*, 2, 385–406.

Huysmans, J., 2000. The European Union and the securitization of migration. *JCMS: Journal of Common Market Studies*, 38(5), pp. 751–77.

Huysmans, J., 2006. *The Politics of Insecurity: Fear, Migration and Asylum in the EU*, London; New York: Routledge.

IDMC, 2016. Global Report on Internal Displacement, Geneva. http://www.internal-displacement.org/globalreport2016/.

IDMC, 2017. Recovery Postponed, Geneva. http://www.internal-displacement.org/assets/publications/2017/20170206-idmc-japan-case-study.pdf.

Joppke, C., 1999. *Immigration and the Nation-State: The United States, Germany, and Great Britain*, Oxford: Oxford University Press.

Jupp, J., 2007. *From White Australia to Woomera: The Story of Australian Immigration*, Cambridge: Cambridge University Press.

Koch, A., 2014. The politics and discourse of migrant return: The role of UNHCR and IOM in the governance of return. *Journal of Ethnic and Migration Studies*, 40(6), pp. 905–23.

Koslowski, R., 2004. *International Cooperation on Electronic Advanced Passenger Information Transfer and Passport Biometrics.* Montreal: International Studies Association.

Koslowski, R., 2008. Global mobility and the quest for an international migration regime. *Center for Migration Studies Special Issues*, 21(1), pp. 103–43.

Lauterpacht, E., & Bethlehem, D., 2003. *The Scope and Content of the Principle of Non-refoulement. Refugee Protection in International Law: UNHCR's Global Consultations on International Protection*, Geneva: UNHCR.
Lehav, G., & Messina, A., 2006. *The Migration Reader: Exploring Politics and Policies*, Boulder, CO: Lynne Rienner Publishers.
Lehr, S., 2017. *Migrant Protection – Personal Interview*, Victoria, BC, September 2017.
Levitt, P., & de la Dehesa, R., 2003. Transnational migration and the redefinition of the state: Variations and explanations. *Ethnic and Racial Studies*, 26(4), pp. 587–611.
MacCannell, D., 1999. *The Tourist*, 2nd ed., Berkeley: University of California Press.
Malkki, L.H., 1996. Speechless emissaries: Refugees, humanitarianism, and dehistoricization. *Cultural Anthropology*, 11(3), pp. 377–404.
Massey, D., Arango, J., Hugo, G., Kouaouci, A., Pellegrino, A., & Edward Taylor, J. (eds.). 2005. *Worlds in Motion: Understanding International Migration at the End of the Millennium*, Oxford: Oxford University Press.
McNamara, K.E., 2007. Conceptualizing discourses on environmental refugees at the United Nations. *Population and Environment*, 29(1), pp. 12–24.
Morrissey, J.W., 2013. Understanding the relationship between environmental change and migration: The development of an effects framework based on the case of northern Ethiopia. *Global Environmental Change*, 23(6), pp. 1501–10.
Nansen Initiative, 2015. *Agenda for the Protection of Cross-Border Displaced Persons in the Context of Disasters and Climate Change*, Geneva.
Neuberger, B., 1977. State and nation in African thought. *Journal of African Studies*, 4(2), pp. 198–205.
Neumann, I.B., & Wigen, E., 2013. The importance of the Eurasian Steppe to the study of international relations. *Journal of International Relations and Development*, 16(3), pp. 311–30.
Neumayer, E., 2005. Asylum recognition rates in Western Europe. *Journal of Conflict Resolution*, 49(1), pp. 43–66.
Newland, K., 2010. The governance of international migration: Mechanisms, processes, and institutions. *Global Governance*, 16(3), pp. 331–43.
New York Declaration | Addressing Large Movements of Refugees and Migrants. 2016. http://refugeesmigrants.un.org/declaration (Accessed May 26, 2017).
Piguet, E., 2013. From "primitive migration" to "climate refugees": The curious fate of the natural environment in migration studies. *Annals of the Association of American Geographers*, 103(1), pp. 148–62.
Ransan-Cooper, H., Farbotko, C., McNamara, K., Thornton, F., & Chevalier, E., 2015. Being(s) framed: The means and ends of framing environmental migrants. *Global Environmental Change*, 35, pp. 106–15.
Roberts, D., 2017. We have no system to deal with escalating climate damages. It's time to build one. *Vox*. https://www.vox.com/energy-and-environment/2017/9/8/16272538/what-is-worth-saving.
Robinson, C., 2018. Making migration knowable and governable: Benchmarking practices as technologies of global migration governance. *International Political Sociology*, 12(4), pp. 418–37.
Rydgren, J., 2008. Immigration sceptics, xenophobes or racists? Radical right-wing voting in six West European countries. *European Journal of Political Research*, 47(6), pp. 737–65.
Salter, M., 2003. *Rights of Passage: The Passport in International Relations*, Boulder, CO: Lynne Rienner Publishers.
Scheel, S., & Ratfisch, P., 2014. Refugee protection meets migration management: UNHCR as a global police of populations. *Journal of Ethnic and Migration Studies*, 40(6), pp. 924–41.

Schlesinger, P., 1994. Europeanness: A New Cultural Battlefield. In J. Hutchinson & A.D. Smith (eds.). *Nationalism*, Oxford: Oxford University Press.

Scott, J.C., 1998. *Seeing Like a State: How Certain Schemes to Improve the Human Condition Have Failed*, New Haven, CT: Yale University Press.

Shacknove, A., 1993. From asylum to containment. *International Journal of Refugee Law*, 5(4), pp. 516–33.

Smith, A.D., 1991. *National Identity*, Reno: University of Nevada Press.

Soguk, N., 1999. *States and Strangers: Refugees and Displacements of Statecraft*, Minneapolis: University of Minnesota Press.

Taran, P., 2009. The need for a rights based approach to migration in the age of globalization. In R. Cholewinski, P. De Guchteneire, & A. Pécoud (eds.). *Migration and Human Rights: The United Nations Convention on Migrant Workers*, Cambridge: Cambridge University Press.

Thränhardt, D., 1995. The political uses of xenophobia in England, France and Germany. *Party Politics*, 1(3), pp. 323–45.

Todorov, T., 2010. *The Fear of Barbarians: Beyond the Clash of Civilizations*, Chicago, IL: University of Chicago Press.

Torpey, J.C., 2000. *The Invention of the Passport: Surveillance, Citizenship, and the State*, Cambridge: Cambridge University Press.

UNHCR, 2015a. Global Trends: Forced Displacement in 2015, Geneva. http://www.unhcr.org/576408cd7.pdf.

UNHCR, 2015b. Global Trends 2015, Geneva.

UNHCR, 2016a. UNHCR Figures at a Glance, Geneva.

UNHCR, 2016b. UNHCR Global Appeal 2016–2017 – Needs and funding requirements, Geneva. http://www.unhcr.org/publications/fundraising/564da0e20/unhcr-global-appeal-2016-2017-needs-funding-requirements.html.

United Nations General Assembly, 1951. Convention Relating to the Status of Refugees. Resolution 2198 (XXI).

United Nations General Assembly, 1990. International Convention on the Protection of the Rights of All Migrant Workers and Members of Their Families. GA Resolution 45/158, December 18, 1990.

Walzer, M., 1983. *Spheres of Justice: A Defense of Pluralism and Equality*, New York: Basic Books, Inc.

Watson, S.D., 2009. *The Securitization of Humanitarian Migration: Digging Moats and Sinking Boats*, London: Routledge.

Weil, P., 2001. Access to Citizenship: A Comparison of Twenty-Five Nationality Laws. In T. Alexander Aleinikoff & D. Klusmeyer (eds.). *Citizenship Today: Global Perspectives and Practices*, Washington, DC: Brookings Institution Press.

Weiner, M., 1993. *International Migration and Security*, Oxford: Westview Press.

Weiner, M., 2006. On international migration and international relations. In A. Messina & G. Lahav (eds.). *The Migration Reader: Exploring Politics and Policies*, Boulder, CO: Lynne Rienner Publishers.

Wickramasekara, P., 2008. Globalisation, international labour migration and the rights of migrant workers. *Third World Quarterly*, 29(7), pp. 1247–64.

Williams, P., 2002. Transnational organized crime and the state. *Cambridge Studies in International Relations*, 85(1), pp. 161–82.

Zolberg, A., 2006. International migration in political perspective. In A. Messina & G. Lahav (eds.). *The Migration Reader: Exploring Politics and Policies*, Boulder, CO: Lynne Rienner Publishers, pp. 63–88.

SECTION 2

Disaster Diplomacy and Relief in Practice

Until now in this book, I have focused on how foundational belief systems of international order shape international disaster management regimes, such as disaster relief, disaster risk reduction, and disaster migration. These regimes, as currently structured, reflect contingent resolutions to the ongoing tensions and inconsistencies in a heterogenous international order. One of the core claims developed in preceding chapters is that states respond to the failures of existing regimes by altering those regimes and by creating new supplementary regimes to deal with developments on the ground that were originally excluded from consideration or considered inappropriate for managing disasters. International efforts to create a disaster risk reduction regime emerged from the demonstrated and perceived failure of the disaster relief regime to produce the outcomes it had originally promised – namely, to reduce human suffering from disasters. The same process has occurred with respect to the management of international migration.

At the start of this book, I argued that, in addition to shaping international regimes of disaster management, foundational ideologies of international order also constitute the identity of states and consequently how political leaders and civil society actors respond to disaster and offers of aid. In a destabilized post-disaster environment, offers of assistance from external actors introduce additional potentially destabilizing factors and may further disrupt existing beliefs about the international order and about responsibility for preventing and responding to disasters. For instance, offers of aid from adversaries may call into question the depiction of the international system as relentlessly competitive and hostile, and may introduce an ambiguity into existing antagonistic narratives. Similarly, an offer of aid from an impoverished or less technologically advanced state may disrupt claims about the affected state's identity as modern and technologically advanced. As noted in the introduction to this book, disaster and offers of assistance necessarily implicate a number of ideologies that sustain the current global sociopolitical order.

Governments and civil society actors construct state identity in terms of common foundational belief systems – modernism, nation-statism, economic liberalism, and humanitarianism – and in terms of their power in regional and international dynamics. The group's identity narrative draws on these frameworks in the construction of the group's "myth-symbol complexes" – the narratives that define the group (Kaufman, 2015). Consequently, decision makers implement policies to preserve identity constructs that emerge from these belief systems when actors call these identities into question. Developments that destabilize these identities, such as disasters, require responses that shore up the credibility of the identity claim and, by extension, ideologies of order.

To draw once again on the case of nation-statism, most political elites and civil society actors within a state believe that their states are the ultimate authority within their territory and assert that as a pertinent identity during deliberations on disaster relief, although there may be significant disagreement on the source of the state's authority. This does not mean that this identity is the only, or even the dominant, identity in operation in deliberations on aid or that all states will understand offers of aid as a threat to their sovereignty, but the foundational belief in nation-state sovereignty does have important implications. The first is that governments deliberate on whether or not to accept aid, or on what type of aid to accept, as though such decisions are rightfully theirs to make as sovereign states. They publicly perform this identity by consenting (or not) to the delivery of aid from other states and non-state entities. A second implication of nation-statist beliefs is that it opens up the possibility of civil society or opposition actors framing aid as a threat to state sovereignty. The ideology of nation-state sovereignty operates as a condition of possibility for the framing of aid as a threat to the authority of the state.

Whether a particular offer of aid is represented as a threat to state sovereignty (or the state's economy, or its international position) depends on whether actors frame it as such, which then gives way to the contested process of framing. The framing of aid depends on and reproduces identity constructs and how decision makers attempt to preserve multiple identities in response to a development that calls into question the ideological foundations of those identities. Disasters and the need for outside assistance present opportunities for actors to call into question the state's identity as modern, sovereign, liberal, humanitarian, or as a relevant international or regional power.

In this section, I demonstrate this process by examining a number of cases of aid acceptance and refusal, paying particular attention to how these offers are framed by decision makers. In all cases, I focus on how policy makers activate certain beliefs and identities to inform and legitimize their decisions on aid. These cases demonstrate how ideological belief systems shape decisions in concrete ways in developed and major international powers like the United States and in developing, regional powers like China and Egypt. As I noted in the introduction, I do not focus on less-developed countries such as Haiti and Myanmar, in part because the existing literature already focuses disproportionately on these cases. My aim is to demonstrate that decision making on aid in developed and developing states does not drastically differ from those of less-developed states, although the configurations of belief systems and their relative importance in specific instances can differ drastically. The goal here is not to provide a predictive account of how decision makers will respond to particular offers of aid; rather, my purpose is to demonstrate the dominance of particular ideological belief systems and how they shape decision making in differing contexts.

6

THE UNITED STATES AND HURRICANE KATRINA

The provision of assistance in cases of disaster and crisis is a meaningful order-sustaining activity in international relations – hence the extensive (if insufficient) international disaster relief regime. The acceptance and rejection of aid is also meaningful and has important consequences for bilateral relations and for the disaster-affected population. The burst of literature on Myanmar's rejection of international aid following Cyclone Nargis convincingly shows that refusal of aid can exacerbate widespread suffering in disaster-affected areas and even lead to international crises (Selth, 2008; Seekins, 2009; Bellamy, 2010; Nelson, 2010b; Paik, 2011; Streich and Mislan, 2014). The impact on bilateral relations is less clear, although the existing literature on disaster diplomacy suggests that, while offers of disaster aid seldom alter existing relations between states on their own, they can bolster existing processes of rapprochement (Kelman, 2007, 2011). While Kelman's work on disaster diplomacy is a useful corrective to the overly optimistic accounts of the positive impact of disaster aid, the exclusive emphasis on the potential to alter bilateral state relations misses much of the political purposes of international aid. Indeed, this chapter demonstrates that most states do not offer aid to convert enemies to friends, and aid is seldom accepted or rejected for that purpose. This chapter shows that aid is strongly tied to realist considerations of meeting foreign policy objectives, including strengthening partnerships, delegitimizing opponents, and boosting the state's legitimacy and soft power.

Nelson (2010a), who is among the only scholars to study the phenomenon of aid refusal more broadly, suggests that in developing states, aid is accepted or rejected based on its perceived threat to regime legitimacy. Nelson provides a compelling starting point for understanding why states may reject disaster assistance, although his analysis simply raises more questions: is the reasoning behind aid rejection categorically distinct from aid acceptance? Do the same considerations prevail in developed and developing countries? Perhaps more fundamentally, how do actors produce legitimacy, and how do states situate international aid in that process?

These questions are at the forefront of this analysis of the United States' response to international assistance in the aftermath of Hurricane Katrina. The United States' response is an important case study because of its position in the international system and in the provision of disaster assistance. This chapter explores how the United States used offers of aid to stabilize and legitimize American identities and to support key foreign policy objectives, protect domestic corporations, and enable the neo-liberal restructuring of domestic economic relations in disaster-affected areas.

Hurricane Katrina struck the southern coast of the United States on August 29, 2005. At landfall, it was a category 4 hurricane with wind gusts of 125 mph. Although stronger storms have hit the United States before and since, Katrina was one of the most catastrophic disasters in US history, in both human and economic terms. The hurricane caused tidal surges and flooding that killed almost 1,900 people; affected another 500,000; and caused close to $125 billion in economic damage, mostly concentrated in New Orleans, Louisiana (CRED, 2015). Contributing to the devastation was the poor response by federal, state, and municipal officials, leading to the characterization of Hurricane Katrina as one of the "largest breakdowns of public administration in the history of the United States" (Daniels et al., 2006; Walters and Kettl, 2006). Numerous scholars have noted how the response to Katrina reflected a number of beliefs about race and crime – notably the thoroughly liberal economic concern with protecting private property over people (Elliott and Pais, 2006; Hartman and Squires, 2006).

The focus here is not on the administrative breakdown at the domestic level but rather on the United States' response to international offers of assistance –which exhibited similar administrative confusion that characterized the domestic relief efforts. There were conflicting statements on whether the United States would accept aid and which offers were accepted or rejected, as well as significant waste of accepted aid. Offers of aid from even close allies were selectively accepted – Ferris and Petz (2012) report that of the 77 different offers of aid from "three of its staunchest allies: Canada, Britain, and Israel", the United States accepted only 23 (42). They also report that the United States claimed less than 5% of the cash and oil offered by other states and that a significant portion of the aid that was accepted was never distributed and/or left to rot in warehouses.

It is clear that there were a number of bureaucratic and/or administrative failures in the American response to international aid (Walker, 2006; Van der Linde, 2008); however, this chapter argues that these failures were the product of ideological tensions and contradictions exposed by the disaster and by international offers of aid to the United States. The administrative chaos that characterized the international response reflects the inconsistencies between belief systems that shaped how American officials understood offers of aid. The lack of pre-existing policies and procedures for accepting and distributing international aid reflected beliefs about the United States that rendered the possibility of needing international assistance unthinkable and an administrative capacity to distribute that aid as unnecessary. The scope of the Katrina disaster and the need for international assistance was a significant challenge to key beliefs about international order and the place of the United States in that order.

The United States is generally understood as a technologically advanced superpower that is, among many other things, the global leader in providing humanitarian disaster assistance. American humanitarian identity is supported by its (relatively) large financial contributions to such activities; in 2013, the United States contributed over $4.7 billion in emergency humanitarian aid, along with an additional $27 billion in other forms of development assistance (Swithern, 2015). Notably, the United States spends around 2% of its development assistance budget on disaster risk reduction, and over 6.5% on disaster relief (Swithern, 2015). The United States played a seminal role in shaping the international disaster relief regime, the disaster risk reduction framework, and the migration regimes outlined in Chapters 3 through 5, and until recently contributed substantially to the preservation of these regimes. As with other international regimes heavily shaped and financed by the United States, the United States is, now and historically, the largest financial contributor to international disaster relief.

Since the inception of institutionalized international disaster relief in the late 1800s, the United States has identified exclusively as an aid donor. Despite the fact that the United States experiences a large number of disasters relative to other states and in recent years has spent close to $30 billion per year on domestic disaster relief (Zabarenko, 2013), the United States had never been a recipient of major international disaster relief efforts prior to 2005. The United States' position as the world's leading disaster relief donor supported its identity as the exemplar of a modern, liberal, sovereign, and humanitarian superpower that is crucial to the maintenance of the current international order. The devastation caused by Hurricane Katrina disrupted these identity constructs, not least of which was the shift from aid donor to aid recipient. This shift contradicted key beliefs about how the international system is ordered and about the relative place of the United States, its allies, and its enemies in that order. Consequently, the Katrina disaster required extensive ideological, as well as material, reparation.

Given the large amount of aid offered to the United States and the various factors that influenced the American decision-making process, the pattern of aid acceptance and rejection by the United States in the aftermath of Hurricane Katrina defies simple categorization. It is an exemplary case for analysis because the State Department maintained a detailed list of aid offers and responses and released (or, more accurately, was forced to release) detailed documentation of internal deliberations over international disaster assistance.[1] What this material reveals is that, after initially signalling that it did not need assistance, which is typical of most disaster-affected states asserting their sovereignty, the United States implemented a generic policy of accepting all offers of aid in principle while in practice rejecting aid from many allies and partners as well as adversaries. The response of the international community to the disaster forced American policy makers to reframe offers of aid in ways that reinforced beliefs about how the international system operates and the United States' place in that order.

The documentation also shows that while the State Department and the Federal Emergency Management Agency (FEMA) played the lead role in making decisions over which offers of aid to accept, a number of other government departments participated in the assessment of offers of disaster relief. These included the United States Agency for International Development (USAID), US Department of Agriculture, US Department of Homeland Security, National Security Council, the Department of Health and Human Services, the states of Louisiana and Missouri, and the American Red Cross. In addition to demonstrating the institutional complexity of handling international aid in modern states, the documents reveal that several considerations came to the fore in shaping decisions on aid: the desire to maximize the number of offers accepted, limitations imposed by state and national regulatory frameworks, the difficulty of assessing humanitarian need and the perceived intent of donor states, convergence of standards in donor states, and legitimizing concrete foreign policy goals. Yet simply listing the various factors that influenced American decision makers fails to explain why and when these factors were important. Key to understanding and explaining the American response in the aftermath of Hurricane Katrina and the factors considered relevant to this decision requires an understanding of the identities and belief systems that decision makers reinforced in a period of potential destabilization.

We'll Take Care of it on Our Own

As with all disasters that prompt offers of international assistance, Hurricane Katrina thrust several contradictions of nation-state sovereignty to the forefront. First, it showed the inability of the state, even the most powerful state in the world, to ensure the security of all its citizens and to deliver basic public goods in the event of major disaster. Second, it demonstrated that the everyday practices of nation-states to regulate various aspects of modern life to protect citizens limits the ability of the executive to override the state's existing regulations. For instance, the expectation that the American state ensures the food that its citizens consume is safe and protects domestic food producers resulted in regulations that severely limited the ability of the American state to accept food aid from other states. Last, the Katrina disaster and international offers of aid demonstrated the disjuncture between the belief in a single, ultimate source of authority and the actual division of responsibility within modern states between federal and sub-federal units, between different federal agencies, and between the state and civil society. Given these sources of ambiguity and tension exposed through the Katrina disaster, American political elites devoted considerable effort to repair the cracks in the ideology of nation-state sovereignty.

Having emphasized the centrality of state sovereignty in the Katrina response, I argue that in contrast to other cases of aid acceptance or rejection, sovereignty was not the dominant framework through which decision makers understood foreign offers of aid. The preponderance of military and economic power that the United States possesses meant that it was unlikely that American decision makers would interpret offers of aid as undermining the state's authority in the provision of public goods. American leaders never expressed concern that accepting international aid would invite either invasion, foreign norms, or regime change, as happened in the paradigmatic case of Myanmar after Cyclone Nargis. As with nearly all other cases of international disaster assistance, however, among the first public statements from the federal government of the United States was an assertion of nation-state sovereignty.

On a nationwide television program on September 1, 2005, three days after the initial disaster, President Bush stated, "I'm not expecting much from foreign nations because we hadn't asked for it… this country is going to rise up and take care of it" (Bush, quoted in Pleming, 2005). Bush's statement was not a categorical rejection of aid; in fact, at the time of his statement, the State Department was already in the process of accepting some foreign offers of assistance. Rather, it was a statement to restore belief in American sovereignty and its ability to provide public goods essential to social order within its territory. At the same time, the president noted that he expected other states to offer help in the form of "sympathy… and cash dollars" (Bush, quoted in Pleming, 2005). The primary message was that the American state was fully sovereign and capable of responding to the crisis without foreign assistance and thus was not, in the traditional sense, an aid recipient.

Within days of Bush's statement, the US government reversed its position, with Secretary of State Condoleezza Rice announcing that the United States would accept all offers of aid, opening up other tensions in the statist narrative. Announcing that all offers of aid would be accepted reinforced American sovereignty simply by virtue of claiming the authority to make such a decision; however, the practice of consenting to international assistance rests on an embedded understanding that receiving aid represents a temporary diminishment or abdication of sovereign responsibility, understood in terms of performing its basic responsibility to protect its citizens. To counter this, the American government reframed its acceptance of aid as demonstrative of American power and leadership rather than a diminishment of sovereignty, which I will delve into more in the next section.

Another tension exposed by the decision to accept aid was to demonstrate the fragmentation of state sovereignty. Once the federal government decided to accept international offers of aid, it revealed complications over how sovereignty is distributed and practiced in the United States. This occurred in two ways: first, through discrepancies between offers of aid violating existing laws and regulations on the importation of goods, services, and people, and second, by offers of aid accepted by the federal government violating the division of power between national and sub-national authorities. As noted in an earlier chapter, the ideology of statism places responsibility for the protection of individuals in the hands of the state and places the national government at the top of that hierarchy, with the assumption that it can override domestic restraints.

The subsequent decision to accept as much foreign assistance as possible meant that the federal government confronted existing laws and statutes, and the long-standing division of power in the United States whereby federal and state authorities divide responsibility for the provision of public goods. This means that different agents normally have differentiated responsibilities in the provision of public goods, but in some emergencies, the sovereign may override existing arrangements and distributions of authority. In practice, however, the division of power and authority in a federal state may disrupt this understanding of sovereignty, and there may be no clear mechanism to override the division of power and responsibility and to change the asserted network or hierarchy of authority. This is particularly the case when there is no history or precedent for disrupting the current division of power in the event of a natural disaster. The most relevant law relating to foreign aid was the Stafford Act, which "establishes the programs and processes for the federal government to provide major disaster and emergency assistance" to sub-national authorities through FEMA (Government Accountability Office (GAO), 2007:8). As it pertains to international assistance, the Stafford Act empowers the president or his designate, in this case, FEMA, to "accept and use gifts or donations" for disaster response, but the act requires that these gifts are processed through the agency empowered to deal with it (GAO, 2007) Consequently, FEMA had to direct offers of assistance through various departments: the State Department handled cash donations, the Food and Drug Administration (FDA) approved most food and medicine, the US Department of Agriculture (USDA) handled some animal-derived food products, and the Department of Defense (DOD) assessed military personnel and materiel. Some aspects, such as the qualifications for medical professional, fell outside the jurisdiction of the federal government entirely, requiring state approval. As the United States had not received international disaster aid in the past, the executive lacked the capability to process and distribute such aid, as well as a mechanism to override the authority of states to regulate certain aspects of the economy.

The United States, as with many other states, has an extensive and well-administered set of national and sub-national laws and regulations that govern the movement of goods and people, making the expedited importation of large amounts of disaster assistance difficult. To take one example, medical teams from (most) foreign states were rejected on the grounds that licensing medical professionals falls to the individual American states, and the federal government cannot supersede their authority (Smith, 2005). In this case, the federal government opted to turn down most offers of medical personnel because the process for licensing them was too onerous to do quickly and on a massive scale.

The United States has faced this constraint in providing aid to other disaster-stricken communities and has advocated for measures to expedite disaster relief by reducing domestic regulations. Through the International Law Commission (ILC) and the International Federation of Red Cross and Red Crescent Societies (IFRC), donor states such as the United States have negotiated to expedite disaster relief by reducing customs rules, visa requirements, and domestic regulations on certain goods that impede effective disaster relief (Fisher, 2007:13–15). The stated concern of the ILC and IFRC is that states may enforce national regulations to restrict the entry of goods and people in an intentional effort to delay or refuse aid. This objection reinforces a unified view of state sovereignty that states can and should violate national regulations in the event of emergencies – whereas the American case demonstrates that, in some instances, the division of authority between executive departments, and between national and sub-national units, restricts the ability of national governments to override the existing distribution of responsibility and authority. Consequently, the federal government did not distribute much of the aid it accepted and left it to waste away in spaces belonging to the federal government or redistributed it to other parts of the world (Solomon and Hsu, 2007).

National and sub-national regulations played a key role in decisions on aid. FEMA rejected all offers of pharmaceuticals and medicines, with the exception of medicine from Canada, and asked donors to remove them from all donated medical or survival kits. FEMA rejected offers of foreign medical teams due to lack of recognition of foreign credentials and concerns about possible legal implications and insurance coverage for practicing medicine in the United States. In addition, the division of power between state and federal jurisdiction limited how much the State Department and FEMA could do without state approval. FEMA rejected numerous goods – foodstuffs, cleaning supplies, and water treatment chemicals – because they failed to meet American regulations on labelling, violated trade agreements, or qualified as banned products in the United States. More so than assessed need or meeting foreign policy objectives, national regulations and federal-state division of responsibility shaped the acceptance and rejection of international aid. In essence, the everyday practices of state sovereignty – regulating the movement of people and goods, production of food and medicine, the provision of healthcare – frames aid in terms of the existing rules and limits the types of aid that can be accepted.

The American response to disaster assistance demonstrates the ambiguities in the ideology of nation-state sovereignty and its impact on how states respond to disaster assistance. American leadership performed its sovereignty, first by refusing aid and subsequently by accepting all offers of aid. This performance introduced secondary tensions, namely, (1) that accepting aid signified a temporary diminishment of sovereignty understood as the capacity to provide security and basic public goods and (2) the division of sovereignty in the United States. To alleviate some of these tensions, American decision makers reframed offers of aid in terms of the other ideologies of order, one of which was an emphasis on American power and leadership.

Aid and Foreign Policy Objectives

Realism depicts aid as a tool of foreign policy designed to exert influence on other states and affect their position with respect to a larger political or economic strategy (Schraeder et al., 1998; Hattori, 2001). In the common form of this narrative, receiving aid typically signifies a subordinate position in the bilateral relationship, wherein a state accepts assistance with the understanding that this implies a certain level of compliance with the donor state's larger strategic objective. The American decision to accept offers of aid had the potential to challenge its identity as the superordinate state in the system and exclusively as an aid donor. The devastation caused by Katrina, the administration's inept response, and the provision of international aid all potentially challenged American identity as the most powerful state in the world and the understanding of aid as a tool used by powerful states to curry favour with recipients.

What is fascinating about the case of Hurricane Katrina is how the United States reframed the meaning of aid acceptance within the larger realist narrative by depicting aid acceptance (and refusal) as useful for meeting foreign policy objectives. Rather than providing aid in exchange for other states supporting American foreign policy interests, the United States understood its response to Katrina as accepting aid in exchange for other states supporting American foreign policy interests. The United States reinforced aid as a tool of foreign policy by emphasizing America's history of contributing to disaster assistance globally, by promoting key institutions (North Atlantic Treaty Organization [NATO]) in the provision of disaster aid, delegitimizing offers from those who challenged the American-led international order, and by legitimizing controversial American foreign interventions.

American decision makers reinforced the aid/foreign policy nexus by casting American acceptance of aid as demonstrative of American power and leadership. For the US government, offers of assistance demonstrated that the United States was reaping the benefits of its position in the international order and its successful application of soft power. American contributions to disaster relief in other regions became a fixture in media coverage and official diplomatic communication. The US government and US media outlets routinely recounted instances where the US government provided disaster assistance to countries that donated aid; the 2004 tsunami in Asia, hurricanes in the Caribbean, and earthquakes in China figured prominently (Morse, 2005). In response to a question on why the government initially seemed reluctant to accept foreign aid, Ambassador Harry Thomas, who was placed in charge of coordinating the response to foreign disaster relief, stated all offers were being considered and that "the American people should know that others are helping us ... in the way we help others" (Thomas, 2005). The various departments kept extensive lists on offers of aid to (1) demonstrate widespread support for the United States and (2) ensure that it publicly recognized and thanked each donor state.

American decision makers explicitly acknowledged that they should accept aid to meet foreign policy objectives and separated offers of aid into two streams: USAID and the State Department handled international offers of aid and worked with FEMA to determine which offers could be used, and which could not. Meanwhile, FEMA was the lead agency on the ground assessing need. The relationship between the two streams of disaster aid were depicted in terms of push-and-pull forces: FEMA pulled assistance from domestic and international sources by identifying areas of need, while USAID/State Department pushed assistance on FEMA by trying to find uses for foreign offers, even those of little use in fulfilling assessed need. Internal documents show that the acceptance of international aid was not concerned with "filling gaps in FEMA's operational capacity" but to "satisfy diplomatic objectives" (Stitt, 2005). Internally, FEMA viewed international aid as disruptive of national relief efforts, because it brought in goods that were not needed and introduced interests and objectives besides the efficient and fair distribution of aid. The resulting recommendations from the Office for Foreign Disaster Assistance (OFDA) suggested the development of three lists in future disasters: a pull list that included items the United States would request in future emergencies, a push-list that included items FEMA could take if "required to meet a foreign policy directive", and a no-go list that included items that would not be accepted (Stitt, 2005).

Second, the US administration reinforced the connection between aid and foreign policy objectives through the prioritization of NATO. While most offers of aid were quite small and mostly bilateral (and directed through the American Red Cross), larger, coordinated, multilateral aid occurred through NATO rather than the UN. In doing so, the United States reinforced the importance of military power and alliances as the bedrock of international order, even in the provision of humanitarian assistance. The internal discussions show that American officials spent much of their time working with NATO in the delivery of aid. It was not just greater attention given to NATO allies and allies in the war on terror, although this is also evident; rather, the discussions reveal a concerted effort to privilege NATO as the institution through which international aid would be collected and transported to the United States.

In the aftermath of the Cold War, NATO has attempted to "convert its military capability into political capital" to promote itself as a legitimate humanitarian actor and to ensure regional security (Huysmans, 2002:600). One way it has done so is by expanding into the realm of disaster relief; it possesses military transport abilities that are the most significant resource in effective disaster relief (Hanning, 1978; Walker, 1992). Since 1998, NATO and NATO-affiliated groups such as the Euro-Atlantic Partnership Council (EAPC) and the Euro-Atlantic Disaster Response Coordination Centre (EADRCC) have played a lead role in coordinating, transporting, and providing disaster relief in Europe, including Eastern Europe and former Warsaw Pact states. The functional expansion of NATO into disaster relief brought non-member states in closer contact and cooperation with NATO, often heightening Russian suspicions (Ratti, 2009; Pouliot, 2010). Although these NATO-affiliated organizations cooperate with the UN, in many instances, they operate as a competitor to UN agencies (Huysmans, 2002:604). Privileging NATO in the provision of humanitarian assistance is a political act that promotes a particular vision of international order in which security organizations such as NATO perform the key task of ensuring regional, and in some cases, global, stability rather than the UN (Huysmans, 2002:616) or other institutions.

In the aftermath of Hurricane Katrina, the United States turned to NATO as the lead agency in the coordination and delivery of disaster relief to the United States from EAPC and EADRCC countries. Over 39 states offered and channelled their aid through NATO; in all, 12 NATO flights delivered 190 tons of relief (Department of State, 2005; EADRCC, 2005). There were no recorded instances of non-NATO delivery of relief, and most states outside NATO's sphere of influence were encouraged to provide cash rather than physical material. While the United States accepted some UN assistance from the UN Disaster Assistance Team, the World Health Organization (WHO), and UNICEF, these entities played a much less significant role than did NATO. Much of the internal correspondence released pertained to organizing NATO flights and ensuring that they had priority landing at the main reception centre in Little Rock, Arkansas.

The importance of NATO and Eastern expansion of NATO is evident in discussions over offers of aid from countries in Eastern Europe, in particular Ukraine. In the declassified American documents, Ukraine was depicted as "anxious to provide" aid to demonstrate a shift toward the West and Europe (Hultman, 2005). While the US government decided that it did not need most of the aid from Ukraine, it did accept "token" Ukrainian aid in the form of aircraft transportation for NATO supplies and a "symbolic" amount of supplies. At this time, the government of Ukraine initiated a shift toward Europe away from Russia, and pushed the EU and NATO for recognition as a market economy, integration with the EU, and membership in NATO (Kuzio, 2003, 2006). Although the non-redacted US cables do not directly reference this shift, they do note that Ukraine's new president Yushchenko was scheduled to visit the United States on September 14, that the government of Ukraine was facing a "political crisis", and that the decision on Ukrainian aid should take into "important political considerations" (Bula-Duane, 2005). The special consideration of the Ukrainian offer of aid and the focus on NATO demonstrates how the acceptance of aid supported American foreign policy objectives and bolstered American-led security institutions.

A third illustration of the aid/foreign policy nexus concerned the treatment of aid offers from states that opposed the United States' international position and its use of power to achieve international outcomes. Officials in the State Department, USAID, and FEMA were aware of the political implications of rejecting offers of aid, and sought to accept as many offers as possible, and to reassure most donors that their aid was rejected based on the humanitarian principle of need rather than as a statement on the nature of their bilateral relations with the United States. In other cases, however, the rejection of aid was justified by calling into question the humanitarian intent of the donor. Consequently, there were two forms of aid rejection: (1) rejection based on recipient need and (2) rejection based on donor identity. Only three states were singled out in such a way, and all three were key opponents of the exercise of US power internationally: Iran, Cuba, and Venezuela.

The released internal documents demonstrate that the State Department was unsure from the outset how to handle the offers of assistance from these three states in particular. Relations between Iran and the United States in 2005 were hostile. Just three years earlier, President Bush had included Iran as one of three states in his "axis of evil" speech and earlier in 2005 had extended sanctions against Iran over its nuclear program. Despite a few voices to the contrary, the United States and Israel viewed Iran's pursuit of nuclear weapons as a threat to international order, and as a threat to Israeli security and to the nuclear weapons regime (Waltz, 2012; Clarke, 2013). Iran was a frequent and outspoken critic of American foreign policy; during his UN General Assembly speech on September 17, 2005 (less than three weeks after Katrina and in the latter stages of the international relief operation), President Ahmadinejad excoriated US foreign policy in Iraq and Afghanistan and for targeting Iran's nuclear power program. He further suggested that Hurricane Katrina demonstrated that the US government acted in the interests of a small powerful elite rather than for the American population as a whole (Ahmadinejad, 2005).

Despite (or perhaps because of) the high level of hostility between the two states and the politicization of Hurricane Katrina, Iran offered disaster relief to the United States. Kelman (2007) reports that the offer consisted of "20 million barrels of oil and assistance for rescue efforts" (Kelman, 2007:297). Participating in bilateral humanitarian assistance was at odds with the understanding of Iran as a pariah state intent on challenging the foundations of American-led international order. Unlike all other offers, the United States never officially recorded or acknowledged the Iranian offer, although internal documents and a few public statements imply that Iran did make an offer, which the State Department categorically rejected. In one State Department press release on the Iranian offer of aid, the US government claimed that the media had reported an Iranian offer but that the State Department had not received an official offer through regular diplomatic channels (Sullivan, 2005). Questioned on Iran's offer, another representative for the State Department simply stated that it had rejected Iran's offer because it was "conditional. And that speaks for itself" (Thomas, 2005). Thomas did not elaborate on the Iranian conditions, but Van der Linde reports that it involved lifting sanctions on Iran (2008:9). Accepting Iran's offer of aid and lowering sanctions would have reinforced the traditional realist understanding of aid that would have placed the United States in the subordinate position and disrupted the modified realist position adopted by the US administration. By revealing that Iran's offer was conditional, it undermined the ostensibly humanitarian nature of disaster aid, reinforcing the depiction of Iran as a state outside the existing international order and thus legitimately subject to the application of US power and coercion. By rejecting the offer of aid, the United States reasserted its dominant position internationally.

The United States also received an offer of assistance from Cuba – a state long in conflict with the United States and subject to a prolonged period of American sanctions. In the aftermath of Katrina, the Cuban government offered "1,585 doctors and 36 tons of unspecified material" (Regan, 2005). From the outset, the State Department handled the offer outside the established, albeit ad hoc, procedure as it understood the offer to be politically contentious. Internal documentation noted the "sensitivities involved" and issued special instructions for handling this offer: directing all correspondence through the Cuba desk (Regan, 2005) and not including the offer "in any White House acknowledgement" (Rojas, 2005).

In the end, the State Department publicly rejected the offer on the grounds that Cubans were in need of the assistance as a result of Hurricane Dennis, which had struck Cuba earlier in July 2005 (Tumminia, 2005). The press guide from the State Department "encouraged Cuba to instead use the resources it offered to the US to help the Cuban people". The same document noted that Cuba had rejected official US aid in the aftermath of Hurricane Dennis, but that the United States had directed assistance to Cuba through non-governmental organizations (Tumminia, 2005).

The US government accepted aid from other states recently damaged by major disasters and who would have benefitted from using those resources domestically, yet in those cases, the offer of aid was presented as generous and an important demonstration of support and solidarity given their own need. The case of Kenya stands as a prime example. The State Department inadvertently publicized Kenya's offer of $100,000 as $500 million, a sum that the State Department saw as "damaging to the credibility of their government" and given that Kenya relies heavily on foreign aid, it expressed concern that the mistake could reduce "the amount of international humanitarian support that it receives" (Marshall, 2005). Rather than publicly reject Kenya's offer and suggest it would be best spent "mitigating the catastrophic impact of HIV/AIDS ... and the huge population of orphans and street children", as was mentioned in internal discussions (Marshall, 2005), the State Department accepted the offer and publicly thanked the Kenyan government and people for their generosity. In rejecting Cuba's offer, the United States called into question the legitimacy of the Cuban regime and its humanitarian identity, while asserting its own humanitarian credentials. The depiction of the Cuban government as more concerned with scoring political points that ensuring the health and well-being of its own citizens legitimated longstanding American antagonism toward Cuba.

The United States also received an offer of aid from Venezuela, one of the most activist states at the time in terms of altering the American-led international order and with whom the United States had a strained relationship. In his speech before the UN General Assembly in 2005 (just 17 days after Katrina), President Chavez called for the establishment of a new international order and the relocation of the UN to "the South", and accused the United States of being the source of "great evils" and sponsoring state terrorism. Chavez used the Katrina disaster to criticize the US government and show that Americans "do not have a government that protects them from … disasters" (Chavez, 2005). From the perspective of the United States, the offer from Venezuela was similar to the offer from Cuba – part of a political game. In fact, the State Department documentation often conflated or referred to the offers as a single problem – in terms of what to do about the offers from "Cuba and Venezuela".

Despite the conflation of the Venezuelan and Cuban offers and the relatively high level of hostility between Venezuela and the United States, the two states maintained significant economic relations, particularly in the trade of oil. Thus, while the Venezuelan government presented itself as a leader in challenging the American international order, it was also keen to reinforce its identity as a reliable trade partner with the United States. In its official offer of gasoline, the Venezuelan ambassador stressed Venezuela's record as "a consistent and reliable supplier of energy to the United States" whose offer of aid would reach the United States more quickly, and thus would be more effective, than oil products from other American oil suppliers (Herrara, 2005). Thus, both the United States and Venezuela faced a difficult situation; Venezuela was assisting a state whose international leadership it publicly condemned, and the United States had an interest in rejecting the aid to send a message to states critical of the current American-led international order.

In addition to gasoline, the government of Venezuela offered $6 million to the Red Cross, along with generators, water, and food (Department of State, 2005). Unlike other offers of aid, however, Venezuela offered portions of its aid to the affected states directly, Mississippi and Louisiana, rather than the US government, and through CITGO, the Venezuelan-owned oil and gas corporation based in the United States that owns several refineries in Louisiana. The State of Louisiana, rather than FEMA or the State Department, accepted the offer of aid, and the State Department listed the CITGO portion of the aid as a private donation to the state of Louisiana from a corporation with local ties to the affected community. The State Department accepted the cash offer to the Red Cross and rejected the other aspects of the offered aid.

As with the offers from Iran and Cuba, internal communications in the State Department suggest that the administration treated the offer with suspicion and consulted with the National Security Council on whether to acknowledge the offer. The officer responsible for receiving and responding to the offer thought that the Venezuelan government was "playing with us", by repeatedly changing the content of the offer and demanding immediate shipment (O'Herron, 2005a). In the end, the preservation of trade relations presaged on the trade of oil and localized through the connections between CITGO and the state of Louisiana seemed to trump foreign policy objectives and the hostile relations between the two states. While the US federal government did not receive aid from Venezuela, the American Red Cross and the state of Louisiana benefitted from Venezuelan largesse, and the State Department publicly accepted and thanked the Venezuelan government.

Last, the United States used international offers of assistance to legitimize the war on terror and its interventions in Iraq and Afghanistan. The United States treated offers of relief from states that publicly and continuously opposed US intervention in Afghanistan and Iraq with suspicion, but did not always reject them. Rather, the State Department used offers of humanitarian assistance to bolster the legitimacy of its foreign interventions. Indeed, the US government publicly depicted offers of disaster relief from the governments of Afghanistan and Iraq as demonstrative signs that US military intervention abroad was a positive development that was conducive to goodwill toward the United States. The US ambassador to Iraq, in his statement of thanks, stated that the United States' purpose in Iraq was to "help the Iraqi nation to stand on its own feet after decades of dictatorship" and that the disaster relief provided by Iraq demonstrated "the United States and the new Iraq share common values" (Khalilzad, 2005).

The State Department tied other offers of aid directly to support for the war on terror. For instance, the official statement of thanks to Egypt placed its offer of assistance for victims of Katrina alongside Egypt's humanitarian aid to Afghanistan in support of Operation Enduring Freedom (Bagby, 2005). Internal documents also note that American officials were eager to "get the message out" that Arab and Muslim countries had contributed aid (Lawrence, 2005). The message was to demonstrate support from the Arab and Muslim world and in turn show that the war on terror was not a war on Arabs and Muslims and had not alienated these groups. Declassified documents also show that the United States gave Latin American states that provided disaster relief and contributed troops in Iraq priority in representation at the Hispanic Heritage Month event in 2005. While the United States did not reject offers of aid based on whether states supported or participated in their foreign interventions, the State Department used offers of assistance to legitimize these interventions.

This episode reveals that almost all states viewed disaster aid to the United States as a way to achieve foreign policy objectives: from currying favour with the United States to discrediting US leadership, to attempting to change US foreign policy. Rather than seeing aid donation as a challenge to US leadership and soft power, based on the conventional understanding of aid donation as a form of soft power, the US government inverted that understanding. In this revised realist framework, offers of aid legitimized US leadership on disaster relief, strengthened key US-led institutions, delegitimized states that oppose US leadership internationally, and reinforced its role as the world's enforcer of international order. The US response to disaster aid reflected and reinforced realist beliefs about the use of aid in achieving foreign policy objectives.

Modern Standards for Providing Aid

Modernist beliefs about disasters and foreign aid also shaped the US government's response to international offers of aid. The devastation caused by Katrina raised some ambiguity over the construction of US identity as modern, technologically sophisticated, and effectively governed. Numerous national and international commentators expressed surprise that such a disaster could befall the most advanced nation on earth (Forero and Weisman, 2005), and many allocated responsibility for the disaster to the persistence of "non-modern" characteristics of US society: corruption, racism, apathy. A few rare elements in US society, such as evangelist Pat Robertson, even framed Hurricane Katrina as divine punishment for various sins.

The US government's response to offers of aid clearly reflected a modernist belief system and the importance of modernity as a standard of civilization. As experts in disaster management identified needs and created lists, and administrators counted, tracked, and distributed aid, they essentially employed a modern "standard of civilization" in the acceptance and rejection of aid. The State Department expected donating states to meet certain basic criteria to have their aid accepted: all potential donor states were required to identify "an English speaking contact" with sufficient expertise to discuss and relay information about their aid offer, and donating states were instructed to wire funds in US dollars. The relatively recent promotion of English as the international language of diplomacy and business has been a core element in the expansion of the US international order (Phillipson, 2008) and signals which states are capable of maintaining "diplomatic exchange and communication", which operates as a basic standard of civilization (Gong, 2002:80). Similarly, participation in the global financial system has become a defining standard of assessing states by dividing the international community based on its incorporation of market exchange and participation in a global economic exchange, including the use of the basic technologies (Gong, 2002:78). Requiring English contacts and using wire transfers as the primary method of providing humanitarian assistance reproduced a standard expectation of modern states and reinforced these institutions and practices as central to international order.

This modern "standard of civilization" also applied to other aspects of donated aid. The State Department, USAID, and FEMA subjected certain forms of aid (food, medical personnel, and technical experts) to scrutiny through these standards. With few exceptions, the Katrina taskforce rejected all offers of medical teams, pharmaceuticals, and medical devices, and subjected offers of foodstuffs, such as meals that are ready to eat (MREs) to extra inspection. The reasons for not accepting this type of aid reflected a combination of federal/state jurisdiction and standards on quality. Foodstuffs and medical supplies from most states required FDA approval. Gaining this approval was possible, but FEMA would only pursue approval in instances where it perceived the donating state as "having similar standards of care" (Bass, 2005b). The FDA sent out its regulations and guidelines for bottled water, food products, and supplies, and all donated materials were required to meet FDA regulations, including labelling requirements on language, ingredients, and nutritional information.

As for personnel, the Centers for Medicare and Medicaid Services (CMS) Taskforce requested that all offers of medical personnel include specifics about their training and skills, as well as the equipment specifications for any equipment they would bring (O'Herron, 2005b). Although the US government redacted much of the discussion on which states met this standard from the documents, examining the states from whom medical professionals and disaster management and technical experts were accepted reveals a preference for states perceived as having similar standard of civilization. Forty-nine states offered expertise of various kinds (medical, forensic, engineering, search and rescue [SAR]), with the United States accepting only eleven offers,[2] mostly from other members of the Group of 8 (G8). The internal documents suggest that need and speed played the primary role in the rejection of Russian and Japanese experts. The Russians offered SAR teams described in State Department memos as experienced and capable; however, the US government met the demand for SAR teams domestically and with support from Canada and Mexico, whose SAR teams arrived within hours due, in part, to the North American Free Trade Agreement (NAFTA) trade deal that had standardized many of these areas across the three countries. In addition to the elite G8 club of modern states, the United States accepted medical teams from states with specialized expertise: engineers and pumping teams from Belgium and the Netherlands, and Spanish-speaking mental health experts from Argentina, Colombia, and Mexico.

The acceptance of mental health experts from these three states was due to FEMA's assessment that the United States lacked the domestic capacity to assist "thousands of Spanish speaking people" affected by Katrina (Drucker, 2005). Nonetheless, prior to accepting these offers, FEMA and the State Department assessed their qualifications. In all cases, they required proof of certification, relevant experience for each member of the health teams, and letters of recommendation. In addition to this professional documentation, the process of legitimization for the Argentine mental health experts entailed listing their "considerable field experience" in close to 30 countries, and the organizations they had worked with, which included "UNESCO, WFP [World Food Programme] and UNHCR"; as well as letters of commendation from Kofi Annan (Schiffer, 2005). FEMA only accepted aid based on a combination of assessed need and assurance that the personnel met the basic standard of care expected in the United States.

Many states explicitly acknowledged a standard of modernity, including qualifications, elite training, and work experience of those whose skills they had offered. In particular, donating states used experience in previous international disasters, education and accreditation from an American university, and/or previous training with American professionals as indicative of the highest standard of professionalization. States were eager to have their offers accepted in part as recognition of their status as modern states. These statements, in essence, act as an assertion of modernity, or of meeting the standard of the United States, which many states understood as the epitome of modernity and civilization in terms of its technological expertise in responding to disaster.

The response of the United States to offers of MREs demonstrates both the administrative failure associated with Hurricane Katrina and the application of modernist standards of civilization to offers of aid. FEMA initially requested MREs from NATO-affiliated states, and these were among the first relief provisions to arrive. However, the US Department of Agriculture rejected MREs from European Union (EU) countries due to trade restrictions on certain agricultural products. Later, MREs from most EU states were accepted after the State Department concluded that rejecting MREs from the EU while accepting them from Egypt and Russia amounted to "the worst possible PR story" (Bass, 2005a). The negative "PR" reflected a belief that the United States should accept food from close allies with a standard of care similar to the United States.

To alleviate concerns over public perception of food aid, the State Department, working with USDA, accepted MREs from a number of European states, although it still blocked some MREs from the United Kingdom because it contained meat deemed inadmissible due to food safety regulations – USDA had blocked British meat and poultry due to bovine spongiform encephalopathy (BSE), or mad cow, concerns. The United States then redistributed British MREs through the Food for the Poor charity to Haiti (Chan, 2005). In this case, the modernist standard operated in declaring the food unfit for consumption in the United States, a modern society, but fit for Haiti, understood as a pre-modern or less-than-modern state with greater need, fewer regulations, and presumably lower standards on quality of food.

Of course, the assessment of some states and their offers of aid as failing to meet a modernist standard of civilization represented a diplomatic dilemma for those concerned with maintaining the United States' international influence, hence the creation of lists based on assessed need by technical experts and lists based on diplomatic objectives. The most diplomatic way to achieve both goals was to present all refusals of aid as based on assessed need and to encourage states to offer cash rather than in-kind donations of expertise and materiel. However, even here we see the standard of civilization at work, with the United States creating the expectation that all states have English-speaking contacts, American currency and the capacity to wire funds electronically. US officials accepted expertise from states generally perceived as equally modern, namely, their advanced Western partners in the G8, as well as a few states with niche expertise, such as water-pumping expertise and English-Spanish bilingual, post-disaster, mental health teams. Modernist beliefs framed offers of aid in terms of their technological sophistication, as measured against the United States' standards. Much of the aid that was rejected, unused, or redonated was due to its failure to meet US standards.

Foreign Aid and the Domestic Economy

Disasters typically destroy infrastructure and disrupt existing economic relations, operating as a form of creative destruction in which inefficient allocations of capital can be replaced by more productive ones (Klein, 2007; Mutter, 2015). This means that the rebuilding process following disaster offers many opportunities for the generation of profit and economic restructuring; consequently, beliefs about economic order (and the previous capital investments) are paramount in shaping disaster relief and rebuilding. Indeed, there is a significant amount of scholarship on Hurricane Katrina demonstrating how neo-liberal economic beliefs shaped the disaster and, the actions of local responders and profoundly influenced the rebuilding efforts. The downsizing and privatization of FEMA under the Bush administration severely restricted its ability to respond effectively (Mutter, 2015:174). During the disaster, local officials often demonstrated greater concern about protecting property rather than protecting vulnerable people, focusing inordinately on the problem of looting and leading to a militarized police response (Tierney et al., 2006; Dynes and Rodriguez, 2010). In the rebuilding process, the federal government funded support for the private sector to rebuild while reducing welfare programs (Davis, 2005). Damaged property left behind by evacuees was quickly sold off to entrepreneurs to build the city "back better", which resulted in the dispossession of poorer, mostly black, residents (Zack, 2014:60–1). The state and city also used the mass displacement and destruction of schools to reconstruct the New Orleans school system along neo-liberal principles of privatized education and charter schools (Saltman, 2009; Gotham, 2012). New Orleans has become a classic case demonstrating the influence of liberal and neo-liberal economic principles in disaster management (Kates et al., 2006; Gladstone and Préau, 2008).

The international picture is less straightforward. The embedded liberal compromise that continues to shape the international economic order encourages both open trade policies offset by protections for some domestic producers and workers. Disasters often demonstrate the benefits of free-trade policies in the form of fewer restrictions on the movement of goods and people. In the case of Katrina, America's NAFTA partners were able to provide expedited relief; some portions of Canadian and Mexican aid and personnel arrived within 24 hours and before many FEMA personnel (Mutter, 2015:175).

However, because disasters represent an opportunity to generate profit and stimulate local economic activity, states have incentives to promote/protect domestic industries rather than freer movement of goods and personnel. The response to the Katrina disaster suggests that, in cases of disaster, the protectionist component of the embedded liberal compromise is particularly attractive. The US government sought to protect its corporations from foreign competition in a number of ways. As a starting point, the Bush administration's emphasis on cash donations as the preferred form of aid limited the extent to which aid would threaten domestic producers. In the case of the Netherlands, Van der Linde notes that the United States readily accepted Dutch aid and expertise in water management during the crisis, but it excluded Dutch private corporations from the rebuilding process due to American protection of domestic shipping and dredging industries (Van der Linde, 2008:14). Similarly, in the case of Greece, who offered two cruise ships to house those left homeless by the disaster, the US government rejected this offer in favour of a no-bid contract to an American cruise line with strong political connections to the Bush family (Van der Linde, 2008:15). In both cases, Van der Linde suggests that the US government's priority was with "providing a market opportunity for domestic firms" at the expense of foreign firms and US taxpayers (16).

There were other instances of aid rejection due to concerns about its effect on US domestic industries. FEMA turned down offers of wood, for instance, due to fears that it would negatively affect wood producers and suppliers in the United States. The released documents show that FEMA identified materials not needed because they were locally available. In most cases, the United States turned down offers of aid that included material easily acquired in the United States. In other cases, the US government accepted material for political purposes but did not distribute it and thus did not affect local producers. While the Katrina case amply shows how policy makers may use disasters to reshape local economies and produce profit for private actors, the US response to international offers of aid shows that the government sought to ensure that aid did not displace US producers and that the disaster profits were distributed primarily to US corporations. From the embedded liberal framework, the US government approached most offers of aid based on its impact on the domestic economy. Consequently, the response was domestically (neo)-liberal (market-oriented) and internationally protectionist.

Can Dictators Be Humanitarian?

While the United States and donating states clearly prioritized foreign policy objectives in the provision of aid, it would be misleading to exclude the impact of humanitarian considerations. Domestically, Rooney (2017) estimates that Americans donated $3.7 billion in relief. International non-governmental organizations (INGOs) collected over $3.3 billion in donations to assist the victims of Hurricane Katrina, largely through the Red Cross (Eikenberry et al., 2007).

Internationally, close to 150 states offered direct assistance to the US government. While many of these states had "political" motivations for offering aid, the pattern of charitable giving globally and within the United States suggest that the humanitarian frame was also influential in the international response to Hurricane Katrina.

The media framing of the disaster contributed to the humanitarian response, although humanitarian frames were not dominant. Dynes and Rodriguez (2010) suggest that the media relied on five frames (damage, death, help, authority, crime) in their coverage of Katrina, and three of these draw on the larger humanitarian frame that relies on a victim narrative of helpless, passive, depoliticized individuals. Dynes and Rodriguez observe that the use of these particular frames distorted developments on the ground, disproportionately focusing on looting and crime, and minimizing some aspects of suffering associated with the disaster and the government's response. The disproportionate focus on looting and crime may have even contributed to the prevalence of military and police personnel in international offers of aid. Most problematically, Dynes and Rodriguez conclude, the frames that dominated coverage ignored the myriad instances of pro-social activity in which local residents cooperated to prevent damage and to actively assist one another (2010:35). Internationally, humanitarian frames performed as designed in the case of disaster: to elicit widespread shock and support.

Rather than address how humanitarian frames shaped offers of aid, in this chapter, I am more interested in how humanitarian beliefs shaped the US response to offers of aid. As is the case in other instances of disaster assistance, humanitarianism functioned as a basic framework for understanding international offers, which in turn reinforced the understanding of disaster relief as humanitarian (rather than political), voluntary (rather than obligatory), and bilateral (rather than multilateral). The State Department kept a detailed list of aid offers, in part to ensure public acknowledgement and gratitude for each offer. Indeed, a significant proportion of the internal correspondence related to confirming offers of international aid and coordinating the announcement with the donating state. For instance, the State Department communicated to its officers "we must ensure that our appreciation for their generosity and continued support continue to be expressed clearly and publicly. As delivery … is confirmed, embassies are to make public statements to the government, in the media, and in other appropriate venues thanking donors for their help at this critical time" (Gisvold, 2006). While there is no doubt that this fulfilled political objectives and diplomatic protocol, it also reinforced the humanitarian framing of aid as intending to relieve human suffering rather than fulfilling political objectives. Of the three offers of aid that were framed as political rather than humanitarian, which I addressed in a previous section, they were either rejected (Cuba and Iran) or reluctantly accepted through alternative routes (Venezuela).

The maintenance of a humanitarian/political binary not only influenced US decision makers on whether to accept or reject foreign aid; it also identified the types of political considerations that would disqualify an offer as humanitarian. While many states lobbied the US government to accept their offers of aid for political reasons, the conditionality of Iran's offer symbolized the politicization of aid, legitimizing its rejection. Similarly, the government and media depicted Cuba as non-humanitarian for offering aid that it could use to help the Cuban people and for earlier rejecting US humanitarian assistance. While the United States accepted some of Venezuela's offer, in large part due to trade considerations, it channelled that aid through a corporation, the state of Louisiana, and a non-governmental organization (NGO) – the Red Cross.

At the same time, singling out each state for public acknowledgement reinforced nation state–nation state solidarity as well as the bilateral benefits of the humanitarian system. For many states, it was important to donate aid to the United States in exchange for recognition and acknowledgement. Numerous correspondences in the released material emphasized that states were pressuring the State Department (who in turn pressured FEMA) for their aid to be accepted. In addition to solidifying their identity as humanitarian, having their aid accepted would increase their international profile and potentially secure an audience with US representatives at the upcoming United Nations General Assembly (UNGA) meetings.

Last, the response to Hurricane Katrina emphasized the voluntary nature of humanitarian assistance. At the outset, the US government indicated that any offers of aid were voluntary – states were under no obligation to provide aid and indeed were not expected to provide any. As offers of aid came in, the US government directed much of the international aid through IFRC and, to a lesser extent, the Bush-Clinton Katrina Fund. The Red Cross in particular enjoyed a privileged position in this particular case, which reveals a great deal about how the United States uses humanitarianism in the construction of international order. The United States stressed that NGOs would take the lead role in distributing relief and explicitly requested that states donate aid to the IFRC or other NGOs. This reinforced the existing system of disaster relief within the United States, in which NGOs take the lead role in emergency relief. Most states that donated cash did so through the Red Cross, and the State Department redirected the bulk of disaster relief it received to the American Red Cross (Department of State, 2005).

US reliance on IFRC and other NGOs is not overly surprising, as the IFRC is a voluntary quasi-NGO that enjoys a uniquely privileged position internationally in the provision of humanitarian assistance, dating back to its inclusion in the League of Nations. The emphasis on the IFRC and other NGOs reinforces the US view of limited state intervention into the economy and of humanitarianism as a voluntary, charitable, bilateral activity rather than a compulsory, multilateral obligation (Hutchinson, 1996; Katoch, 2003, Watson, 2014). Not only has the voluntary system exemplified in the IFRC curtailed efforts to create stronger obligations on states to contribute to disaster prevention and relief or to channel those funds through multilateral agencies, but the IFRC and the ARC's prominent place in the system has also enhanced US influence and soft power through the provision of bilateral humanitarian assistance (Kissinger quoted in Kent, 1987:81; Hutchinson, 2001). Prioritizing

NGOs, and the IFRC in particular, in the provision of disaster relief served to reproduce the international humanitarian order in line with a longstanding US vision of its duty to other states as voluntary, charitable, and bilateral.

Recent Developments

The last contribution of this chapter is to place the response to Katrina in contrast with the current Trump administration. The success of the current US administration's challenge to many of the ideologies of order explored in this book suggests that a significant portion of the US population no longer places much faith in the ideologies outlined here – although it is impossible to ascertain what role Katrina played in this development. Trump's election signalled a significant alteration in, and in some cases rejection of, the beliefs that had sustained US involvement in the international order. Trump's presidency represents a reaction against the embedded liberal compromise in favour of protectionism (evidenced by protectionist policies and changing existing trade agreements), against modernism in favour of traditionalism (evidenced by the rejection of science, expertise, and facts), and against humanitarianism in favour of communal interest (evidenced by the cutting of aid programs and refugee admissions). As it pertains to nation-statism, the Trump administration still emphasizes state sovereignty, but it privileges nationalism over statism, and indeed privileges white nationalism over civic nationalism, combined with a deep mistrust of "the state" and regulatory oversight. While the Trump administration appears to adhere to a form of realism – indicated by the expansion of military spending, belief in military action to solve international problems, and an overriding concern with freeriding in NATO – the Trump administration has also alienated some of its closest allies while courting or relying on support from adversaries, such as Russia. These policies suggest a modified and incoherent realist orientation, shorn of realism's characteristic pragmatism and prudence.

Although a direct comparison is not possible, the ideological shift under the Trump administration appears to have affected the international response to hurricanes Harvey and Maria. The most obvious implication is the lack of international offers of aid: while close to 150 states offered aid after Katrina, the publicized offers after Harvey to date number less than a dozen (Erickson, 2017; Toosi, 2017), and there appears to have been two offers of international assistance for Puerto Rico after Hurricane Maria. Both Toosi and Erikson speculate that this is the result of the Trump administration's foreign policies, which include a reduction in foreign aid and migration, a hostile position to freer trade, and an unwillingness to engage in international efforts to combat climate change. Kounalakis speculates that states have withheld offers of aid to signal opposition to Trump's foreign policy, his intentional alienation of traditional allies, cutting of USAID programs, and the America First policy that places emphasis on every country for itself (quoted in Erickson, 2017). The Trump administration's policies and orientations signal a decline in the relative importance the United States places on humanitarian, liberal, and even modernist beliefs in maintaining international order. Indeed, these shifts and the international response to them may have contributed to changes in perceptions regarding the importance of the United States in the maintenance of international order and consequently the importance of the United States in achieving its foreign policy goals and the value of offering aid to the United States. While it is beyond the scope of this book to explore the international response to Hurricane Harvey and the Trump administration more broadly, these developments lend support to the arguments developed in this chapter, which is that donor and recipient states understand disasters and offers of disaster aid in terms of the dominant ideologies of order. Changes to the contingent resolutions within, and the relative balance between these belief systems, affect decisions on offering and accepting disaster aid.

In line with this ideological shift, the Trump administration appears unwilling to accept the few offers it has received, apart from a few supplies from Canada and the use of satellites from the EU (Erikson, 2017). Although the two disasters are not directly comparable and there is insufficient information on Hurricane Harvey to draw firm conclusions, there are indications that ideological change in the United States has altered its approach to offers of aid. The GAO recommendation after the Katrina failure was to strengthen FEMA and the State Department, and communication between them in handling donations. Under the current administration, the State Department is significantly weaker, with attrition reducing its capacity to process offers of aid (Kounalakis, cited in Erickson, 2017). The American Foreign Service Association (AFSA) reports that the State Department "has lost 60 per cent of its Career Ambassadors since January [2017]", cautioning that "the rapid loss of so many senior officers has a serious, immediate, and tangible effect on the capacity of the United States to shape world events" (Stephenson, 2017). Combined with a hiring freeze on entry-level positions, the AFSA suggests that the Trump administration is "weakening the American Foreign Service" and "forfeit[ing] the game to our adversaries" (Stephenson, 2017). Former secretary of state Madeleine Albright views the severe weakening of the State Department under Trump as a national emergency (Albright, 2017). I argue that this clearly reflects a shift in the ideology of nation-statism in the United States under Trump, in which the president undermines the bureaucratic legitimization of state authority and its ability to provide basic public goods in favour of an ethno-nationalist legitimization of state authority.

The failure of the Trump administration to respond to the devastation in Puerto Rico after Hurricane Maria further supports this interpretation. UN experts reported "a dissimilar urgency and priority given to the emergency response in Puerto Rico compared to the U.S. states affected by hurricanes in recent months" (Farha, quoted in Cumming-Bruce and Robles, 2017). FEMA deployed far fewer personnel to Puerto Rico, while President Trump delayed requesting additional funds from Congress for Puerto Rico. In addition to downplaying the devastation of Maria by suggesting it "was not a real catastrophe, like Katrina", Trump worried that aiding Puerto Rico was "throwing our budget out of whack", concerns not expressed in the relief efforts in Texas and Florida after hurricanes Harvey and Irma (Gibson, 2017). Building on racial stereotypes, Trump claimed that Puerto Rico suffered from poor leadership and "wanted everything done for them" (Baker, 2017). Trump further claimed that the economic crisis facing the island was "largely of their own making" and that "all infrastructure was a disaster before hurricanes" (Baker and Dickerson, 2017). In doing so, Trump raised the issue of responsibility for poverty and vulnerability and the worthiness of aid recipients, drawing on racial stereotypes and existing levels of poverty and infrastructure to legitimize the restrictions placed on disaster relief in Puerto Rico.

Furthermore, the Trump administration refused initially to waive the Jones Act to facilitate aid delivery to Puerto Rico. The Jones Act, a protectionist relic from the 1920s, "decrees that only American ships could carry goods and passengers from one U.S. port to another ... every ship must be built, crewed and owned by American citizens" (Denis, 2017). In addition to preventing the arrival of foreign aid, the Jones Act has increased the cost of living on Puerto Rico, which the disaster exacerbated (Denis, 2017). Trump's rationale for refusing to waive the act was that "we have a lot of shippers and a lot of people ... who work in the shipping industry that don't want the Jones Act lifted" (Mortimer, 2017). In contrast, the administration waived the Jones Act after hurricanes Harvey and Irma to facilitate the shipment of oil after pipelines were damaged, which the administration claimed was a matter of "national security" (Nixon and Stevens, 2017). Politicians from both parties called on the president to waive the Jones Act to speed up aid delivery and reduce the cost of supplies on Puerto Rico. While eventually Trump did so, his initial reluctance demonstrates his administration's twin concerns of privileging US domestic shipping and oil industry and his ethno-nationalist conception of US identity in which Puerto Rico is less than fully American, buttressed by longstanding racist stereotypes of Puerto Ricans and other non-white Americans (see Smith, 1997; Burnett and Marshall, 2001). The differential treatment of Puerto Rico is clearly not simply the product of the Trump administration; there is a long history of colonial attitudes and polices toward Puerto Rico. However, the slow and hesitant response of the federal government to assist Puerto Ricans in the aftermath of Hurricane Maria highlights this historical exclusion and its exacerbation under the Trump administration.

In response to offers of aid for Puerto Rico from Cuba and Venezuela, the United States reportedly, and not at all unexpectedly, rejected these offers – although the federal government did not provide an official statement regarding these offers (Pulley, 2017). In both cases, the US government treated the offers of aid as they had after Katrina – through a realist lens. Although it was not a prominent media story, some media speculated that Venezuela's campaign of aiding disaster-affected states in the Caribbean region after the devastating 2017 hurricane season was a tool to "wield political influence in the region", aimed at undermining support for sanctions against the Maduro regime (Charles, 2017). This approach to the offers of aid from Cuba and Venezuela bear a strong similarity to the Katrina episode outlined earlier, in which states offer aid to gain political influence or to meet a specific foreign policy objective, and the United States turned down these offers of aid to deny them this influence. The *New York Times* reported that the mayor of San Juan was directly negotiating with Caracas and would accept aid from Venezuela, although it is not clear if Venezuela delivered its offer of diesel fuel (Alexander, 2017).

To the extent that international aid reached Puerto Rico, it came through a number of international NGOs who sent teams to provide aid. Oxfam reported that it "rarely responds to humanitarian emergencies in the U.S. … but as the situation in Puerto Rico worsens and the federal government's response continues to falter, Oxfam has decided to step in" (Maxman, 2017). Save the Children similarly sent aid to Puerto Rico, noting that it never previously "had to send staff into the United States with satellite phones and other disaster equipment typically used for remote relief operations" (Ramm, quoted in Liautaud, 2017). The responses of these agencies are revealing in a number of respects: first, they demonstrate the widespread perception that the US government had failed in its duty to provide disaster assistance to Puerto Rico. More important for this book, these groups reproduced a number of narratives and identities that derive from the ideologies of order explored here: a statist narrative reasserting the primary duty of the state to provide disaster assistance; a modernist narrative reinforcing the United States as the prototypical modern state and the necessity of technological solutions to post-disaster recovery; as well as a humanitarian narrative of voluntary, charitable assistance for the relief of human suffering for those deemed worthy of such assistance. In this instance, INGOs performed the role of stabilizing existing order and performing ideological work to repair cracks in the ideological framework underpinning the management of disaster.

Conclusion

This chapter illustrated well a number of claims developed throughout this book. The first is the ongoing influence of ideologies of order in shaping decisions on disaster relief. These are not simply abstract belief systems that rarely inform pragmatic action or that only surface in international negotiations to establish basic principles or aspirational goals for international regimes. These ideologies of order shape how policy makers understand disasters and offers of aid, which in turn strongly influence how they respond to these offers. US decision makers after Hurricane Katrina, faced with unprecedented offers of aid, relied on existing belief systems about disaster, foreign aid, and US leadership in the international order to shape their response. Given the urgency and short time frame to act, it is not particularly surprising that these actors would draw on immediately accessible frameworks to understand and respond to offers of aid.

These ideologies did not simply constrain actors to behave in a predictable manner, however; they also create space for agents to act. This was evident through the creative use of frames (use of humanitarian frames to explain decisions reached on realist, protectionist, or modernist bases), as well as innovation within dominant ideologies. The US government acted innovatively within a realist framework to subvert the dominant understanding of the relationship between aid and soft power. As subsequent chapters will show, the United States is not the only government to pursue foreign policy objectives through aid acceptance, although this chapter (when placed in contrast with subsequent chapters) does demonstrate the importance of power and the privileged position that the United States enjoys in shaping the narrative around its aid acceptance and rejection.

This chapter also revealed the operation of multiple belief systems, even as particular ideologies of order are dominant. In this case, realist beliefs were dominant in US thinking – the US government consistently prioritized foreign policy objectives in the consideration of offers of aid. My argument is not that each of these ideological frameworks are equally prominent in shaping the US response; instead, I show that even when certain frameworks are dominant, the others are so deeply intertwined and necessary to the legitimation of international order that they continue to shape policy makers' and bureaucrats' understandings of international aid and consequently decisions on aid. At the same time, the ideologically heterogenous nature of the international order provides actors with significant latitude in framing international developments.

Ultimately, the prioritization of foreign policy considerations led to a number of problematic outcomes. Consistent with the literature on disaster diplomacy, it is not clear that the US response actually helped achieve any of its foreign policy objectives. The US response to some offers of assistance was inadequate, and the US government failed to use much of the aid that it did accept (Mayer, 2011). Solomon and Hsu (2007) report that of the \$454 million offered in cash and \$400 million in oil (to be sold for cash), the United States ended up receiving only \$126 million in cash. The discrepancy was due to various factors: lack of follow-up by the State Department, states channelling their donation to the Red Cross or Bush-Clinton Katrina Fund, or deciding that the United States no longer needed the in-kind donation by the time it actually responded to the offer (Solomon and Hsu, 2007). The Government Accountability Office (GAO) recognized these administrative failures and recommended a number of procedural, regulatory, and legislative changes to facilitate more efficient use of international aid in the future (GAO, 2006). In essence, the GAO sought to institutionalize and streamline the ad hoc response of FEMA, the State Department, Department of Defense, and USAID. I should note that the call to expand and improve state bureaucracies represents a thoroughly statist and modernist response to a perceived administrative failure.

In the next chapter, I will further explore the impact that ideologies of order have on policy makers' decisions on disaster assistance. By drawing on a wider range of cases outside the Global North, I show that these belief systems have a strong influence on decisions on aid even in quite different contexts: culturally, ideologically, and politically. While the internal variations on these belief systems and the relative weight afforded each strongly influence understandings of disasters and disaster aid in each country, the centrality of these ideologies of order to international order makes it difficult to understand and respond to disasters outside these frameworks.

Notes

1 The primary material for this chapter is from the large body of declassified material (157 documents, some in excess of 200 pages) released by the US State Department on international offers of aid. This material was released through access to information requests and legal action conducted by Citizens for Responsibility and Ethics in Washington (CREW).

2 Expertise was declined from the following states: Albania, Armenia, Australia, Austria, Bangladesh, Bosnia/Herzegovina, Bulgaria, Croatia, Czechoslovakia, Denmark, Estonia, Greece, Guatemala, Honduras, Hungary, Iceland, India, Indonesia, Japan, Jordan, Korea, Latvia, Lithuania, New Zealand, Nicaragua, Norway, Pakistan, Paraguay, Philippines, Poland, Portugal, Romania, Russia, Serbia, Switzerland, Taiwan, Venezuela, Vietnam.

References

Ahmadinejad, M., 2005. Address of Mahmood Ahmadinejad, President of the Islamic Republic of Iran before the Sixtieth Session of the United Nations General Assembly, U.N.G. Assembly, September 2006.

Albright, M., 2017. The national security emergency we're not talking about. *The Washington Post*. 29 November 2017

Alexander, H., 2017. Let him tweet. I don't mind. I'm too busy, says target of president's anger. *The Daily Telegraph*. October 7, 2017.

Bagby, M.G.B., 2005. Remarks of Major General Byron S. Bagby, Chief of the Office of Military Cooperation, American Embassy Cairo E. Embassy, ed.

Baker, P., 2017. Trump lashes out at Mayor of San Juan. *New York Times*. October 1, 2017.

Baker, P., & Dickerson, C., 2017. Aid won't last, Trump warns Puerto Ricans. *New York Times*. October 13, 2017.

Bass, J., 2005a. Notification to Cairo Re: MREs C. M. S. Taskforce1, Washington, DC: CREW and Department of State.

Bass, J., 2005b. Supplies in Support of Katrina C. M. S. Taskforce1, Washington, DC: CREW and Department of State.

Bellamy, A.J., 2010. The responsibility to protect – five years on. *Ethics & International Affairs*, 24(2), pp. 143–69.

Bula-Duane, M., 2005. Tuesday's update on Ukrainian assistance: Ukraine eager to send supplies C. M. S. Taskforce1, Washington, DC: CREW and Department of State.

Burnett, C.D., & Marshall, B., 2001. *Foreign in a Domestic Sense: Puerto Rico, American Expansion, and the Constitution*, Durham, NC: Duke University Press.

Chan, C., 2005. MREs – summary of this morning's discussion C. M. S.Taskforce1, CREW and Department of State.

Charles, J., 2017. As Venezuelans starve, their government offers hurricane aid to Caribbean islands. *Miami Herald*. October 4, 2017.

Chavez, H., 2005. Statement by Hugo Chavez, President of the Bolivarian Republic of Venezuela. New York: United Nations.

Clarke, M., 2013. Iran as a "pariah" nuclear aspirant. *Australian Journal of International Affairs*, 67(4), pp. 491–510.

CRED, 2015. EM-DAT the International Disaster Database. International Disaster Database. https://www.emdat.be/.

Cumming-Bruce, N., & Robles, F., 2017. U.S. response to storm-hit Puerto Rico is criticized by U.N. experts. *New York Times*. October 30, 2017.

Daniels, R., Kettl, D., & Kunreuther, H. (eds.) 2006. *On Risk and Disaster: Lessons from Hurricane Katrina*, Philadelphia: University of Pennsylvania Press.

Davis, M., 2005. Gentrifying disaster, *Mother Jones*. http://www.motherjones.com/politics/2005/10/gentrifying-disaster.

Denis, N., 2017. The law that is strangling Puerto Rico. *New York Times*. September 26, 2017.

Department of State, 2005. *Matrix Supplement*, Washington, DC: Department of State.

Drucker, M., 2005. Mexican and Colombian mental health professionals. The faster we get the info, the faster we can accept. September 12, 2005. CREW and Department of State, Washington DC

Dynes, R., & Rodriguez, H., 2010. Finding and framing Katrina. In D. Brunsma, D Overfelt, & J.S. Picou (eds.). *The Sociology of Katrina: Perspectives on a Modern Catastrophe*, Plymouth: Rowman & Littlefield.

EADRCC. 2005. EADRCC Final Report Hurricane Katrina - USA, Brussels: NATO - Euro-Atlantic Disaster Response Coordination Centre.

Eikenberry, A.M., Arroyave, V., & Cooper, T., 2007. Administrative failure and the international NGO response to Hurricane Katrina. *Public Administration Review*, 67(Suppl 1), pp. 160–70.

Elliott, J.R., & Pais, J., 2006. Race, class, and Hurricane Katrina: Social differences in human responses to disaster. *Social Science Research*, 35(2), pp. 295–321.

Erickson, A., 2017. Dozens of countries offered help after Hurricane Katrina. After Harvey, not so much. *Washington Post*. September 1, 2017.

Ferris, E., & Petz, D., 2012. *The Year That Shook the Rich: A Review of Natural Disasters in 2011*, London: The Brookings Institution – LSE Project on Internal Displacement.

Fisher, D., 2007. *Law and Legal Issues in International Disaster Response: A Desk Study*, Geneva: IFRC.

Forero, J., & Weisman, S., 2005. U.S. Allies and others send offers of assistance. *New York Times*, p. 31. https://www.nytimes.com/2005/09/04/us/storm-and-crisis-foreign-aid-us-allies-and-others-send-offers-of.html.

GAO, 2006. *Hurricane Katrina: Comprehensive Policies and Procedures are needed to ensure appropriate use and accountability for International Assistance*, Washington, DC.

GAO, 2007. *Hurricanes Katrina and Rita: Federal Actions Could Enhance Preparedness of Certain State-Administered Federal Support Programs*, Washington, DC.

Gibson, C., 2017. How colonialism and racism explain the inept US response to Hurricane Maria. *Vox News*. https://www.vox.com/the-big-idea/2017/10/5/16426082/colonialism-racism-american-response-puerto-rico-maria.

Gisvold, L., 2006. FWT: Hurricane Katrina, Recognizing Foreign Assistance. CREW and Department of State, Washington, DC.

Gladstone, D., & Préau, J., 2008. Gentrification in tourist cities: Evidence from New Orleans before and after Hurricane Katrina. *Housing Policy Debate*, 19(1), pp. 137–75.

Gong, G.W., 2002. Standards of civilization today. In M. Mozaffari (ed.). *Globalization and Civilizations*, London: Routledge.

Gotham, K.F., 2012. Disaster, Inc.: Privatization and post-Katrina rebuilding in New Orleans. *Perspectives on Politics*, 10(03), pp. 633–46.

Hanning, H., 1978. NATO and Disaster relief. *Disasters*, 2(2–3), pp. 101–104.

Hartman, C., & Squires, G., 2006. *There Is No Such Thing as a Natural Disaster: Race, Class, and Hurricane Katrina*, New York: Taylor & Francis Group.

Hattori, T., 2001. Reconceptualizing foreign aid. *Review of International Political Economy*, 8(4), pp. 633–660.

Herrara, B.A., 2005. *Offer of Assistance: Embassy of Bolivarian Republic of Venezuela H. C. R. S. of State*, Washington, DC: CREW and Department of State.

Hultman, D., 2005. *GOU Requests Confirmation No Assistance Needed for Katrina*. Washington, DC: CREW and Department of State.

Hutchinson, J.F., 1996. *Champions of Charity: War and the Rise of the Red Cross*, Boulder, CO: Westview Press.

Hutchinson, J.F., 2001. Disasters and the international order-II: The International Rrelief Union. *The International History Review*, 23(2), pp. 253–98.

Huysmans, J., 2002. Shape-shifting NATO: Humanitarian action and the Kosovo refugee crisis. *Review of International Studies*, 28(3), pp. 599–18.

Katoch, A., 2003. International natural disaster response and the United Nations. In International Federation of the Red Cross and Red Crescent Societies. *International Disaster Response Laws, Principles and Practice*, Geneva: IFRC, pp. 47–56.

Kates, R.W., et al., 2006. Reconstruction of New Orleans after Hurricane Katrina: A research perspective. *Proceedings of the National Academy of Sciences of the United States of America*, 103(40), pp. 14653–60.

Kelman, I., 2007. Hurricane Katrina disaster diplomacy. *Disasters*, 31(3), pp. 288–309.

Kelman, I., 2011. *Disaster Diplomacy*, London: Routledge.

Kent, R.C., 1987. *Anatomy of Disaster Relief: The International Network in Action*, Sussex: Pinter Publishers.

Khalilzad, Z., 2005. Statement by Ambassador Zalmay Khalilzad: Hurricane Katrina Relief. CREW and Department of State.

Klein, N., 2007. *The Shock Doctrine: The Rise of Disaster Capitalism*, New York: Metropolitan Books/Henry Holt.

Kuzio, T., 2003. *EU and Ukraine: A Turning Point in 2004?* Institute for Security Studies, Paris: European Union.

Kuzio, T., 2006. Is Ukraine part of Europe's future? *Washington Quarterly*, 29(3), pp. 89–108.

Lawrence, W., 2005. Incoming Call on Tunisian Aid to land in Little Rock. CREW and Department of State.

Liautaud, A., 2017. Puerto Rico is in such dire shape international charities are sending aid. *Vice News*. https://news.vice.com/story/puerto-rico-is-in-such-dire-shape-international-charities-are-sending-aid.

Marshall, B., 2005. AOL News – International Aid for Katrina Recovery Misrepresented C. M. S. Taskforce1, Washington, DC: CREW and Department of State.

Mayer, M., 2011. Accepting disaster relief from other nations: Lessons from Katrina and the Gulf oil spill, Washington. https://www.heritage.org/homeland-security/report/accepting-disaster-relief-other-nations-lessons-katrina-and-the-gulf-oil.

Maxman, A., 2017. Statement by Oxfam America president regarding Puerto Rico. Oxfam Statement. https://www.oxfamamerica.org/press/statement-by-oxfam-america-president-abby-maxman-regarding-puerto-rico-hurricane-response/.

Morse, J., 2005. Asia-Pacific opens hearts, wallets to US victims of Katrina. US Department of State. http://news.findlaw.com/wash/s/20050908/20050908163814.html.

Mortimer, C., 2017. Trump backs down and allows ships to deliver aid from US to Puerto Rico. *The Independent*. September 28, 2017.

Mutter, J., 2015. *The Disaster Profiteers: How Natural Disasters Make the Rich Richer and the Poor Even Poorer*, Toronto: Macmillan.

Nelson, T., 2010a. Rejecting the gift horse: International politics of disaster aid refusal. *Conflict, Security & Development*, 10(3), pp. 379–402.

Nelson, T., 2010b. When disaster strikes: On the relationship between natural disaster and interstate conflict. *Global Change, Peace & Security*, 22(2), pp. 155–174.

Nixon, R., & Stevens, M., 2017. Harvey, Irma, Maria: Trump administration's response compared. *The New York Times*. September 27, 2017.

O'Herron, P., 2005a. *Forensic Experts*, Washington, DC: CREW and Department of State.

O'Herron, P., 2005b. *Katrina Task Force: WHA End of Shift Report*. Washington, DC: CREW and Department of State.

Paik, W., 2011. Authoritarianism and humanitarian aid: Regime stability and external relief in China and Myanmar. *The Pacific Review*, 24(4), pp. 439–62.

Phillipson, R., 2008. The linguistic imperialism of neoliberal empire. *Critical Inquiry in Language Studies*, 5(1), pp. 1–43.

Pleming, S., 2005. Foreign governments line up to help after Katrina. *Reuters*. https://reliefweb.int/report/united-states-america/foreign-governments-line-help-after-katrina.

Pouliot, V., 2010. *International Security in Practice: The Politics of NATO-Russia Diplomacy*, Cambridge: Cambridge University Press.

Pulley, R., 2017. Wake of devastation. *New York Times*. October 14, 2017.

Ratti, L., 2009. Back to the future? International relations theory and NATO-Russia relations since the end of the Cold War. *International Journal*, 64pp. 399–422.

Regan, J., 2005. *Official Cuban Offer: Pass to the Task Force*, Washington, DC: CREW and Department of State.

Rojas, F., 2005. *Memo Re: Anything on the Hurricane Relief Latin American Countries?* Washington, DC: CREW and Department of State.

Rooney, P. 2017. "American generosity after disasters: 4 questions answered". The Conversation, September 11, 2017. http://theconversation.com/american-generosity-after-disasters-4-questions-answered-83277

Saltman, K.J., 2009. Schooling in disaster capitalism: How the political right is using disaster to privatize public schooling. In S. Macrine (ed.). *Critical Pedagogy in Uncertain Times: Education, Politics and Public Life*, New York: Palgrave MacMillan

Schiffer, G., 2005. *Mexican and Columbian Mental Health C. M. S. Taskforce1*, Washington, DC: CREW and Department of State.

Schraeder, P.J., Hook, S.W., & Taylor, B., 1998. Clarifying the foreign aid puzzle: A comparison of American, Japanese, French, and Swedish aid flows. *World Politics*, 50(2), pp. 294–323.

Seekins, D.M., 2009. State, society and natural disaster: Cyclone Nargis in Myanmar (Burma). *Asian Journal of Social Science*, 37(5), pp. 717–37.

Selth, A., 2008. Even paranoids have enemies: Cyclone Nargis and Myanmar's fears of invasion. *Contemporary Southeast Asia: A Journal of International and Strategic Affairs*, 30(3), pp. 379–402.

Smith, D.B., 2005. *FW: Supplies in Support of Katrina*, Washington, DC: CREW and Department of State.

Smith, R.M., 1997. *Civic Ideals: Conflicting Visions of Citizenship in U.S. History*, New Haven, CT: Yale University Press.

Solomon, J., & Hsu, S., 2007. Most Katrina aid from overseas went unclaimed. *Washington Post*, p. A01. April 29, 2007.

Stephenson, B., 2017. President's views. *The Foreign Service Journal*. http://www.afsa.org/time-ask-why.

Stitt, T., 2005. *HSC LL Summary*, Washington, DC: CREW and Department of State.

Streich, P.A., & Mislan, D.B., 2014. What follows the storm?: Research on the effect of disasters on conflict and cooperation. *Global Change, Peace & Security*, 26(1), pp. 55–70.

Sullivan, G.W., 2005. *NEA Press Guidance: Iran Offer of Oil and Relief for Hurricane Katrina*, Washington, DC: CREW and Department of State.

Swithern, S., 2015. United States – Global Humanitarian Assistance. http://www.global-humanitarianassistance.org/countryprofile/united-states#tab-types-of-aid.

Thomas, H. Executive Secretary of the Department of State, 2005. Special Press Briefing on International Offers of Support following Hurricane Katrina, Washington, DC: CREW and Department of State.

Tierney, K., Bevc, C., & Kuligowski, E., 2006. Metaphors matter: Disaster myths, media frames, and their consequences in Hurricane Katrina. *The Annals of the American Academy of Political and Social Science*, 604(1), pp. 57–81.

Toosi, N., 2017. World in no rush to offer Trump help post-Harvey. *Politico*, https://www.politico.com/hurricane-harvey-world-reaction-mexico.

Tumminia, F., 2005. WHA Press Guidance: Cuba: Response to offer of aid for Hurricane Katrina victims, Washington, DC: CREW and Department of State.

Van der Linde, D., 2008. The politics of disaster: A study into post-Katrina international aid. *Disaster Diplomacy*, http://www.disasterdiplomacy.org/pb/vanderlinde2008.pdf.

Walker, D.M., 2006. *Hurricane Katrina – GAO's Preliminary Observations Regarding Preparedness, Response, and Recovery*, Washington, DC.

Walker, P., 1992. Foreign military resources for disaster relief: An NGO perspective. *Disasters*, 16(2), pp. 152–9.

Walters, J., & Kettl, D., 2006. The Katrina breakdown. In R. Daniels, D. Kettl, & H. Kunreuther (eds.). *On Risk and Disaster: Lessons from Hurricane Katrina*, Philadelphia: University of Pennsylvania Press.

Waltz, K.N., 2012. Why Iran should get the bomb: Nuclear balancing would mean stability. *Foreign Affairs*, 91, p.6.

Watson, S.D., 2014. Institutionalizing and implementing the disaster relief norm: The League of Red Cross societies and the International Relief Union. In A. Betts & P. Orchard (eds.). *The Diffusion of International Norms: Explaining the Institutionalization: Implementation Gap*, Oxford: Oxford University Press.

Zabarenko, D., 2013. U.S. spent $62 billion on disaster relief in fiscal 2011 and 2012. Reuters. September 11, 2013.

Zack, N., 2014. The ethics of disaster and Hurricane Katrina: Human security, homeland security, and women's groups. In C. Hobson, P Bacon, & R. Cameron (eds.). *Human Security and Natural Disasters*, London: Routledge.

7

EGYPT AND THE CAIRO EARTHQUAKE

One of the central claims that I put forward in this book is that the foundational ideologies of order explored thus far are *international* ideologies: they provide a foundation for all states in the international system either as a guiding belief system or as a set of principles through which relevant actors are constituted and build legitimacy. This is most prominently exemplified in the context of decolonization, which drastically increased the number of states in the system and also reproduced, with varying degrees of contestation and alteration, these dominant ideologies of order. Through the process of building sovereign nation states, decolonizing leaders asserted and reinvented their self-identities, producing a set of identities that were simultaneously set in opposition to but also mimicked the identities of the former colonial powers, creating a form of hybrid identity (Bhabha, 1984; Berenskötter, 2017). New states in the system adopted basic behavioural expectations of "Westphalian states" because a principal normative force behind their push for independence was that they shared with Western states the basic characteristics of self-determining nations capable of self-governing (Chan, 2017:96). In short, the call to independence was rooted in a claim to participate on an equal footing in the international order on the existing terms of that order, with slight alterations.

Many decolonial nationalist leaders were educated in Western institutions and pursued industrialization, incorporating select aspects of Western thought and identity formation while rejecting others (Matossian, 1958; Kedouri, 1971). Consequently, the leaders of newly decolonized states approached the construction of international order through a syncretic mix of dominant ideologies of international order with components of domestic or regional values and goals. This meant that the tensions and contradictions inherent to social order were (and are) particularly acute in many former colonial states and are subject to disruption during crises and disasters by domestic or regional actors who resisted the incorporation of Western ideologies of order, but also by powerful states who resisted any deviation from the international order that sprung from the decolonization process. The ongoing pressure placed by Western states on decolonizing states to adopt certain forms of order further exacerbates these tensions.

Some commentators suggest that ideological foundations within several newly decolonized states, particularly in the Middle East, were so contradictory that these states could not survive for more than a few years (Halliday, 2005; Chan, 2017). While a full explication of the persistence of states suffering from severe ideological fragmentation is beyond the scope of this book, two findings from this book, explored more fully in this chapter, help shed light on such persistence. First, the existence of multiple legitimizing ideological foundations within a heterogenous international order provides regimes with a sufficient number of frameworks to narrate material developments in a number of ways to reduce dissonance and to resist challengers. Second, international aid (and state consent to such aid) after disasters has operated as a way of normalizing the dominant ideologies of order and reducing deviations from those belief systems. To be sure, disasters potentially expose the failure of contingent resolutions to ideological contradictions and ambiguities and thus become windows of opportunity to challenge or modify the existing order. In the case of former colonial states, international assistance has the potential to destabilize because it closely approximates the colonial relationship and signifies dependence on external actors, yet it also serves a stabilizing function by providing material goods necessary to maintain state legitimacy and/or its coercive capacity.

In this chapter, I examine Egypt's response to the 1992 Cairo earthquake to reveal the influence of the ideological belief systems explored in the previous chapters and in particular how they interact with local contexts and belief systems. This case provides an interesting and illustrative comparison to the United States. As with many decolonizing states, Egypt embarked on intense modernizing processes following incorporation into the international state system in the early twentieth century. After gaining independence from Britain and building on longstanding historical claims as an independent political community, Egyptian leaders positioned the country as a regional power in the contentious Middle East and North Africa (MENA) region and as a leader of the non-aligned movement (NAM) that fully embraced a modernist project while opposing key aspects of the liberal economic order. In response to a series of economic and political crises, Egypt embarked on liberal economic reforms that in turn produced significant social disruption and exacerbated tensions with Islamist groups that continue to exert significant influence.

In short, Egypt's very recent history illustrates the ongoing influence of variations of realism, modernism, economic liberalism, nation-statism, and humanitarianism that represent a unique set of contingent resolutions to the tensions within and between these belief systems. The Egyptian case reinforces the central claim of this book – that the modern, liberal, sovereign-state-making project suffers ideological inconsistencies that disasters often exacerbate and that international aid is meant to rectify. However, the provision of international assistance designed to maintain the existing international order opens up new tensions and contradictions. The Egyptian case reveals a great deal about the prevalence of these ideologies of order and their fragility. First, it reveals the increasing dominance of realist and liberal economic beliefs. The evidence suggests that policy makers in Egypt did not interpret the Cairo disaster as a failure of economic liberalism and responded by continuing liberalizing reforms. Indeed, as occurred in the United States, the disaster created opportunities for further liberalizing reforms. Similarly, policy makers in both states understood disaster aid in terms of power dynamics and meeting strategic foreign policy goals, although there was disagreement about what those foreign policy objectives ought to be. Both cases also demonstrate the influence of modernism and the belief in scientific and technological solutions to natural disaster; in Egypt it is combined with the ongoing challenge that Islamist political thought poses to the secular values associated with modernism. The disaster response in Egypt also demonstrated the tension between nationalist and bureaucratic legitimization of state authority. Last, this case also demonstrates that humanitarian beliefs about alleviating human suffering are widespread, as is the usefulness of humanitarianism as a standard of civilization to delegitimize assistance from some actors.

I should note at the outset some of the limitations of the Egyptian case, particularly in comparison with the previous chapter's examination of the US response to Hurricane Katrina, in which large amounts of internal documentation were available. As Shama notes with regard to researching on Egypt: there are few academic sources, and government policy making is shrouded in secrecy (2014:2). Use of media sources, while useful, present their own challenges – most notably government control. For the purposes of this research project, I rely primarily on *Al-Ahram*, which is the most prominent Egyptian media source but one that has, historically, been quite closely tied to the government (Shama, 2014). Consequently, the media narrative should not be understood as a reasonably neutral and objective account of developments. This also has methodological benefits, in that the account presented can be understood as closely approximating official government position and thus offers insight into the beliefs of the regime and how they counter challenges.

At the time of the 1992 Cairo earthquake, Egypt was in the midst of particularly turbulent period. Beset by an Islamist insurgency, a stalled Palestinian peace plan, participation in an US-led war against another Arab state, and a highly disruptive economic reform program, the ideological contradictions and material failures of Egyptian social order were openly evident. In this environment, the 1992 Cairo earthquake presented an opportunity for the Mubarak regime and their opponents to bolster or disrupt the existing social order – as became evident in the provision of disaster relief.

To understand the disruption associated with the earthquake and the post-earthquake recovery, it is important to understand the ideological tensions in Egypt at that time. Upon independence from Britain and the overthrow of King Farouk, the Free Officers under Gamal Abdel Nasser pursued a vision for Egypt and the Arab world that simultaneously adopted, altered, and diverged from the foundations of international order espoused by the dominant powers, and this has had lasting influence.

The Modernist Project in Egypt

As with many decolonial leaders, Nasser was a high modernist, seeing industrialization and technological advancement as necessary to resist Western imperialism and to improve the lives of Egyptians and Arabs. Nasser sought to remake Egyptian society through a variety of far-reaching reforms and policy initiatives that included secularism, large-scale industrialization, ousting the monarchy, agrarian reform, and major development projects. In many respects, modernization and the expected economic development (and independence) that would accompany that process became a key framework of legitimization of Nasser's regime and the Egyptian state. While Sadat and Mubarak abandoned several core components of Nasser's vision for Egypt, they maintained his modernist beliefs. In many respects, their partial and incremental adoption of market-led economic development emerged from a perceived failure of Nasser's state-led economic development to modernize Egyptian society and economy.

The modernist project in Egypt also entailed an endorsement of secularism, albeit significantly altered. As Agrama (2010) notes, Egypt's secularism exhibits "peculiar ambiguities" and is frequently (and problematically) labelled as "partially or precariously secular" (499). It is characterized as such because its "legal system is derived from European, mostly French-based law" even as its constitution identifies sharia law as its primary legal foundation (Agrama, 2010:497). Islamic institutions and movements are placed under the authority of the Egyptian state, or they are banned. Yet religious authority continuously operates in the political and legal spheres. As with all secular states, Nasser's modernization efforts and state-building project necessarily entailed an entanglement with secularism's insistence on demarcating the line between religious and secular and/or scientific authority. In Egypt, this led to the identification of Islamists as a threat and the strategic use of Islam to support foreign policy choices (Shama, 2014:135). The continual use of Islam to justify the Egyptian state and its foreign and domestic policies, combined with deep distrust of Islamist movements, contributed to ideological contradictions that presented a serious challenge to successive regimes. Indeed, the threat of Islamism to secular leaders in Egypt was so central that it became a key factor in shaping Egyptian foreign policy.

Throughout Egypt, Islamist ideology appeals to religious sentiments to promote unity across a wide range of social cleavages (Onis, 2001). Drawing from the work of influential intellectual Sayyid Qutb, Islamism in Egypt stressed the failures of key Western ideologies and practices: capitalism, communism, nationalism, colonialism, and scientific advancement; it identified Islam as the only credible belief system (Bayat, 2017:33). Despite this general orientation, Islamist groups in Egypt (and other countries) varied significantly on the degree to which Islam was compatible or not, with these ideologies (see Zubaida, 2000). Consequently, Islamism in Egypt is not a single, coherent movement and encompasses various strains and positions on key questions regarding Islam, the state, the nation, ethnic identity, scientific advancement, and liberal economic policies (although they share general agreement on Israeli/Egyptian relations). Brumberg differentiates the various perspectives into three strains: radical, reformist, and liberal, and even these strains contain a fair amount of heterogeneity (Brumberg, 2002). Nonetheless, there are key shared components of Islamism in Egypt. Brumberg (2002) suggests all three strands of Islamism share a belief that the secular state has failed to solve the twin problems of poverty and inequality. This, according to Gumuscu (2010), was the result of how economic reforms in Egypt failed to create a "strong and devout bourgeoisie" that would embrace a pragmatic, moderate Islamist position. In Egypt, the dominant Islamist groups promote an alternative vision to the modern, statist, liberal economic order that prevails currently in much of the world (Berman, 2003:257). In principle, Egyptian Islamist groups oppose secularism and liberal economic policies, and promote an alternative to state sovereignty in the form of a pan-Muslim political community.

More recently, however, Bayat notes that Egypt has witnessed the rise of "neo-Islamism", which combines Islamist thought with pro-market, laissez-faire, neo-liberalism, not unlike the rise of Pentacostalism and the prosperity doctrine in regions previously dominated by liberation theology (Bayat, 2017:84). For Bayat, these developments reflect the perceived failures of socialism and its influence in the developing world, as well as the growing hegemony of neo-liberalism. It also demonstrates how ideological systems can mutate and combine to create various hybrid belief systems.

This is evident in the relationship between Islamist movements and modernity as well. Many Islamist movements contain strains that tend toward anti-modernism in their promotion of a religious, anti-secular cosmology. Various developments, including natural phenomena, are depicted in religious terms – as punishment from Allah. However, most Islamist groups embrace scientific training and technological solutions (Onis, 2001:282). Islamism in Egypt has an ambiguous relationship with key tenets of modernist ideology. Secularism as a key aspect of modernist ideology does not exclude religion. Rather, secularism places religion as a cultural object that must be managed to fit within modern, state-based principles (Agrama, 2010:499). Islamism rejects this governance of religion, even as it struggles to contain or govern non-religious truth claims based on science and reason. Consequently, the Islamist movement is varied based on different groups and traditions of thought that attempt to resolve the ambiguity between religious cosmology and science and technology. In Egypt, the moderate strand of Islamism asserts that scientific inquiry is intrinsic to and an obligation contained in sharia law and that the perceived opposition between "religion and science is a Catholic-European phenomenon" (Agrama, 2010:8). This version of Islamism is contested by both extremist Islamists and secularists. Nonetheless, moderate Islamist groups in Egypt, such as the Muslim Brotherhood, are often associated with professional and scientific societies. Many of these societies possess more modern equipment and provide a higher standard of care than that of the Egyptian government, indicating that Islamism does not necessarily entail a complete rejection of modernist enterprises, such as modern medicine or building codes (Clark, 2004).

Other aspects of Nasser's sociopolitical order were also in tension with the modernist foundation of the Egyptian state and economy. Egypt's closed economy and state control over universities and civil society limited the inflow of foreign expertise and technology (Weiss and Wurzel, 1998). As a result, Egyptian firms and researchers were closed off from the rest of the world and many technological developments. The impact on Egyptian society was evident in the Cairo earthquake: outdated seismic equipment, building codes, and materials; lack of search and rescue (SAR) technology; and insufficient education for the general population on how to act in the event of an earthquake. It also undermined Egypt's privatization efforts in the 1990s. It was difficult for the Egyptian state to sell many state-owned enterprises in part due to their outdated technology (Weiss and Wurzel, 1998).

One potential source of modernization occurred through technology transfers from sponsor states, but this was directed almost exclusively toward military modernization. Soviet aid was primarily military, as was the US aid that supplanted it. Of the $3 billion per year that the United States provides in aid to Egypt, over $2 billion is military aid (Weiss and Wurzel, 1998:164). Consequently, while the Egyptian state had access to relatively modern military technology, it lagged behind in most other sectors. The failure of Egypt's modernization program was not due to a rejection of modernist beliefs but of the sociopolitical systems put in place to pursue other values and goals.

Nationalism, Pan-Arabism, and the Centralized State

Anti-colonial nationalism emerged as one of the most powerful ideological motivators of early Egyptian independence (Bayat, 2017:170). However, the initial anti-colonial resistance movements in Egypt to Ottoman and then British control combined two distinct nationalisms: an ethnonational pan-Arabism (*qawmiyya*) and territorial state-based nationalism (*wataniyya*) (Abulof, 2015:667; Dawisha, 2016). Directed at liberating Egypt from foreign control, the ideological tensions between the two nationalisms did not manifest immediately, although the centrality of the Palestinian struggle and Nasser's endorsement of pan-Arabism upon taking control in 1952 thrust these tensions to the forefront of Egyptian politics. In addition to these competing nationalist frameworks, Nasser's adoption of state-led development, often referred to as Arab socialism, was ideologically statist (in its territorial and bureaucratic sense) and thus at odds with pan-Arab nationalism that challenged existing state borders. In the Arab socialist program, the legitimacy of the state derived primarily from the distribution of social goods made possible by an industrial economy driven by "large-scale and capital-intensive public sector companies" (Weiss and Wurzel, 1998:56,73). This resulted in a highly centralized political system in which the government, and the president in particular, exerted a high degree of control over most aspects of political and civil life in Egypt, including approving political parties, trade unions, universities, and civil society actors (Weiss and Wurzel, 1998:141). The Egyptian state model did not envision the state residing atop a hierarchy of public goods providers justified by consent of the population. Rather, the Egyptian state sought a monopoly on the provision of these goods, co-opting many entities and completely excluding others, mostly Islamists and Communists (Weiss and Wurzel, 140).

Given Egyptian reliance on the state and state-owned corporations for employment and other basic goods, the shift to a liberal market economy under Sadat, and more fully under Mubarak, was highly destabilizing. Mubarak argued that "turning away from our public sector heritage is absolutely necessary", but he consistently moderated and delayed this shift to minimize social unrest (Weiss and Wurzel, 1998:79). Until the early 1990s (and to a certain extent still today), social order in Egypt was highly dependent on the state for maintaining low unemployment and subsidizing the cost of basic goods (Weiss and Wurzel, 1998:105). Consequently, economic reforms occurred in areas where it was least likely to damage state authority and its ability to provide goods, such as privatizing the tourist industry or implementing reforms most likely to affect the rural poor rather than the urban rich or middle class (Weiss and Wurzel, 1998:105). The Egyptian state's legitimacy rested on the direct provision of public goods, including jobs and cheap food and fuel.

Nasser coupled the bureaucratic legitimation of the Egyptian state focused on provision of public goods with a pan-Arab nationalism that challenged and transcended existing state boundaries. For a brief period, this led to a union of Egypt and Syria as the United Arab Republic (Abulof, 2015:668). This union ideologically complicated the legitimization of the Egyptian state. In many respects, pan-Arab nationalism represents a rather conventional adoption of nation-statist ideology, wherein state boundaries should coincide with the national community – in this case, the Arab nation. In this context, Nasser's pan-Arabism exposed the ideological contradictions of nation-statism as an ideology of international order, namely, the existence of two legitimizing narratives for state authority (nationalist and bureaucratic), as well as the existence of multiple "nationalized" identities (qawmiyya and wataniyya) – which in turn exposed the incongruence between state borders and the national community. This was (and is) particularly acute in the case of Arab nationalism that crosses the boundaries of numerous states: Egypt, Syria, Libya, Saudi Arabia, Iraq, and Sudan, among others.

By the time of the 1992 earthquake, pan-Arabism was no longer a strong political force in Egypt, replaced by the tension between Egyptian nationalism and Islamism. Pan-Arabism suffered several setbacks: resistance from competing Arab states; the dissolution of the United Arab Republic; Egypt's involvement in Yemen; and, perhaps most damaging, defeat in the Six Day War with Israel in 1967 (Ajami, 1978; Abulof, 2015). The transition from Nasser to Sadat, combined with the rise of Arafat and the Palestine Liberation Organization (PLO) and Egypt's success in the 1973 confrontation with Israel, further signified the demise of pan-Arab ideology in Egypt and an embrace of Egyptian nationalism (Ajami, 1978). Under Sadat and later Mubarak, the primary nationalist narrative in Egypt was wataniyya. This influenced Egyptian foreign policy, shifting further toward a realist orientation concerned with competing with and balancing the power of other states in the region rather than uniting Arabs under a single political entity. Despite these many changes, pan-Arabism did not completely disappear with the death of Nasser; it continues to find expression in Egyptian society, particularly with respect to the Palestine-Israel question and Egypt's desire to be seen as a leader among Arab states.

In short, at the time of the 1992 earthquake, the legitimacy of the Egyptian state rested heavily on the provision of basic public goods and state-based nationalism. However, its legitimacy was challenged by the inefficient provision of public goods and a relatively weak Egyptian nationalism that competes with both pan-Arabism and Islamism. The Cairo earthquake fully exposed these tensions, demonstrating the weakness of state institutions, the robustness of Islamic groups in providing public goods, and continued pan-Arab sentiment that complicated Egypt's official position regarding peace with Israel and participation in the Iraq War.

From Socialism to Economic Liberalism

Among the most radical components of Nasser's movement was his adoption of Arab socialism that rejected both the economic liberalism endorsed by many Western states and the communist alternative offered by the Soviet Union. A key economic component of Nasser's program was to nationalize key sectors of the economy to reduce or eliminate foreign ownership (Chan, 2017:39). As with many post-colonial states, Egypt under Nasser "admired Western achievements" and understood the pragmatic benefits of cooperating with the United States and its allies, while simultaneously harbouring deep resentment of "Western imperialism" (Podeh, 1996:161). Combined with a fear of communism that foreclosed full alignment with the Union of Soviet Socialist Republics (USSR), Nasser promoted an ideological commitment to neutrality and non-alignment and carved a path of economic development that diverged in important ways from both the economic liberalism of the West and the state-led communism of the USSR.

The most glaring tension of this economic ideology arose due to Egypt's material need for external financial support. Nasser's modernization program required significant capital investment and thus great power patronage as a source of external funding and technology. Subsequent Egyptian governments sought to balance their strong opposition to imperialism and external control with their financial dependence on external actors necessitated by the pursuit of modernization and industrialization. Nasser relied on Soviet support, with Sadat radically shifting to the United States in the early 1970s, driven by the belief that the United States offered a better opportunity for foreign investment and technological modernization as well as a resolution of Egypt's conflict with Israel (Shama, 2014:34). Allying with the United States, however, necessitated a modest liberalizing shift away from the state-directed socialism of Nasser. Given the centrality of the state in the Egyptian economy, the shift toward a private-sector-based economy was ideologically and materially disruptive. Indeed, one effect of Sadat's liberalizing program was an increase in the cost of food staples, which resulted in food riots across Egypt in the late 1970s (Shama, 2014).

As would become the standard practice for both Sadat and Mubarak when liberalizing reforms appeared to undermine the legitimacy of the state, Sadat backtracked and attempted to delay further reforms. Rather than abandon their commitment to economic liberalization, Sadat and Mubarak slowly and incrementally moved toward liberal economic reforms in order to reduce opposition and maintain social order. While economic liberalism was only gradually and incrementally implemented, it had several effects that made completely abandoning economic liberalism an increasingly difficult proposition: greater dependence on key liberal institutions for regime survival (US aid and International Monetary Fund [IMF] and World Bank loans), and greater influence for business groups in Egypt. Shama (2014) notes that Egypt became so dependent on US financial assistance that several commentators described it as an addiction or referred to Egypt as a US colony. Shama also notes that, in the three decades after Sadat's shift, there was a dramatic increase in the power of formal business lobbying groups, as well as the number of businesspeople in cabinet positions and in the number of high-level politicians who joined the private sector after retiring from public service (Shama, 2014:89–90). Over time, supporters of socialism and isolation from global markets enjoyed less influence or were forced out of high-level positions. Not all business organizations in Egypt favoured freer international trade; indeed, many benefitted from protectionist measures, however, the growing influence of business indicated a gradual erosion of the socialist alternative and the internalization of economic liberalism.

As Egypt's commitment to economic liberalism grew, the primary economic debates were largely framed in terms of the embedded liberal compromise, namely, how to open Egypt's economy to the international economic system while protecting local businesses and workers. As with many countries around the world, after the late 1980s, Egypt was subject to greater neo-liberalizing pressures that sought to challenge the embedded liberalism it had only recently, and gradually, adopted. Of course, in Egypt, neo-liberalizing pressures were not particularly strong and were slow to materialize, as Mubarak frequently resisted neo-liberal measures in order to preserve social order and his regime. This stalled implementation was supported by the United States and the IMF due to Egypt's geostrategic importance in the MENA region.

The adoption of economic liberalism and the move to a more open economy did not shore up the Egyptian economy, which continued to face severe economic turmoil throughout the late 1990s and to the present time, evidenced by recurrent balance of payments crises and labour unrest. By the late 1980s, the government was running large deficits and faced difficulties in providing basic public goods, such as education and healthcare (Bromley and Bush, 1994). By the early 1990s, the economic situation in Egypt was so dire it influenced Egypt's decision to participate in the US coalition against Saddam Hussein. In return for their participation, Egypt received close to $30 billion in debt relief (Bromley and Bush, 1994:202), including $6.7 billion in military debt (Shama, 2014:164).

In an attempt to rectify Egypt's recurrent economic problems, Mubarak embarked on a renewed commitment to liberalizing reforms in 1991, which entailed structural adjustment characterized by privatization of state-owned enterprises (Assaad, 2007:2). Assisted by the World Bank and IMF, Egypt began to liberalize its economy by strengthening the market, increasing private property rights, and reducing the role of the state as the primary employer (Assaad and Arntz, 2005:434), as well as cutting public subsidies on energy (Bromley and Bush, 1994:201). To implement these reforms and limit opposition, the Egyptian state also embarked on a process of political "deliberalization": altering voting arrangements; limiting civil society organizations; restricting voting in unions and professional syndicates; and strengthening penalties for those who "undermine social peace", including those who strike, demonstrate, or oppose economic reforms (Kienle, 1998:221).

From this perspective, the 1992 earthquake represented both a potential danger and opportunity. One potential danger of the disaster was to demonstrate the weakness of the Egyptian state in providing basic relief and the failure of its modest liberalizing reforms to modernize Egyptian society. However, the stalled implementation of economic liberalization in Egypt meant that there were two credible economic frames – as a failure of liberalization to modernize Egypt and reduce Egyptian's vulnerability, and alternatively, as a failure of Egyptian society and institutions to accept liberal economic reform. As I will show later in the chapter on the narrative around the earthquake, the Mubarak regime strategically employed both frames: using the earthquake to convince international actors to delay liberalizing reforms while simultaneously using the recovery process to implement a number of reforms that had faced domestic opposition prior to the earthquake.

A Realist Foreign Policy?

While it is impossible and unnecessary here to provide a full account of the shifts and nuances of Egyptian foreign policy from Nasser to the 1992 earthquake, it is important to provide a broad sketch of Egypt's understanding of its place in the international and regional order to help understand Mubarak's response to the Cairo earthquake and offers of international aid. As with other features of the Egyptian state, it is perhaps most accurate to describe its foreign policy as ideologically fragmented, combining realist beliefs about the need to balance power/threat with commitments to pan-Arabism and non-alignment. This is not to suggest that either pan-Arabism or non-alignment signify a rejection of realist principles; rather, they demonstrate the contradictions that arise from the application of realist beliefs. In one respect, both signify a belief in the importance of power and the necessity to balance against greater powers – Israel regionally, and the "West" and the Soviet Union globally (Chan, 2017:41). Similarly, it is important to recognize that its leaders viewed Egypt as a regional power and leader – and both pan-Arabism and non-alignment increased its standing and power in the region, which is symptomatic of a realist belief system.

In other respects, pan-Arabism and non-alignment reflect a modification of realist beliefs in response to regional and local contexts. The overlapping of power and threats in the region complicate the application of realist principles – as David (1991) captures through the concept of omni-balancing. Egypt faced a balancing environment that included global powers/threats (United States and USSR), regional power/threats (Israel, Saudi Arabia, Iran), and domestic powers/threats (Islamist organizations). The application of realist beliefs about the importance of accumulating and balancing power in such an environment exposes the contradictions in realist beliefs. Strategies to balance against threat at one level may have implications at other levels, some of which pose a more immediate threat.

Egypt's early commitment to pan-Arabism is a case in point. The pursuit of a unified Arab state contravened a foundational realist principle, namely, the survival of the sovereign state as the primary value of all states. And yet the unification of Arabs under Egyptian leadership both increased Egypt's status in the region and expanded the power of Egypt to balance against Israel. To a certain extent, pan-Arabism reflected a nation-building exercise that neglected important differences between the various Arab states and societies and set Egypt in tension with Arab and non-Arab states with whom it might have cooperated to balance against greater powers and threats. As Shama (2014) notes, there were (and are) clear divisions among Arab states concerning the appropriate form of government (monarchy/republic), the role of Islam in the state (secular/theocratic), as well as confessional differences (Sunni/Shi'a). Egypt's commitment to pan-Arabism led to the Arab "Cold War" (Kerr, 1967) and an ill-advised military adventure in Yemen that had little strategic value for Egypt. Perhaps most damaging from a realist perspective, Nasser's focus on the pan-Arab project detracted attention from Egypt's primary threat at the time – Israel, which contributed to the military defeat in 1967 (Barnett and Levy, 1991:379–80). Rather than uniting Arab states against common threats and increasing Egyptian power, pan-Arabism contributed to intense intra-Arab rivalry, Egyptian intervention in

Yemen, economic hardship, and ultimately military defeat (Barnett and Levy, 1991:379–80). These developments contributed to the waning of pan-Arabism, and Nasser and Sadat came to adopt a more traditional realist approach – with Egypt providing for its own security by unilaterally balancing against Israel and other regional competitors through military modernization and strategic international alliances. Sadat's radical shift from an alliance with the USSR to the United States reflects a stronger emphasis being placed in basic realist principles, combined with the other ideological shifts already mentioned.

The other key internationally oriented belief system adopted by Nasser – non-alignment – incorporated several positions that eschew or alter core realist principles. According to Babbili, NAM rejected traditional "power relations" in favour of principled positions on decolonization, economic equity between North and South, reduced tensions of the Cold War (Babbili, 1990:319, 322), and a commitment to the Palestinian cause (Morphet, 2004:518). In place of realpolitik and competition between newly decolonized states, NAM identified common values for all states of the "third world", particularly around resistance to colonialism, imperialism, and racism (Gupta, 1992:66). Consistent with its belief in cooperation between states, NAM also placed great emphasis on international organizations such as the UN rather than relying on traditional sources of power to achieve political goals. Gupta (1992) suggests that the adoption of NAM reflected an effort on the part of newly independent nation states to continue participation in the global political and economic system, but on different terms, namely, by developing a new position in the global economy and a new relationship with former colonial powers.

In doing so, NAM did not entail a wholesale rejection of realist beliefs. Certain realist foundations, such as the centrality of the nation state, and the importance of sovereignty and independence, were central to NAM. It also reflects a recognition of the need for weaker powers to work together to counter disparity in the global distribution of power – a central tenet of realist belief. By working together, countries supporting NAM quite successfully used their numeric power in the UN General Assembly to shape a wide range of global initiatives and UN reforms, often in direct opposition to the United States and its allies (Morphet, 2004:525). In many ways, the emergent North-South divide that was made manifest through NAM was just as central to this period as was the East-West divide often associated with the Cold War period, despite much more emphasis being placed on the latter. Furthermore, the North-South divide persists well beyond the Cold War, even if NAM is not the force it once was. In short, the commitment to non-alignment represented an alteration of realist beliefs rather than wholesale rejection and operated alongside more conventional realist beliefs and practices.

Egypt is a prime example of this ideological fracture; even under Nasser, Egypt's commitment to non-alignment was from the outset buttressed by conventional realist power politics and engagement in rather classic, pragmatic alliance considerations to provide financial and military aid to achieve its security and economic goals (Barnett and Levy, 1991:371). While Egypt was formally non-aligned, pursued principled positions through NAM, and generally eschewed any alliance structure that could be perceived as renewed colonialism, successive regimes pursued "limited" alliances with the two great powers and sought to balance against regional powers – even those within NAM.

This "omnibalancing" behaviour reflected the blurred line between internal and external threats (David, 1991). David suggests that, rather than attempting to accrue power and form alliances to protect against external threats, as is commonly asserted in realism, states accrue power and create alliances to respond to both internal and external threats (David, 1991:236). While David and others present the concept of omnibalancing as explaining how realism (as an academic theory) can explain the behaviour of these states, I suggest that this behaviour reflects a modified application of realist ideology by these leaders. Realist ideology, as I have presented it in this book, reinforces a belief that the domestic and international realms operate according to distinct logics and patterns of behaviour. Because the international realm is anarchic, foreign policy decisions are understood in terms of competitive power politics. In contrast, the domestic realm reflects other values, such as the pursuit of the good life or justice, and to the extent that force is still central, that force operates in the name of these other principles.

The modified realism adopted in many states, particularly post-colonial states, refutes such a distinction and applies realist principles of power accumulation and alliance formation to both internal and external threats. Omnibalancing reflects key aspects of realist ideology – belief that the international system is anarchic and conflictual and that states act rationally in the pursuit of power to ensure their survival (David, 1991). For David, this is most evident in the case of Sadat's abrupt switch from aligning with the Soviet Union to aligning with the United States after the 1973 October War against Israel. Rather than signalling an ideological reversal from socialism to capitalism, Sadat's change of great power patron reflected an ideological shift from the ideational aspects of pan-Arabism and non-alignment to realism. After 1967, Israel remained the dominant international threat to Egypt, not because it threatened to take more Egyptian territory but because Israeli occupation of Egyptian territory empowered Islamist actors (and others) to challenge the authority of the Egyptian state. For Sadat, it was imperative to work with the great power most likely to help Egypt regain territory lost to Israel, which had become a significant source of domestic unrest (David, 1991:249). The obvious tension that emerged from this adoption of modified realist ideology is that it simultaneously requires peace with Israel to sustain US support and to prevent a military threat from Israel, and continued hostility toward Israel to prevent a domestic challenge to the Egyptian state and to assert regional and Arab leadership. This had led, as many commentators have observed, to a "cold peace" (Stein, 1997; Podeh, 2007).

Realist beliefs also contributed to Egypt's hostile relationship with Iran, for instance. Although a number of factors have contributed to hostility between the two states, from regional competition to Iran's memorialization of Sadat's assassin, to perceived Iranian aggression toward Arab states, Shama suggests that the primary factor arises from the perceived threat of Islamist groups in Egypt and the perception that Iran was actively supporting these groups (2014:146). Egyptian hostility to Iran is firmly rooted in realist ideology, slightly altered to include internal and external threats, given that Iran does not pose a direct military threat to Egypt. Egypt's orientation toward other states in the region reflects not only consideration of external threats from other states but just as significantly from internal threats, such as those posed by Islamic militants. At the time of the 1992 earthquake, Egypt was in the midst of dealing with Islamist violence and also a non-violent challenge from the Muslim brotherhood.

In short, at the time of the 1992 earthquake, realist beliefs were influential in shaping Egyptian foreign policy, although there were still strong residual currents of pan-Arabism (particularly regarding Israel) and non-alignment solidarity (with regard to anti-colonialism and the deeply unequal international economic structure). This played out most notably in terms of how the Mubarak regime responded to offers of aid from other states (Israel in particular), and from internal Islamist movements.

Disaster and Response

On Monday October 12, 1992, an earthquake measuring 5.9 on the Richter scale struck northern Egypt, crumbling housing structures in and around Cairo and damaging numerous antiquities in surrounding areas. While not particularly powerful in terms of its magnitude, the impact of the earthquake was quite devastating. Over 550 people perished in the quake, a further 92,000 were affected or rendered homeless, and the total economic damage of the quake exceeded US $1.2 billion (Centre for Research on the Epidemiology of Disasters). Although Egypt is surrounded by a number of active tectonic plates, it is not located on a major fault line and is not particularly prone to intensive earthquakes. It does have a long and well-documented history of seismic activity (Degg, 1993), although the last earthquake to have caused significant damage in Egypt occurred close to 150 years previously, in 1847, which killed close to 100 people around Cairo and destroyed several thousand dwellings (Ambraseys et al., 1994). The relative infrequency of earthquakes in Egypt had two important consequences that operated at cross-purposes. First, it contributed to the low prioritization of earthquake readiness, which exacerbated the impact of the earthquake. Egypt had insufficient and outdated seismic equipment, lax enforcement of building standards, and minimal capacity to respond rapidly to the extensive damage caused by the quake. One result of this is that the damage caused by the earthquake was more extensive than the magnitude of the quake itself would suggest. Second, the infrequency of earthquakes reduced expectations that earthquake preparedness was an essential public good that the Egyptian state should have developed.

As the previous sections have demonstrated, in the fall of 1992, there were numerous cracks in the ideological foundations of the Egyptian state and the Mubarak regime. Contrasting beliefs about the source of social order at the domestic, regional, and international level, combined with repeated demonstration that policies based on these belief systems had failed, threatened to upend the Mubarak regime. In this situation, a massive disaster had the potential to contribute to further ideological uncertainty and potential regime change. This outcome was not desirable for the Mubarak regime, Mubarak's major power patron (the United States) or the other regional powers whose own regimes faced pressure for reform. Consequently, the response of the Egyptian state to the disaster and to offers of disaster aid reflected a concern with sustaining regime legitimacy with respect to its primary challengers: Israel and domestic Islamist groups.

Internal Legitimacy and Islamism

In the period prior to the earthquake, the Egyptian government was facing a severe challenge from a range of Islamist organizations. Militants from Gama'a al-Islamiyya had for a brief period of time "liberated" part of Cairo and declared an "Islamist state within the state" (Al-Sayyid, 1993:229; Bayat, 2017:188). Militant groups had been targeting foreign tourists at several key tourist sites. Perhaps even more problematical for the Mubarak regime was the moderate Islamist movement in Egypt, which was associated primarily with the Muslim Brotherhood and advocated for a greater role for political Islam. It depicted the state's secularist principles as heretical and atheist and as agents of the West – although it adopted a more ambiguous position with regard to science and technology and was generally supportive of modernization (Najjar, 1996). Indeed, the Muslim Brotherhood had strong support among the most "modern" portions of Egyptian society and controlled the medical and engineering syndicates.

While the Mubarak regime outlawed the Muslim Brotherhood, support for it continued to grow. One reason is that, as the Egyptian state retreated as part of its liberalizing reforms, the Muslim Brotherhood (as well as other Islamist groups) became increasingly involved in providing basic public goods, such as healthcare, education, insurance, and social welfare – goods previously associated with the state (Berman, 2003:260). Islamist groups had started to fulfil the functions expected of the state, simultaneously undermining one of the state's claims to legitimacy and increasing their own.

The response of the Egyptian state to these challenges combined increased repression with co-optation (Berman, 2003). Even as the government attempted to restrict the activities of Islamist groups, it was increasingly turning to Islamic principles to legitimize its rule (Shama, 2014). Consequently, the domestic situation in Egypt was one of serious ideological instability. The earthquake then presented an opportunity to expose the ideological tensions and ambiguities underpinning the existing order by demonstrating the Mubarak regime's "weak" Islamic commitments as well as the failure of the Egyptian state and its liberal economic reforms to modernize Egypt and provide basic social services.

Establishing the existential meaning of the earthquake was a key point of contestation between Islamists and modernists. For Islamist organizations, the earthquake itself had religious meaning – it was more than just a natural event exacerbated by poor building standards and failed modernization. Numerous groups within Egypt, including the Muslim Brotherhood, suggested that the earthquake was punishment from Allah for the sins of the Egyptian people and its government – in essence for adopting a secularist rather than Islamist form of state (Abdel-Fattah, 1992; Higazi, 1992). The religious interpretation of the earthquake was common, particularly in rural areas (Homan, 2001), and reflected the ongoing debate within Egypt over its identity as a modern, secular state or as an Islamic state. Leaders who rejected the religious interpretation as "superstitious" and insisted on modern interpretations were accused of "slandering the Islamic religion" (Anis, 1992c).

In contrast, the government and its supporters did not interpret the disaster as a divine statement on the Egyptian state or as a failure of modernism but as a failure to modernize. In this interpretation, responsibility for the disaster rested with outdated seismic equipment, frequently violated building codes, uninformed citizens, and the lack of modern disaster response organizations (Nafie, 1992a; Degg and Homan, 2005). The primary claim of the modernists was that Egypt had only imported a superficial modernity without the requisite commitment to developing scientific knowledge (Higazi, 1992). The religious interpretation was rejected as symptomatic of an anti-science Islamist movement.

In addition to contestation over the meaning of the disaster, the Egyptian state faced contestation over its rescue and recovery efforts by the Civil Defence Department (Khalil, 1992) and over emergency housing and financial compensation for victims of the disaster (Fattah, 1992). The Egyptian media was divided on the adequacy of the government response.[1] Pro-government papers praised Mubarak's decision to immediately return from his trip to China to manage the disaster and depicted the government's response as competent and generous (Anis, 1992a). Opposition papers were highly critical and accused the government of responding too slowly, of underreporting the amount of damage and the number of dead, and of failing to regulate construction projects (Anis, 1992c; Salama, 1992). Even the pro-government papers reported that many survivors of the disaster felt that government officials were slow to respond and operated in a top-down manner that did not take into consideration community needs. Consequently, people were often housed in apartments that lacked running water and electricity and that were located significant distances away from their home communities, schools, and places of employment, with few or no transportation options (Shehab et al., 1992). There were also reports of people waiting weeks for government building inspectors to assess their homes and workplaces.

In response to these failures, Islamist organizations helped during rescue operations and provided tents, food, and water to survivors. At first, the government and media praised the volunteerism of Egyptians who had donated blood, rescued neighbours, and provided shelter and food to survivors in the aftermath of the disaster (Nafie, 1992b). Once it became clear that the Muslim Brotherhood and the Gama'a al-Islamiyya were leading these rescue efforts, the Egyptian state prohibited any non-state entity from contributing to disaster relief. The government stressed that the state had sole responsibility to provide new housing for those who had lost their homes and financial compensation for those injured or killed. To undermine the Islamist movement and bolster its legitimacy, the government centralized administration of disaster relief – putting it in tension with its ongoing project of privatization and state withdrawal (Abdel-Fattah, 1992).

To the extent that the Egyptian state did encourage civil society participation, it required coordination through the Egyptian Red Crescent Society (ERCS), which Suzanne Mubarak, President Mubarak's wife, supervised (*Al-Ahram*, 1992b; El-Saadani, 1992a). As I detailed in Chapter 2, the international community and individual states endorse and empower the voluntary national societies of the International Federation of Red Cross and Red Crescent Societies (IFRC) as the key agency of disaster relief due to its quasi-non-governmental status. However, the ERCS was not the primary voluntary civil society organization in Egypt and possessed very little specialized capacity. The inability of the Egyptian state to channel civil society response through the safe confines of the ERCS reflects the highly political nature of disaster relief in Egypt, and the weakness of its non-governmental organization (NGO) sector. In the media coverage of the post-disaster response, the ERCS was rarely mentioned; rather, the emphasis was on the inadequacy of the government response and on the contributions of various Islamist groups.

By all accounts, Islamist groups were very effective in responding to the earthquake. Because the Muslim Brotherhood controlled the doctor's and engineer's syndicates, it was able to use its control of these entities to provide assistance. This included providing medical care and building inspections for structural integrity – two highly modern activities often expected of the state rather than civil society (Shehab et al., 1992). The emergency assistance offered by the Muslim Brotherhood paralleled, and sometimes exceeded, that of the government. In response, the Egyptian state introduced penalties for groups providing disaster aid outside the state apparatus (Ghazaleh and Atia, 1992). These penalties included seven-year prison sentences for any individual collecting donations for disaster relief (El-Saadani, 1992b). The government also refused to issue permits for the professional syndicates and other charitable organizations (El-Saadani, 1992b). In defence of this policy, the government claimed that it was necessary to ensure fair distribution of relief funds and to stop "fraudulents who tried to take advantage of the national dilemma to realise personal and unjustified interests" (El-Saadani, 1992a). People in poorer areas that had not yet received government assistance protested, often violently, and were subject to violent reprisals from the Egyptian authorities (El-Saadani, 1992b). Spokespeople for the government accused the Islamist groups of promoting the riots to score political points (Anis, 1992b), simultaneously dismissing the concerns of the protesters and the legitimacy of the Islamist parties.

While it is impossible to assess the impact of the earthquake and the relief effort on the ongoing divide between state and Islamist authority in Egypt, it does appear to have contributed to a cumulative impact. In the short term, the Egyptian state's relief efforts, buttressed by international support, was moderately successful in repairing the ideological rupture exposed by the earthquake and its aftermath. Riots over housing were quite small and contained, the government did not face a serious threat of being removed from power, and the Mubarak regime did not need to significantly alter its political and economic policies to shore up its domestic legitimacy. In the long term, however, the perceived failure of the state combined with the effective relief efforts of Islamist groups contributed to several influential trends.

The first was to contribute to the growing influence of the Muslim Brotherhood and other Islamist groups in Egyptian society, which prompted the Mubarak regime to exert greater control of these organizations and NGOs generally (Weiss and Wurzel, 1998:153–6). Of course, as the Mubarak regime continued its policies of privatization and rolling back the state, NGOs became more central to Egyptian life, which in turn deepened religious divides and supported the shift to "marketization and commodification" (Bayat, 2017:95). Mubarak's repressive measures did not stamp out Islamist influence in the syndicates or the popularity of the Muslim Brotherhood, but they did contribute to the longstanding stifling of civil society in Egypt, which fuelled the protests in 2011 (Shehata, 2011). Many of the political groups that organized the 2011 protests were motivated by the closure of opportunities for political expression and community activism that Mubarak initiated in response to the earthquake relief effort, although few appear to be rooted in Islamist ideology.

While the primary force behind the 2011 revolution seems to have been a number of non-Islamist political groups, such as Kefaya, the April 6 Movement, and the "We are all Khalid Said" group (Rutherford, 2013:38), the Muslim Brotherhood initially capitalized on the regime change, resulting in the election of Mohamed Morsi as president. Morsi had been a long-time member of the Muslim Brotherhood, from the centre right-wing of the party, and upon taking office initiated a series of changes that sought to alter the secular foundation of the Egyptian state, although not as profoundly as that advocated by the more hard-line Islamist groups, such as the al-Nur party. Morsi endorsed a view of Egypt as a "civil, national, constitutional, and modern" state but appealed to his Islamist constituency by promoting "Islamic constitutionalism", the implementation of sharia law, and opposition to women or Copts running for the presidency (Rutherford, 2013:49).

The military's quick ouster of Morsi (with fairly high levels of domestic support) and a new ban on the Muslim Brotherhood indicates that key actors in Egypt continue to espouse a particular version of modernism in which secularism is paramount. Rutherford notes that the Egyptian military views itself as responsible for defending Egypt from Islamist organizations, which was evident in their quick response to the election of a moderate Islamist leader in 2011. Nonetheless, three important legacies of the earthquake continue to shape the influence of modernist beliefs in Egypt: the ongoing project of modernization understood in terms of scientific and technological progress, more space for civil society to participate in public life, and widespread support for the moderate Islamism of the Muslim Brotherhood.

International Aid and Structural Reforms

While the Egyptian state was mostly pre-occupied with the challenge of organizing the state response to the earthquake and limiting the activities of Islamist organizations, it also had to coordinate the international response and the ideological tensions that these introduced. As news of the devastation spread, international offers of assistance began pouring in from surrounding states as well as from several Western states. Exactly how much international disaster relief the Egyptian state received is not clear, even as the Egyptian media frequently reported on how much individual states had donated. The Egyptian media reported at the time that the international community had publicly pledged close to US $200 million. However, reports in 2012 following Mubarak's overthrow suggest international disaster relief for Egypt in 1992 had actually topped US $1.3 billion (or 9 billion Egyptian pounds), with the Mubarak regime spending only a tiny fraction of it on disaster relief and rebuilding (Ahram Online, 2012). In addition to the relief funds sent to the Egyptian state, various international actors channelled relief funding through non-state sources, such as the ERC and the Muslim Brotherhood; it is unknown how much international disaster relief entered Egypt through these agencies.

The Egyptian government proved strongly amenable to international disaster relief from Arab and Western states, as well as the UN. Unlike the United States after Katrina, there were no assertions that Egypt did not need foreign aid and no claims that Egypt could respond adequately on its own. Given Egypt's ongoing reliance on foreign aid from states such as the United States and Saudi Arabia as well as on international institutions such as the IMF and World Bank, receiving disaster aid did not fundamentally challenge Egypt's existing self-image. Indeed, the Egyptian government argued that international aid, in particular for preserving its antiquities, was in fact the responsibility "of all civilized nations" and not Egypt's responsibility alone. Consistent with this position, the United Nations Educational, Scientific and Cultural Organization (UNESCO) played a leading role in the efforts to repair and preserve the numerous damaged antiquities in and around Cairo.

The Egyptian government explicitly requested aid from Japan and accepted aid from nearly all states that offered: the majority of whom were allies with Egypt in the Gulf War, including the United Arab Emirates, Saudi Arabia, Kuwait, France, Greece, Italy, Germany,[2] Spain, and the United States. The large amount of support from states allied with Egypt in the Gulf War and Egypt's acceptance of that aid indicate the geopolitical importance of Egypt in the MENA region and the prevalence of realist ideology operating in the post-disaster response. Egypt was a key ally of the United States and other Western states in the Middle East: balancing Iranian and Russian influence in the region, playing a leading role in the negotiations between Israel and Palestine, ensuring US access to oil in the region, and supporting military action against Saddam Hussein (Brownlee, 2012:7–8). Egypt and its Western allies acted on the belief that disaster recovery was essential to regime stability because of the role that Egypt played in maintaining order in the Middle East and balancing other regional powers. The United States alone provided around US $15 million in disaster relief – all bilaterally (*Al-Ahram*, 1992a) – and in addition to the regular stream of US aid to Egypt.

Notably, several of Egypt's regional competitors and opponents in the Gulf War contributed aid and/or condolences – including Libya, Jordan, and Sudan. In some of these cases, hostility between these states was intense and longstanding. Nonetheless, the Egyptian media and government frequently articulated the importance of Arab solidarity, "brotherhood", or "fraternity" (Nafie, 1992b). Nafie (1992b) suggested that "sympathy was felt all over the Arab world" and that Arab states have "always been able to rise above differences". Indeed, in the coverage of external offers of aid, states were essentially categorized into two groups: Arab countries or others. One program initiated by the Mubarak regime to foster both rebuilding and privatization (more on that below) was to have private corporations or individuals rebuild heavily damaged schools, which would be named after the benefactor. This program was open exclusively to Egyptians and Arabs (Nafie, 1992b).

The centrality of Arab contributions to disaster relief speaks to the ongoing importance of pan-Arabism in Egypt and to Egypt's role as a regional leader. As noted earlier, Egypt had only recently been readmitted to the Arab League in 1989 after having been suspended due to its peace agreement with Israel. Furthermore, the Arab League had been quite divided over the 1991 Gulf War, with several members opposing military action against Hussein (despite all condemning his annexation of Kuwait). Arab unity – and Egypt's place in the Arab world – was under strain. The obvious contradictions in pan-Arabist ideology, which postulates a common Arab identity and dictates that Arab states act in accordance with the common interests of all Arabs (Telhami, 1993:439), were evident in Arab disagreement over the Gulf War, US troops in Saudi Arabia, and peace with Israel. The Egyptian disaster offered an opportunity to reassert pan-Arabism and Egypt's place within the regional order by demonstrating mutual aid in the face of disaster – notwithstanding ongoing divisions.

The Egyptian media also regularly reported on the disaster relief provided by Western states, although they framed Western aid differently than that provided by Arab states. Whereas aid and condolences from Arab states was presented as evidence of intra-Arab cooperation and shared identity, aid from the United States, France, Germany, and Japan (which featured most heavily in Egyptian media coverage) was framed almost exclusively in terms of modernity. *Al-Ahram* coverage frequently praised the expertise and technological sophistication of rescue and recovery operations from these countries and described the advanced technology they possessed for monitoring seismic activity and finding survivors trapped in rubble (e.g. Salama, 1992). This narrative reinforced a binary of modern and non-modern that overlapped with a Western/Arab binary. In short, the West was technologically advanced, and Egypt (being insufficiently modernized and without access to similar technology) ought to modernize in line with the Western example; Arab aid was a demonstration of shared national identity.

Indeed, as noted in the previous section, "modernity" emerged as one of the primary points of contestation in framing the disaster. Modernizing forces in Egypt used the earthquake to stress the importance of the modernization project – understood in terms of gaining access to technology, specialized rescue training, and developing modern building standards (El-Saadani, 1992a). The Egyptian government distributed a booklet explaining the scientific causes of earthquakes and what to do when one occurred. Egyptian political elites and foreign experts stressed the importance of Western assistance in upgrading Egypt's seismic monitoring capacity and improving building regulations, which were frequently cited as reasons for the disaster (Khalil, 1992). The Egyptian government stressed the lack of domestic expertise in handling the repair of damaged monuments and the need to import expertise from Germany and the United States (Murad, 1992). Others speculated that the outdated equipment and poor response "discredited Egypt's scientific credibility" (Mohamed, 1992). As part of its rebuilding plan, the Egyptian government used the disaster and disaster aid to acquire up-to-date seismic equipment and to introduce new, stricter building codes.

Of equal or greater concern to the Mubarak regime than offers of aid from foreign states was the ongoing matter of IMF and World Bank loans. Specifically, Mubarak worried that the earthquake and relief and rebuilding efforts would affect its program of economic liberalization and its ability to meet the conditions set by the IMF and World Bank for financial assistance. On this matter, the government used the earthquake to pursue two divergent goals: to delay the reform process by renegotiating with the IMF and to facilitate these reforms by reducing domestic opposition. As noted earlier, in 1991, Mubarak had embarked on an extensive liberalization program funded by the IMF and World Bank, which had required adjustment policies that potentially threatened the ability of the Egyptian state to provide basic public goods. As had his predecessor, Mubarak often balked at the policies demanded by the IMF and generally proceeded incrementally and cautiously to ensure minimal disruption to the existing order. The 1992 earthquake gave Mubarak a compelling reason to delay a number of reforms.

Outlining their plan in the *Al-Ahram*, the government reiterated that it would be unable to meet the budget deficit targets mandated by the IMF and World Bank and called on these institutions to suspend their requirement for Egypt to reduce its budget deficit to 3.5% of the Egyptian budget (Nafie, 1992c). Given the costs of rebuilding, Egyptian officials suggested there was no way Egypt could meet the mandated budget reduction requirements. The IMF relented and in November 1992 agreed to extend the deadline for Egyptian budget cuts (Weiss and Wurzel, 1998:57).

At the same time, the Egyptian state defended its liberalizing reforms and called for expedited implementation of these reforms to "increase exports, cut imports, achieve growth, cope with stagnation in the labor market... and raise productivity" (Nafie, 1992c). Depicting Egypt as competing in an international economic system in which major powers attempt to "keep small and developing countries dependent on them and the outside world for imports", Nafie reiterated the government's belief that economic liberalism alone would lead to modernization of the Egyptian economy and independence from the major powers. In this narrative, the devastation of the earthquake further demonstrated the weakness of Egypt compared with the major powers on whom it had relied for disaster relief. The poor Egyptian response to the earthquake was not a product of economic liberalization and reduced state capacity but of its relative lack of modernization caused by stalled economic liberalization.

The Mubarak regime demonstrated its embrace of economic liberalism in its rebuilding programs. New homes and financial compensation were given to those who had lost their homes or breadwinners (*Al-Ahram*, 1992b; Nafie, 1992a). There was no compensation for those who did not own property or were not actively employed in the workforce. The value placed on human lives reflected an economic liberal perspective: value (and thus compensation) was determined by contribution to the formal economy. In addition, the Egyptian state sought to ensure that no one would illegitimately benefit from the relief program through harsh punitive measures. Falsely claiming to have lost a home in the disaster, even by those who were homeless before the disaster, was subject to harsh new penalties, ranging from seven years in prison to hard labour. The Egyptian government devoted significant resources to preventing and punishing the "unjust" accumulation of property by the "non-deserving poor".

The Egyptian state also used the rebuilding effort as a further spur to privatization. The government called on private businesses to rebuild damaged schools in exchange for naming the buildings after the corporation or individuals, which had previously been restricted (Ragab, 1992). Private corporations responded on the grounds that "the private sector" could produce higher quality schools at a lower cost than the government could produce (Ragab, 1992). Poorer areas of Cairo hit hardest by the quake received the least amount of assistance. Defending the government's response, Ibrahim Nafie (the powerful editor-in-chief of *Al-Ahram*) suggested that residents in poor areas must be forced to leave in order for reconstruction and restoration of damaged monuments (Nafie, 1992b). Similarly, the Egyptian government used the earthquake to justify the removal of houses and shops around damaged monuments and forcing them into commercial districts and shopping centres, something they had been trying to accomplish for many years (Murad, 1992).

There were undercurrents of opposition to the economic reforms that sought to portray the earthquake as symptomatic of these liberalizing reforms. This took place through two narratives: corruption and greedy developers. One of the main narratives surrounding the earthquake was the extent to which privatization and the adoption of economic liberalism had increased vulnerability in Egypt. This was captured primarily through the narrative of greedy developers whose buildings had collapsed in the earthquake. One of the worst tragedies associated with the earthquake involved the collapse of a 14-story building in Heliopolis. According to reports, the building's owner had added four floors and a water tower to the top of the building without a permit (Khalil, 1992), and numerous residents had filed complaints about the safety of the building (Salama, 1992). Reports in the Egyptian media suggested this was the product of "fast-buck developers" who had been permitted to engage in this activity due to liberalizing reforms and lack of government regulation. In response, the government depicted the owners as criminals and introduced harsh penalties for developers that ignored building codes (Nafie, 1992a).

Corruption was also a dominant theme that spoke to the dangers of economic liberalization. In this context, corruption was not simply a critique of the government but of its privatization process. As Weiss and Wurzel (1998) note, corruption in Egypt referred to the manner in which privatization of state-owned corporations occurred, with members of the government and military purchasing state-owned enterprises at below market value. Consequently, in Egypt, privatization had simply empowered members of the ruling classes to profit at the expense of the general population. The Mubarak regime countered this critique with the assertion that the fault for the earthquake damage lay with developers who were corrupt and did not follow building codes, and with local political actors who did not enforce these building codes. The government introduced a number of punitive measures to address both of these groups, although there is little indication that local politicians and developers were actually punished as a result.

The government framed the poor response of the Egyptian state compared with the contribution of Islamist groups to rescue and recovery efforts in terms of modernization and the threat posed by Islamists in Egypt rather than as an indictment on the retreat of the state induced by its liberalizing reforms. The ability of the Mubarak regime to switch ideological frames in describing the earthquake to reduce opposition supports my contention that, while ideological heterogeneity introduces instabilities that periodically produce crises, it also gives actors considerable manoeuvrability in framing crises and disasters in ways that reinforce the ideological status quo. This is also amply demonstrated in the case of Israeli aid.

The Cold Peace and Israeli Aid

The only international aid that the Egyptian state did not accept was Israeli aid. On the one hand, the prominence of Israel as Egypt's main external threat and the highly controversial nature of Egypt's peace deal with Israel made acceptance of Israeli aid a difficult proposition. There were, however, compelling reasons for Egypt to accept Israeli aid. Israel possessed some of the most modern disaster response equipment and personnel immediately available, and with the first 24 hours in rescue operations being crucially important, the decision to accept rescue teams from Germany and Japan rather than Israel meant that humanitarian considerations were secondary to political considerations. Egypt had positioned itself as a regional leader by concluding a peace deal with Israel, and at the time of the earthquake was leading the latest round of negotiations to finalize a peace agreement between Israel and Palestine. In short, its leadership position in the region was tied up with the ability to cooperate with both Palestine and Israel.

The Palestinian position on Israel's offer of disaster aid was clear. Immediately following Israel's announcement of condolences and aid, the PLO publicly criticized the Israeli offer of aid. Broadcast on the Voice of Palestine radio and published by the British Broadcasting Corporation (BBC), the PLO sent condolences to Egypt and praised the offers of aid from Arab states before describing Israel's offer as "amazing and ridiculous" given "their indescribable repressive and savage measures" against the Palestinians. The PLO called on the Egyptian people to refuse the aid so that Israel could not "shroud themselves in a humanitarian character" or present themselves as "rescuers and do-gooders at a time when they do not give any importance to humanitarian values and principles in occupied territories" (Palestine Liberation Organization, 1992). The PLO drew a clear line based on humanitarian considerations – and placed Israel outside the bounds of civilized nations capable of acting in accordance with humanitarian values. Aware of the potential benefits to Israel of providing aid to Egypt, the PLO took a clear stance to ensure that Egypt would not contribute to the legitimization of Israel.

From this perspective, accepting humanitarian aid from Israel was a betrayal of the Palestinians. The Palestinian cause is central to every Arab political movement, including in Egypt (Telhami, 1993:441). In general terms, friends and enemies in the MENA region are defined through their position vis-à-vis the Palestinians and Israel. Even in cases where Arab states fundamentally disagree, such as joining the US-led coalition against Iraq, this is couched in terms of what was best for Palestinians (Telhami, 1993). Egypt had already felt the effects once of being seen to betray the Palestinians when it made peace with Israel and was subsequently removed from the Arab League. Egypt had only been reinstated, and the headquarters of the Arab League moved back to Cairo, in 1987.

At the same time, peace with Israel had been a significant aspect of Egyptian foreign policy, had allowed it to reclaim lost territory, and was a key condition of its aid deal with the United States. As part of terminating the 1973 war between the two states, the treaty arranged for Israel's withdrawal from the Sinai, ensured Israeli navigation and shipping through the Suez Canal, and called for the normalization of relations (Cook, 2012:149). Following its peace with Israel, Egypt had played a key role in negotiating peace between Israel and the Palestinians and other Arab states (Gerges, 1995). Yet this placed Egypt in a difficult position – pushing too hard for peace and normalization with Israel would undermine its leadership in the region.

The awkward peace that emerged between the two states did not produce normalization. Relations between the states have hardly changed, and the amount of trade and cultural interactions between the two states are minimal. Highly antagonistic rhetoric is common in both states, and there have been numerous "incidents" that have strained Egyptian-Israeli relations. Among the more notable incidents are the Israel-Lebanon War in 1982 (in which Egypt recalled its ambassador to Israel), the first and second intifadas, and notable instances of violence in Egypt against Israeli citizens and diplomats (Stein, 1997; Podeh, 2007:115; Cook, 2012:163).

Rather than dramatically altering Egyptian-Israeli relations, the peace treaty ushered in an era of what has been characterized as "non-belligerency" or "hostile peace" (Pine, 1997:332), or "cold peace" (Stein, 1997; Podeh, 2007). Podeh identifies a number of aspects of the post-treaty period that have contributed to the "cold peace": the primary source of tension and conflict (Palestinian statehood) remains unresolved, prominent movements in each state call for the abrogation or alteration of the peace agreement, relations have not progressed from governmental to civil-society interaction, and both maintained contingency plans for war with the other (Podeh, 2007:108). Pine suggests that, for most people in Egypt, Israel remains the prominent existential threat and is commonly labelled as such in the public discourse and in military planning exercises (Pine, 1997:324–5). Despite the existence of a peace treaty, neither Egypt's nor Israel's long-term goals in the region have changed nor has the hostile and untrusting attitude toward each other (Stein, 1997). Indeed, some media commentators suggested that the Egyptian solidarity seen in response to the earthquake replicated the solidarity achieved by opposition to Israel prior to the peace agreement (Sid-Ahmed, 1992).

Egypt's stance toward Israel signals a pragmatic, realist approach that attempts to maximize its regional standing and maintain domestic stability while minimizing the threat from its primary opponent. Egypt benefits most (politically, economically, and militarily) in maintaining the cold peace rather than increasing hostility or normalizing relations. Without having to devote resources to a war against Israel, the Egyptian state has been able to invest in domestic development programs designed to keep the government in power (Stein, 1997). In addition to bringing about the return of Sinai to Egypt, the peace treaty has brought with it substantial aid annually from the United States (Pine, 1997). These incentives are not undermined by the ongoing failure to normalize relations.

Egypt faces a number of positive incentives to sustain cold relations with Israel. Domestically, peace with Israel was, and is, deeply unpopular, and the Egyptian government has been careful not to do more than was "minimally required" by the treaty (Cook, 2012:163). Pine contends that the Egyptian state had no interest in normalizing relations, and many in powerful positions view the treaty as a temporary ceasefire (1997:325). From the time of the signing, through 1992 to the present day, the Egyptian state has made very little effort to counter the negative portrayal of Israel by the media and civil society groups and in the education system. At times, the government has even been complicit in presenting Israel as an ongoing threat (Stein, 1997). Thus, the official peace was not matched by a shift in the domestic identity construction of Israel as the radically threatening other (Stein, 1997; Podeh, 2007:109).

Keeping relations with Israel cold also fits with Egypt's regional or international identity as an Arab state, a Muslim state, and a regional leader (Podeh, 2007). As an Arab state, Egypt was tied to the pre-eminent pan-Arab cause, support for Palestine. Roundly criticized by other Arab states for concluding a peace with Israel, successive Egyptian regimes have attempted to assuage such concerns by taking a leadership role in condemning Israel's land and settlement policies and actively undermining attempts by other Arab states to normalize relations with Israel (Podeh, 2007). As a Muslim state, Egypt supports control of Islamic holy sites in Palestine, again putting it at odds with Israel. And as a regional leader, Israel's economic integration into the region was portrayed as a potential threat to Egypt's leadership, again putting Egypt at odds with Israel (Gerges, 1995). In short, the existing cold peace with Israel was conducive to the maintenance of regional and domestic order.

The offer of aid from Israel was a potentially destabilizing move in a period and location of deep instability. In theory, the offer could have been framed as a gesture of goodwill and a move toward warmer peace, as occurred in the Turkish-Greece context. In contrast to the Turkish case, there were no domestic actors challenging the longstanding narrative of Israel as a threat and few international pressures to improve relations, and the government stood to gain little from doing so. At the time of the 1992 earthquake, Egypt was involved in leading the seventh round of peace talks between Israel and Palestine, which were generally perceived as not going well. Although Israel had made minor concessions, the Egyptian press speculated that Israel was not interested in peace and accused the Israeli negotiators of trying to divide Arab unity on the Palestinian issue by negotiating separate bilateral settlements with individual Arab states (El-Bay and Adham, 1992). Had the offer of Israeli aid occurred at a moment of significant Israeli concessions toward a Palestinian peace agreement, the effect of Israel's disaster diplomacy may have been different. In the context of October 1992 and Egypt's realist approach to regional security, Egyptian bilateral cooperation with Israel was politically dangerous.

Given this context and beliefs about Egypt's place in the Middle East, it is not particularly surprising that the Egyptian state did not accept Israel's offer of assistance. What was surprising was that it was not a highly publicized refusal with a representation of Israel's offer as threatening or as politically motivated rather than humanitarian – as the PLO had depicted it, and as we've seen in the case of the United States with respect to Cuban and Iranian offers of aid after Hurricane Katrina. The Egyptian state did not tie the refusal to the stalled peace negotiations; rather, the offer of aid was acknowledged by the Egyptian government and subsequently ignored. On October 13, the day after the earthquake, the Israeli prime minister Yitzhak Rabin offered condolences and unspecified assistance to Egypt, announcing on Israeli radio that "it is their decision what kind of assistance. Whatever kind of assistance, we are more than ready to offer it" (Rabin, quoted in the *New York Times*, 1992). At the time, Israel had a specialized disaster rescue team that it had previously dispatched to Mexico and Armenia following earthquakes in those states (*New York Times*, 1992). One week after the earthquake and offer of aid, the Egyptian government had yet to respond publicly – to either reject or accept the offer.

Questioned on the matter of Israeli aid on October 19, six days after the offer was made, a spokesperson for the Egyptian foreign ministry confirmed that it had rejected the offer but claimed that there were "no political aspects" to the refusal, that Egypt has "normal relations with Israel and we treat it as any other country" (Ghatrifi, quoted in UPI, 1992). The Israeli government was similarly diplomatic in response. The Israeli chargé d'affaires at the embassy in Cairo noted that the Egyptian government had "expressed thanks" for the offer and speculated that "right now they don't need our assistance" (Shaked, quoted in UPI, 1992). Perhaps a week after the earthquake this was true, although even at this point the Egyptian government was struggling to provide assistance to all Egyptians affected by the earthquake. But in the immediate aftermath, Israel was the state most able to offer assistance, given its geographic proximity and expertise. Instead, Egypt accepted technical expertise and relief teams from states further afield: France, Greece, Algeria, Germany, Italy, Spain, and Japan – indicating that need was not the primary factor in refusing Israeli assistance and that there were political aspects to the aid decision (UPI, 1992).

As was the case with the United States after Hurricane Katrina, the refusal of aid was handled to avoid jeopardizing the disaster-affected state's existing foreign policy. Egypt benefited from its cold peace with Israel, was deeply invested as the key peacemaker in the ongoing Israel-Palestine peace negotiations; relied heavily on US aid; and was embarking on a series of economic reforms, the success of which depended on Western institutions. These factors made it less likely that Egyptian officials would publicly condemn the Israeli offer of aid and frame Israel as incapable of being humanitarian – as the PLO had. On the other hand, Egypt's efforts to maintain domestic stability and advance its regional and Arab leadership meant it had to be seen as supporting the Palestinians, or at the very least not acting in direct contrast to the expressed Palestinian position.

Conclusion

The response of the Egyptian state to the earthquake and offers of assistance in the aftermath of the earthquake shed further light on the arguments I have developed earlier in this book. The ideological foundations of modern international order are on full display in this case, as the responses of the Egyptian government reflected the profound influence of and contradictions within modernism, nation-statism, economic liberalism, and realism. Humanitarianism, to the extent it figured in this case, was used in reference to the relief provided by Western states and as a discursive tool by Palestinians to place Israel outside civilized international society. As demonstrated in the US case, humanitarianism operates in two distinct ways: the first is to draw offers of aid, the second is to delegitimize some offers of aid and justify the rejection of that aid. Egyptian authorities did not use the humanitarian discourse for this purpose; the Israeli offer of aid was largely treated with silence. In addition to further demonstrating the importance of understanding silence as a form of agency and action (in contrast to discursive or practice-oriented approaches), silence in this particular case served multiple purposes. By remaining silent, Egypt could reject Israeli aid in line with the "Arab" position but without explicitly denying Israel's place within the civilized international order, which would in turn further call into question Egypt's peace treaty with Israel and aid arrangement with the United States.

Yet this case also demonstrates the importance of local contexts and how they interact with the foundational ideologies of international order; the Islamic and pan-Arab ideologies prevalent in Egypt demonstrate how the ideological foundations promoted and enshrined in international agreements on disaster management merge with and confront ideologies particular to specific times and places. In this case, realist ideology that places great importance on patterns of enmity and amity regionally did not fit with Arab solidarity contained in pan-Arab ideology. The empowerment of civil society associated with liberal economic beliefs stood in tension with Egyptian concerns with state sovereignty due to the tensions between secularist and Islamist ideology.

The Egyptian case also exposes the difficulty of implementing a universalist approach to disaster management. The local context in Egypt, for instance, demonstrates that promoting an independent and robust civil society as an integral tool of disaster management may undermine belief in the nation state as the legitimate political entity. It may also challenge the veracity of secularism as the modernist accommodation of religious belief. The ideological tensions and ambiguities within and between the foundational beliefs of the international disaster regime are further compounded when applied in settings where these belief systems face serious challenges from competing ideological frameworks. Attempts to build state competence in the provision of public goods, including disaster preparedness and relief, may be in tension with economic liberalism's insistence on reducing government expenditures and privatizing national industries.

In many decolonizing states, attempts to adapt local belief systems with international ideologies of order, which are spread and enforced by international institutions and powerful states, can lead to profound ideological uncertainty and recurrent crises. The hybrid belief systems and identities that have emerged in these contexts, as demonstrated in the Egyptian case, require persistent ideological work and often coercive enforcement. At the same time, the infusion of resources from international actors can assist governmental actors to temporarily address these contradictions and tensions. States that are highly dependent on internationalized belief systems and material assistance to maintain their own domestic order then are limited in their ability to alter international order. This is not true of all former colonized states. Chapter 8 will examine a case where a developing, decolonized state has been able to offer an alternative to the existing international order and introduces a number of disruptions to the dominant belief systems.

Notes

1 At the time, the Egyptian media was highly partisan, and individual papers were identified as pro-government or oppositional, even as the media was subject to government control and repression (Khamis, 2011).
2 Although not an official participant in the Gulf War, Germany provided funding for the war.

References

Abdel-Fattah, N., 1992. The political struggle for hearts and minds. *Al-Ahram*, p. 7.
Abulof, U., 2015. "The people want(s) to bring down the regime": (Positive) nationalism as the Arab Spring's revolution. *Nations and Nationalism*, 21(4), pp. 658–80.
Agrama, H.A., 2010. Secularism, sovereignty, indeterminacy: Is Egypt a secular or a religious state? *Comparative Studies in Society and History*, 52(3), pp. 495–523.
Ahram Online, 2012. Mubarak govt spent just LE70million from LE 9billion earthquake aid fund: Report. Ahram Online. http://english.ahram.org.eg/NewsContent/1/64/37435/Egypt/Politics-/Aboutus.aspx.
Ajami, F., 1978. The end of pan-Arabism. *Foreign Affairs*, 57(2), p. 355.
Al-Ahram Staff, 1992a. Quake's work unfinished. *Al-Ahram*, p. 1.
Al-Ahram Staff, 1992b. The aftermath. *Al-Ahram*, p. 1.
Al-Sayyid, M.K., 1993. A civil society in Egypt? *Middle East Journal*, 47(2), pp. 228–42.
Ambraseys, N.N., Melville, C.P., & Adams, R.D., 1994. *The Seismicity of Egypt, Arabia and the Red Sea: A Historical Review*, Cambridge: Cambridge University Press.
Anis, M., 1992a. Forty seconds that shook Egypt. *Al-Ahram*, p. 7.
Anis, M., 1992b. The wrath of God. *Al-Ahram*, p. 7.
Anis, M., 1992c. Widening cracks. *Al-Ahram*, p. 7.
Assaad, R., 2007. Labor supply, employment and unemployment in the Egyptian economy, 1988–2006. In R. Assaad (ed.). *The Egyptian Labor Market Revisited*. Cairo: American University in Cairo Press.

Assaad, R., & Arntz, M., 2005. Constrained geographical mobility and gendered labor market outcomes under structural adjustment: Evidence from Egypt. *World Development*, 33(3), pp. 431–45.

Babbili, A.S., 1990. Understanding international discourse: Political realism and the non-aligned nations. *Media, Culture and Society*, 12, pp. 309–24.

Barnett, M.N., & Levy, J.S., 1991. Domestic sources of alliances and alignments: The case of Egypt, 1962–73. *International Organization*, 45(3), pp. 369–95.

Bayat, A., 2017. *Revolution without Revolutionaries: Making Sense of the Arab Spring*, Stanford, CA: Stanford University Press.

Berenskötter, F., 2017. Deep theorizing in international relations. *European Journal of International Relations*, 24, pp. 814–840. doi:10.1177%2F1354066117739096.

Berman, S., 2003. Islamism, revolution, and civil society. *Perspective on Politics*, 1(2), pp. 257–72.

Bhabha, H., 1984. Of mimicry and man: The ambivalence of colonial discourse. *Discipleship: A Special Issue on Psychoanalysis*, 28, pp. 125–33.

Bromley, S., & Bush, R., 1994. Adjustment in Egypt?: The political economy of reform. *Review of African Political Economy*, 21(60), pp. 201–13.

Brownlee, J., 2012. *Democracy Prevention: The Politics of the U.S.-Egyptian Alliance*, Cambridge: Cambridge University Press.

Brumberg, D., 2002. Islamists and the politics of consensus. *Journal of Democracy*, 13(3), pp. 109–115.

Chan, S., 2017. *Plural International Relations in a Divided World*, Cambridge: Polity Press.

Clark, J.A., 2004. *Islam, Charity, and Activism: Middle-Class Networks and Social Welfare in Egypt, Jordan, and Yemen*, Bloomington: Indiana University Press.

Cook, S., 2012. *The Struggle For Egypt: From Nasser to Tahrir Square*, Oxford: Oxford University Press.

David, S.R., 1991. Explaining third world alignment. *World Politics*, 43(2), pp. 233–56.

Dawisha, A.I., 2016. *Arab Nationalism in the 20th Century: From Triumph to Despair*, Princeton, NJ: Princeton University Press.

Degg, M., 1993. The 1992 "Cairo Earthquake": Cause, effect and response. *Disasters*, 17(3), pp. 226–38.

Degg, M., & Homan, J., 2005. Earthquake vulnerability in the Middle East. *Geography*, 90(1), pp. 54–66.

El-Bay, D., & Adham, K., 1992. Israel's gesture "insignificant." *Al-Ahram*, p. 3.

El-Saadani, E., 1992a. The earthquake experts. *Al-Ahram*, p. 7.

El-Saadani, E., 1992b. Where does charity begin. *Al-Ahram*, p. 7.

Fattah, N.A., 1992. The political struggle for hearts and minds. *Al-Ahram*, p. 7.

Gerges, F., 1995. Egyptian-Israeli relations turn sour. *Foreign Affairs*, 74(3), pp. 69–78.

Ghazaleh, P., & Atia, T., 1992. Nothing to lean on. *Al-Ahram*, p. 11.

Gumuscu, S., 2010. Class, status, and party: The changing face of political Islam in Turkey and Egypt. *Comparative Political Studies*, 43(7), pp. 835–61.

Gupta, A., 1992. The song of the nonaligned world: Transnational identities and the reinscription of space in late capitalism. *Cultural Anthropology*, 7(1), pp. 63–79.

Halliday, F., 2005. *The Middle East in International Relations: Power, Politics and Ideology*, Cambridge: Cambridge University Press.

Higazi, A.A.-M., 1992. A gap in our knowledge. *Al-Ahram*, p. 8.

Homan, J., 2001. A culturally sensitive approach to risk?: "Natural" hazard perception in Egypt and the UK. *Australian Journal of Emergency Management*, 16(2), pp. 14.

Kedourie, E., 1971. *Nationalism in Asia and Africa*, London: Weidenfeld & Nicholson.

Kerr, M., 1967. *The Arab Cold War*, Oxford: Oxford University Press.

Khalil, N., 1992. When fate and greed conspire. *Al-Ahram*, p. 2.

Khamis, S., 2011. The transformative Egyptian media landscape: Changes, challenges and comparative perspectives. *International Journal of Communication*, 5, pp. 1159–77.

Kienle, E., 1998. More than a response to Islamism: The political deliberalization of Egypt in the 1990s. *Middle East Journal*, 52(2), pp. 219–35.

Matossian, M., 1958. Ideologies of delayed industrialization. *Economic Development and Cultural Change*, 6, pp. 217–28.

Mohamed, A.-A., 1992. Quick to speak, slow to act. *Al-Ahram*, p. 12.

Morphet, S., 2004. Multilateralism and the non-aligned movement: What is the Global South doing and where is it going? *Global Governance*, 10(4), pp. 517–37.

Murad, M., 1992. Neglect is what shook them. *Al-Ahram*, p. 11.

Nafie, I., 1992a. Lessons that must be learned. *Al-Ahram*, p. 6.

Nafie, I., 1992b. Much has already been done. *Al-Ahram*, p. 6.

Nafie, I., 1992c. There is no alternative. *Al-Ahram*, p. 6.

Najjar, F.M., 1996. The debate on Islam and secularism in Egypt. *Arab Studies Quarterly*, 18(2), pp. 1–21.

New York Times Staff, 1992. Israel offering aid to Victims. *New York Times*, p. A14. October 13, 1992.

Onis, Z., 2001. Political Islam at the crossroads: From hegemony to co-existence. *Contemporary Politics*, 7(4), pp. 281–98.

Palestine Liberation Organization, 1992. PLO radio ridicules Israeli aid offer to Egyptian quake victims. Voice of Palestine. October 13.

Pine, S., 1997. Myopic vision: Whither Israeli-Egyptian relations? *Israel Affairs*, 3(3–4), pp. 321–34.

Podeh, E., 1996. The drift towards neutrality: Egyptian foreign policy during the early Nasserist era, 1952–55. *Middle Eastern Studies*, 32(1), pp. 159–78.

Podeh, E., 2007. Normal relations without normalization: The evolution of Egyptian-Israeli relations 1979–2006—The politics of cold peace. In S.M. Gabbay (ed.), *The Search for Israeli-Arab Peace*, Brighton: Sussex Academic Press, pp. 107–29.

Ragab, G., 1992. Private money to build public schools. *Al-Ahram*, p. 4.

Rutherford, B.K., 2013. *Egypt after Mubarak: Liberalism, Islam, and Democracy in the Arab World*, Princeton, NJ: Princeton University Press.

Salama, S., 1992. Beyond the earthquake. *Al-Ahram*, p. 7.

Shama, N., 2014. *Egyptian Foreign Policy from Mubarak to Morsi*, London: Routledge.

Shehab, S., Khalil, N., & Nassar, G., 1992. Quake victims seek new homes. *Al-Ahram*, p. 2.

Shehata, D., 2011. The fall of the Pharaoh: How Hosni Mubarak's reign came to an end. *Foreign Affairs*, 90(3), pp. 26–8.

Sid-Ahmed, M., 1992. The earthquake: Political aftershocks. *Al-Ahram*, p. 6.

Stein, K.W., 1997. Continuity and change in Egyptian-Israeli relations. *Israel Affairs*, 3(3–4), pp. 296–320.

Telhami, S., 1993. Arab public opinion and the Gulf War. *Political Science Quarterly*, 108(3), pp. 437–52.

UPI, 1992. Israel still awaiting word from Cairo on quake aid offer. United Press International. October 19, 1992.

Weiss, D., & Wurzel, U., 1998. *The Economics and Politics of Transition to an Open Market Economy*, Egypt, Paris: OECD.

Zubaida, S., 2000. Trajectories of political Islam: Egypt, Iran and Turkey. *The Political Quarterly*, 71(s1), pp. 60–78.

8

CHINA AND CHANGE IN THE INTERNATIONAL ORDER

There is no shortage of speculation on the potential changes to the international order that will accompany the rise of China (and other developing states). Of course, what qualifies as change depends a great deal on the theoretical grounding as well as the level of analysis (see Gilpin, 1983; Rosenau, 1990; Holsti, 2004; Phillips, 2011; Paul, 2018, for different types of change), but what these accounts reveal is that there are multiple types of change, and given the complex nature of societies, there is no single, universal model or mode of change. Rather than propose such a model, I have focused in this book on a single potential contributor to change – disasters – in a heterogenous international order beset with tensions and contradictions. I have argued that disasters present an opportunity for change by demonstrating the failures of existing belief systems and institutions to deliver on their promised expectations to produce a particular type of order. The post-disaster period is a prime opportunity for developing and promoting new ideas and beliefs. Some of the most foundational aspects of modern international order (mastery of nature, state responsibility for emergency management, conceptions of private property and risk, humanitarian obligations to outsiders) emerged as societies confronted devastating disasters. As a result, disasters are subject to increasing amounts of "management" to shape understandings of disasters and to control their impact and potential for producing change.

The impact of disasters and their potential to foment change is mitigated, however, by distribution of power – within and between states. This book has repeatedly demonstrated that certain actors hold a privileged position in framing disasters, presenting alternatives and alterations to existing beliefs and institutions, and in funding these initiatives. For most of the period under study in this book, the United States and its allies have held a preponderance of power and have demonstrated a capacity for revising existing beliefs and institutions, thus limiting the potential and scope for change presented by disasters. The present period, in which there appears to be a consequential redistribution of power combined with a predicted rise in the devastation caused by disasters, has the potential to usher in significant change. Separating the impact of changes in the distribution of power from ideological change in the belief systems legitimizing the current international order can be difficult and can ultimately overstate the degree to which the two developments are independent of each other. Yet the dominant approaches in international relations (IR) focus more sharply on the former development, arguing that changes in the distribution of power lead rising states to challenge existing hegemons and the international order/institutions they've created, often leading to violent conflict (Organski, 1958; Gilpin, 1983; Kennedy, 1987; Levy, 1987; Ikenberry, 2014).

From these accounts, one would expect that with the current redistribution of power, the norms, rules, and institutions that underpin the existing order, including those that address the environmental degradation and vulnerability to natural hazards, will face significant challenges and change. In this chapter, I assess how a changing distribution of power, with a primary focus on China, could potentially affect international disaster management, which has been heavily shaped by the United States. Although the extent to which the United States is declining and China rising remains debated (see Cox, 2012; Brooks and Wohlforth, 2016), recent developments indicate that we are in the midst of "the relative decline of American power, and the emergence of a risen China" (Layne, 2018:90). Of course, it is a distinct possibility that China will be unable to sustain its current pace of economic development, to maintain domestic political stability, or to close the military gap with the United States. Similarly, it is possible that the United States will be able to reverse its purported decline. Nonetheless, this chapter proceeds on the basis that the distribution of power has already shifted to an extent such that China is already in a position to assert alternative belief systems and institutions in the international order.

Drawing on China's approach to foreign aid, disaster management, and disaster migration, I argue that an increase in China's power and influence has led to modifications of key components of the existing disaster management regimes as China attempts to adjust to expectations of major powers and to address various contradictions in the existing order. In contrast to China's position in the early post–World War II period, when it sought to create an ideological schism and replace the existing system, China's position today is largely pro-systemic. As Krebs notes, China's approach to humanitarian aid at the present time "resembles those of 'traditional donors'" (2016:169). Consequently, China's rise on its own is unlikely to be highly disruptive of the existing order, nor will it usher in dramatically new norms or institutions for dealing with disasters and human vulnerability to natural hazards. In short, changing distribution of power on its own does not produce substantial change in the foundations of international order. China has increasingly adopted existing norms, rules, and institutions concerning the management of disaster, introducing modest and incremental modifications that still maintain the ideological foundations of the existing international order. In many respects, and certainly as it pertains to disaster management, China is reinforcing the foundational norms of nation-state sovereignty, modernization, economic liberalism, and realist power competition, although it is challenging certain humanitarian beliefs and norms.

The reduced role for humanitarian beliefs and values in China's current approach to foreign aid reflects a number of factors. The first two identified by Hanna Krebs focuses on the unique cultural and historical features of China. She notes that humanitarianism has a long history in China but that there are significant differences between humanitarianism in traditional Chinese thought and its Western variation, with the primary difference concerning the centrality of the state (rather than civil society) to prevent human suffering (Krebs, 2016). The second cultural factor is China's suspicion of Western humanitarianism as imperialist that derives mostly from its communist period (Krebs, 2016). In addition to these, I suggest a third factor, namely, China's relatively recent rise to prominence in the current international order. As I outlined in Chapter 3, the major Western powers embraced and shaped international humanitarianism to solve some of the tensions and contradictions in the existing international order – in essence, to preserve the international order and their place in it. Until recently, China has not faced the same pressure to preserve the existing order and help repair the cracks in its foundations and thus had no need to embrace Western-style humanitarianism. As China becomes increasingly integrated into the international order and thus held partially responsible for its failings (to alleviate poverty, inequality, vulnerability to natural hazards), it will likely find humanitarianism as an increasingly useful framework and practice. Indeed, as I show in this chapter, China's humanitarian activities increasingly mirror those of the other great powers, modified by its cultural and historical legacies that emphasize state sovereignty and suspicion of charitable claims.

Of course, failure of existing institutions and belief systems manifest through recurrent disasters would not necessarily lead to change in the absence of other actors having the capacity to create and fund alternatives. As I have shown throughout the book, existing institutions for managing disasters have failed repeatedly but have persisted with minor modifications and through the creation of new institutions to address perceived shortcomings of the existing order. This is most clearly the case with the development of the disaster risk reduction regime as a response to the failure of the disaster relief regime to reduce human vulnerability to disasters. While the creation of new institutions enabled the persistence of the existing order, for much of the past 70 years, there was no viable alternative to the US-led international order. The Soviet Union and the non-aligned movement (NAM) offered partial alternatives, but both lacked the economic ability to develop long-term sustainable alternatives to the key institutions underpinning international order, such as the UN, the International

Monetary Fund (IMF), World Bank, and so on. During much of the Cold War period, China offered development assistance on terms that differed substantially from Western norms, but it was not until China reached a certain level of economic prosperity that it could present a strongly viable alternative to US-led institutions. As outlined in Chapter 7, during the Cold War, key members of NAM were essentially forced to work within the US-led economic order to facilitate their goal of economic modernization. The Soviet Union did not offer an alternative to the Bretton Woods system (Pollard, 1985:274) and was largely isolated from the post-war global economic order by a combination of its own decisions and US resistance (Pollard, 1985; Minkova, 2018).

China now has the economic capability to offer an alternative and presently engages in a "twin-track" approach of operating within existing institutions while simultaneously developing alternatives (Jacques, 2012). The accumulation of tensions and contradictions of the current international order combined with viable alternatives suggests a high probability of revisions to the international order; although China's acceptance of many of the dominant ideologies of order indicate that these revisions will be relatively minor, reflecting a slight reprioritization among modernist, realist, liberal, statist, and humanitarian values.

This chapter maintains that one potentially productive way to assess the impact that a rising power will have on international order is to examine its approach within a particular field of practice to assess (1) whether it prioritizes a different set of values and (2) whether the trend is toward normative divergence or convergence. It is important to note here that the fields of disaster aid, development, and disaster management are not static – they are dynamic and subject to feedback loops – so assessing normative convergence is inherently problematic. This is particularly the case with foreign aid, as Western states continue to alter these fields in response to critiques and perceived failures of the existing system and to rising challengers (Mawdsley, 2012:17, 34). For example, the 2005 Paris Declaration and Accra Agenda for Action set out new principles to enhance aid effectiveness, reduce poverty, and encourage good governance in an attempt to modify and thereby preserve the existing aid system by introducing a number of value shifts within the humanitarian framework. This shift was not just internally driven by aid donors but reflected the input from recipient states, as well as states such as China who straddle the donor/recipient identity. The shift will show that norm convergence in the field of foreign aid and development is not a one-way process of developing and non–Development Assistance Committee (DAC) states adopting fixed Western norms, but of Western states also adopting ideas and norms from these actors. Although disaster management is but one component of the international order, understanding how a change in power distribution has affected this area can provide some telling insights into the more expansive consequences of a changing distribution of power on international order as well as the potential impact that disasters will have on international order.

China and the Existing International Order

There is currently much anxiety over the implications of China's perceived rise. Graham Allison's recent book contends, for instance, that war between a rising China and the United States is almost certain as they vie for hegemony (Allison, 2017). While Allison likely overstates both China's desire for global hegemony and the inevitability of conflict between the two states, increased Chinese economic and military power could signal the return of a bi- or multipolar system – although as Brooks and Wohlforth (2016) argue, that is unlikely to occur in the near future. Proceeding on the basis that this shift is occurring although not imminent, it is crucial to ask the following questions: will a change from US unipolarity to US/China bipolarity or multipolarity be likely to produce changes in the values and goals pursued by the units in the system?

On this question, much of the current discussion boils down to whether China is a revisionist or status quo state. Complicating such an assessment is that the question itself can be problematic: these terms are overly simplistic and inadequately theorized in the IR literature (Johnston, 2003:8), and the foundational norms of the international system are themselves contradictory and heterogenous. In addition, China's (and the United States') orientation toward the system is not fixed, and there is a dearth of information from China that makes it difficult to discern its intentions (Johnston, 2003). Nonetheless, it is still useful to assess the potential impact that the rise of China may have on international order and the management of disaster. For instance, one might ask, is China likely to alter the fundamental values and norms underpinning the disaster relief, risk reduction, and migration regimes or to discard the regimes completely? Will China introduce changes that alter whether and how states engage in disaster diplomacy? For those concerned that current practices reinforce an international order that produces human vulnerability to natural and human-made hazards, is there hope that a new distribution of power will transform the system and produce meaningful action on climate change and disaster management? The analysis in this chapter suggests that the answer to the preceding three questions is: "highly unlikely".

Returning to the question of whether China will pursue a radical reorientation of international order, IR scholars do not agree. A number of scholars expect China's rise to fundamentally reorder the international realm (see Friedberg, 2005; Kurlantzick, 2007; Jacques, 2012). Jacques, for instance, claims that "China will transform the world far more fundamentally than any other new global power in the past two centuries" (2012:579). He envisions changes in the dominant actors (civilization-states rather than nation-states) on norms of domestic political authority, family structure, racial hierarchy, and international relations. Kurlantzick depicts China's rise as a potential future threat to "US political, security, and economic interests" and undermining democratization globally (2006:40). There is widespread concern that China flouts existing principles of international order, most clearly in its state-led market economy that had led to order-challenging policies such as preferential lending and subsidy practices, violations of intellectual property rights, and currency manipulation (Staiger and Sykes, 2010; also Nolke et al., 2014). Writing specifically about China's approach to foreign aid, Hong notes that Western states and scholars often portray China as "an irresponsible" great power for undermining the existing "development paradigms" (2012:162).

On the other hand, there is a growing body of work showing that China, especially in recent years, acts more and more like a status quo power than a revisionist one. China has become much more integrated into the multilateral international order and has made conciliatory efforts to reassure its neighbours that it does not seek significant changes (Christensen, 2006:119; Glosny, 2006). For Christensen, China's membership in the World Trade Organization (WTO) and other regional economic entities signals a commitment to the underlying existing economic foundations of the international order. Glosny notes that China has made significant efforts to assuage concerns among its neighbours in Southeast Asia by committing to basic principles of sovereign equality and non-interference (2006:36). Alistair Johnston observes that China is "more integrated and more cooperative with international institutions than ever before" (2003:49). Breslin concludes that China has been most successful in achieving its goals by "working within existing frameworks and norms" and in some cases has been more liberal than other regional and global actors in the promotion of "regional free trade agreements" (2009:818). China has embraced the importance of multilateral institutions and developing soft power; in pursuing these objectives, China has invested in numerous order-stabilizing initiatives such as arms control agreements, trade deals, and foreign aid projects (Murray, 2012:58). In short, as a rising power, China does not behave as a state that intends to alter the ideological foundations of the existing international order.

To the extent that China does advocate change, its advocacy is (at present) fairly modest. Three foreign policy objectives in particular seem to animate China's government: "reintegration of Taiwan, rectification of disputed borders, and control of the South China Sea" (Friedberg, 2005:27). While such goals may be troubling to neighbouring states, depending on how these goals are pursued and met (diplomacy or military might), they do not represent a fundamental revision of the international order, unless China undermines the freedom of navigation that underpins global trade. Glosny suggests that, while China may have these tangible goals, its primary concern is maintaining regime security and economic modernization. To achieve these goals, Chinese political leaders have made great efforts to "improve China's image to reduce fears" of its rise (Glosny, 2006). Breslin stresses that China needed to "build its reputation as a good global citizen and regional neighbor in an effort to remove the distrust and sense of insecurity among China's neighbors" (2009:822). According to Breslin, China does not threaten the United States by offering an alternative international order, but by abiding by and playing "the rules Westerners themselves have formulated" and "beating them at their own game" (2009:831).

For liberal internationalists, China supports the status quo because it has risen and prospered under the existing order and need not alter that order to "fully thrive within this system" (Ikenberry, 2008:24). According to Ikenberry, the current order differs markedly from previous historical periods in that this order has "a remarkable capacity to accommodate rising powers" (Ikenberry, 2008:31). Thus, traditional theories regarding changing distribution of power no longer hold, and the best strategies for the United States and its allies are to further entrench existing rules and institutions and encourage China to work within them. Zheng Bijian argues that China's political elites embraced economic liberalization as the only way to industrialize, reduce poverty, and gain access to much needed resources, and the success of this approach has cemented China's commitment to economic openness, with some modest innovations (2005:20). Many liberal internationalists are similarly optimistic regarding China's rise due to their belief in the pacifying effect of economic interdependence and the notable increase of Chinese economic interdependence with the United States and other states in the region (Friedberg, 2005:13). Reaching a similar conclusion from a critical perspective, Matthew Stephen argues that the current structure of global capitalism encourages rising powers, such as China, to rely on "existing global institutions, forcing them to collaborate with established powers" rather than to expand territorially at the expense of other powers (2014:914).

In essence, the mode and operation of power differs today from previous eras, indicating that the structural incentives that encouraged rising powers to challenge existing institutions no longer exist. Stephen differs from Ikenberry, however, in the implications of China's rise within the liberal/capitalist system. Stephen suggests that, while the structure of global capitalism has forced rising states to integrate in the global economy and thus to accept "an increasingly dense regulatory framework", the form of capitalism developed in countries such as China is not politically liberal and challenges key liberal aspects of the international order, such as subordination of state to the market (neo-liberalism), democracy, universal human rights, and humanitarian intervention (2014:930). Nolke et al. also show that China's state-led capitalism "is incompatible with international economic institutions" designed to facilitate liberal market capitalism (2014:561).

China's ensconced position in the global capitalist/economic liberal system suggests that it is unlikely to challenge economic liberalism as a foundational international ideology of order, although it may challenge associated politically liberal norms. The problem with exclusively economic accounts is that they rely on reductive conceptions of international order. For Stephen, international order is understood as a capitalist order; which he expects will "lead to abandoning economic nationalism and autarky" (2014:923). In such a global capitalist system, Stephen explains the "continuation of geopolitical rivalries in the form of particular flashpoints among the major powers" as "isolated hangovers from the process of exclusive territorial state formation" (2014:923). The insinuation is that global capitalism is the single determining aspect of international order and that these other idiosyncratic behaviours within the international system (nationalism, geopolitical rivalry) will diminish with the spread of neo-liberal capitalism.

On the contrary, I suggest that it is not clear that neo-liberalism spread as far as is often suggested; indeed, it already appears to be on the wane as China offers a viable alternative to it. Many of these behaviours that Stephen suggests will fade away reflect deep-rooted ideologies of order strongly institutionalized in the international order. While these ideologies of order exist in tension with aspects of economic liberalism, they are not hangovers from a previous order but are constitutive of the current order and are of increasing importance in part due to China's influence. In addition, these beliefs influence policy makers to adopt policies that undermine their economic interests as defined in a capitalist/economically liberal system. As I have demonstrated throughout this book, these belief systems help sustain global capitalism/economic liberalism by allowing actors to frame global developments in terms of these beliefs rather than as failures of economic liberalism.

Ikenberry's conception of international order is similarly problematic. He suggests that the liberal order is based on "rules and norms of nondiscrimination and market openness", "coalition based character of its leadership", and "an unusually dense, encompassing, and broadly endorsed system of rules" (2008:30). These unique aspects of the liberal order allow for rising powers to succeed and achieve their goals without needing to revise the existing order. While certain aspects of this characterization ring true, the international order is far more complex and far less benign: discrimination is common and entrenched in certain belief systems – most clearly in nation-statism and realism but also in the institutions of the liberal economic order and in the standards of civilization that emerge through the universalizing ideologies of modernism and humanitarianism. Markets are not completely open and remain closed or restricted in key areas that continue to privilege developed states, and the current "coalition of leadership" excludes vast swathes of the global population and different perspectives on the proper relationship between the market, state, and society.

The ideological foundations of this order are under serious strain, and while it may be able to accommodate the ambitions of rising powers, the existing order is unlikely to continue unchanged in an era of increased exposure to disasters. To support this contention, I turn to China's approach to foreign aid and development. I start with this area rather than disaster management because China has prioritized this area (more so than disaster management) and there is sufficiently close overlap between practices of foreign aid and disaster aid to identify significant trends that will likely shape the field of disaster management.

Foreign Aid and Economic Development

Since it established foreign aid programs in the 1950s, China's approach to foreign aid has undergone a number of important shifts, slowly changing from an approach openly hostile to both the US-led international order and the Soviet alternative, to its current approach that accepts many of the foundations of the existing aid regime (Hong, 2012; Hirono, 2013). Originally devoted to supporting anti-colonial movements and promoting Chinese socialism, over time China's aid programs have shifted toward a market-oriented form of economic development, emphasizing mutual benefits, equality, and non-conditionality (Mahmoud, 2010:209), although it is important to note that this non-conditionality applies to domestic governance – China's aid programs are frequently conditional on acceptance of the one-China policy and on support for China at the UN (Krebs, 2016:175). Nonetheless, the growing body of work on China's current approach to foreign aid and development suggests that it does offer a distinct approach to aid and development that differs from the practices and principles as established by Western states; however, the principles behind China's aid do not radically alter the foreign aid paradigm. Although much of the literature highlighting the divergence between Chinese aid and that provided by the Organisation for Economic Co-operation and Development (OECD) Development Assistance Committee (DAC) countries may overstate the extent of these differences, the existing evidence does indicate that China has a slightly different set of priorities informing its provision of aid than do most Western donors (see Dreher et al., 2018).

Commonly (and erroneously) referred to as "emerging donors", states such as China, India, and Saudi Arabia have a long history of providing foreign aid that predate the official development aid (ODA) programs of most DAC states (Manning, 2006:372; Woods, 2008:1206). These non-DAC states generally do not conform to the principles established by DAC donors and, despite years of experience in providing foreign aid, have largely been excluded from consultation and international meetings to discuss guidelines and best practices on foreign aid (Manning, 2006:379). While DAC has attempted in recent years to engage with non-DAC donors and incorporate them into the DAC aid regime, there continues to be a general feeling that the Western powers that support the existing order exclude non-Western states from key institutions and negotiations around foreign aid and development. A telling pronouncement related by Mawdsley occurred when the signatories to the 2003 High Level Forum on Aid Effectiveness in Rome, which did not include non-DAC donors, declared that they have achieved "broad agreement" among the "international development community" (2012:9). While subsequent meetings have not been as narrow in their view of the international development community, there is still some concern that Western states continue to view the DAC states, Western nongovernmental organizations (NGOs), and inter-governmental organizations (IGOs) as "the international development community" despite the increased prominence of non-DAC donors in development initiatives. Leadership of the prominent development assistance organizations, such as the IMF, World Bank, and Asian Development Bank (ADB) are also unrepresentative, with decision making in those bodies controlled by the United States, Europe, and Japan (Walz and Ramachandran, 2011:22; Yu, 2017:360).

In addition to increasing its participation in these organizations and pushing for internal reform, China has also created alternative institutions, such as the Asian Infrastructure Investment Bank (AIIB), which gives greater voice and influence to "regional countries and the largest shareholder, China" (Ogawa, 2018:80). This initiative was a significant shock to the system – the United States vehemently opposed the AIIB and unsuccessfully pressured its allies to reject the new bank. The inability of the United States to prevent the establishment of the AIIB and a "parallel – shadow – international order outside the framework of the pax Americana" is clearly indicative of China's new leading role in the international order and its willingness to create alternative institutions to achieve that goal (Ogawa, 2018:100). It is notable, however, that the AIIB largely reinforces and strengthens the existing model of foreign aid and economic development along economically liberal lines.

As a result of these and other initiatives, China is now one of the leading development donors. According to Cheng and Smith, China has provided aid to over 120 countries and is now the sixth largest donor (2016:53). Murray notes that China, through its Import-Export Bank, has surpassed the World Bank as the "largest source of loans to Africa" (2012:55). In the East Asian region, China has surpassed the United States in bilateral aid; Chinese aid to the Philippines is four times that of the United States, and twice as large as US aid to Indonesia (Jacques, 2012:359). Even in Latin America, long considered the United States' sphere of influence, China has become a major source of development funding. China's Belt and Road Initiative (BRI), launched in 2013, is an economic development initiative that will invest billions in hard infrastructure projects linking Asia, Africa, and Europe, cementing the influence of China in the region if successfully implemented (Yu, 2017). For many, BRI is theoretically equivalent to (and empirically larger than) the Marshall Plan that cemented the United States' hegemony in the post–World War II period (Huang, 2016) and has the potential to extend Chinese influence into the "Eurasian heartland" (Layne, 2018:101).

China's growing presence in foreign aid and development assistance on its own does not fundamentally challenge the basic principles of international order. In earlier periods, China did use foreign aid to explicitly challenge capitalism, Western dominance, and Soviet influence (Hong, 2012). Today, China's development assistance and foreign aid policies are economically liberal – rationalizing the liberalization of economies along the modern day "silk road" and promoting free trade, economic development, and increased security as mutually beneficial and proceeding hand in hand (Blanchard and Flint, 2017). Huang describes BRI as "injecting new thinking into international economic development policy", but thinking that is broadly consistent with economic liberal principles of connectivity, free trade, economic stability, and friendly cooperation (2016:320). Of course, alternative accounts of China's development initiatives such as BRI are less sanguine on the beneficial rhetoric used to justify the initiative. Some scholars view it as a mechanism to export China's excess production capacity, which is symptomatic of what other capitalist states have done through trade agreements and development assistance (Huang, 2016:315; Rolland, 2017:130). Regardless of whether one views China's embrace of the market as positive or negative, it is clear that BRI bears a striking similarity to OECD development initiatives in reinforcing economically liberal principles.

As with Western states, the Chinese approach to foreign aid and economic development also contains substantial realist influences; in other words, it believes the primary value of foreign aid is to "enhance the power and interests of the donor in the power play between states" (Mawdsley, 2012:25). From this perspective, China's rising prominence in foreign aid and development assistance is strategic, contributes to Chinese security, and demonstrates how rising powers use economic strength to create regional and potentially global hegemony (Rolland, 2017:133). Melber describes China's aid and development initiatives in Africa as reflecting "appallingly blunt self-interest" (2010:214). This has been a longstanding feature of Chinese (and Western) foreign aid, particularly in Africa. Bartke notes that Mao's early emphasis on economic assistance to other developing states was to "demonstrate China's power as leader of the third world" (1989:7). The Bandung Conference and subsequent establishment of the Afro-Asian People's Solidarity Organization in 1955 led to an aid program for African states that sought to resist colonialism, enhance Chinese leadership, and reverse international recognition of Taiwan (Soderberg, 2010:123; Mawdsley, 2012:55).

Hamrin argues that, during the 1970s, China was strongly engaged in "power politics" and used its development aid program to encourage countries to side with China against the Soviet Union for leadership of the third world (1986:42). While several of China's foreign policy objectives have changed significantly over this period, the use of economic assistance and foreign aid to demonstrate and build power and leadership continues. Yu (2017) observes that BRI is clearly designed to increase Chinese access to resources and markets, expand Chinese influence, and address several contradictions within its domestic economy. China has used aid to develop close relations with India's neighbours to balance against India in South Asia (Jacques, 2012:442). Jacques notes that China played a significant role in ameliorating the Asian financial crisis and gaining support in the region through aid and unconditional concessionary loans (2012:350). According to Nordin and Weissmann, China currently employs a dual narrative of market-based economic growth and nationalism to promote increased Chinese leadership regionally and globally (2018:242).

China has adopted a similar approach in Africa and Latin America, using foreign aid and loans to increase Chinese influence and prosperity. In short, China's approach to foreign aid and development assistance (1) is not tied to a singular objective and set of beliefs – it combines economic liberalism, realism, modernism, nation-statism, and, to a lesser extent, humanitarianism; and (2) is not radically different than the approach of the DAC donors. As Mawdsley notes, the foreign aid policies of any given state reflect a variety of contextual and structural factors and cannot be accurately captured by any single paradigm (2012:29). In keeping with the argument developed in this book, I suggest it is because policy makers do not operate on a single belief system regarding the purpose of aid and international order. Drawing on the available evidence, China is increasingly mirroring the blending of liberal, statist, and realist beliefs that shape foreign aid programs of the other major powers. Nordin and Weissmann similarly find that China's combination of economic liberalism with renewed commitment to sovereignty and the realist objective of maximizing Chinese power "contribute[s] to rather than challenge the existing international order", particularly in the face of Trump's apparent retreat from that order (2018:243).

Similarities notwithstanding, there are some important differences in China's approach to foreign aid. From the outset, China viewed humanitarian assistance, and the conditions that accompanied it, as a tool of imperialist, capitalist states (Krebs, 2015). In this framework, Chinese aid was meant to limit the influence of capitalist states over the developing world (Krebs, 2015:11). This thinking has not completely disappeared from Chinese narratives on aid, as evidenced in its prioritization of mutual benefit and non-conditionality. China continues to view political conditionality as a feature of Western imperialism (Hirono, 2013:211), although it does insist on recognition of the one-China policy as a condition for Chinese development assistance. As Layne notes, China's approach to development assistance means that "norms favoring democracy promotion, humanitarian intervention, and human rights" are likely to recede in importance (Layne, 2018:108). This is already evident in China's development assistance, which emphasizes mutual benefit, non-interference and South-South cooperation (Dreher et al., 2018:191). China's ongoing representation of its economic assistance as mutually beneficial simultaneously assuages concerns that China is acting as an imperialist power that is the primary beneficiary of these investments and distinguishes its initiatives from Western aid and development programs. The prevailing depiction of Western development assistance is that it is humanitarian and charitable – a framing that suggests that the action is apolitical and that donors receive no benefit from the exchange. As noted in the earlier sections on humanitarianism, this representation then allows these states to impose conditions for the purported benefit of recipients. In contrast, China treats foreign aid and development aid as mutually beneficial and respectful of state sovereignty rather than in terms of humanitarian charity; consequently, it imposes fewer political conditions (Jacques, 2012:426; Ogawa, 2018:74).

The conception of development aid as charity has led DAC member states to create an extensive set of principles governing donorship and development assistance, such as those contained in the Paris Declaration of 2005, the Accra Agenda for Action, and the Global Partnership for Effective Development Co-operation (GPEDC). These initiatives establish a set of principles and practices that not only guide the actions of donor states but also in essence establish basic standards for partner countries. The Paris Declaration, for instance, details expectations of both donor and partner (recipient) states, while lending institutions often impose a number of conditions prior to approval. China is a signatory to the Paris Declaration, although as Melber notes, it is not clear if it signed as a donor or recipient and has yet to incorporate these principles into its aid programs (2010:235).

China's priorities when providing foreign aid differs from DAC on other key issues as well. Numerous commentators note that China invests heavily in infrastructure projects, which DAC donors are less willing to fund, and in agriculture, which Western donors have neglected (Mahmoud, 2010:201). Manning suggests that non-DAC states such as China use multilateral channels less often and are more likely to provide in-kind assistance than cash (2006:379). Dreher et al. suggest that non-DAC states focus less on recipient need than do DAC states (2018:1950). Walz and Ramachandran note that Brazil, Russia, India, China, and South Africa (BRICS) view recipient countries as business partners and tend to tie the majority of their aid to recipients purchasing goods, services, and labour from the donating state (2011:19). While China generally ties its aid to ensure recipient countries purchase Chinese goods and services, Woods (2008) observes that they rarely "apply direct conditionalities" or insist on international standards of sustainable development, environmental protection, poverty reduction, or gender equality, which is common for DAC donors. Similarly, Dreher et al. find that China's development assistance programs are strongly connected to foreign policy objectives, although they do stress that there is little evidence that China is gearing its aid programs toward resource extraction, nor are they directed disproportionately to authoritarian or corrupt states (2018:191).

On many key norms promoted by DAC for the provision of foreign aid, China offers an alternative and has articulated these principles explicitly: equality and mutual benefit, non-conditionality, reduced burden on recipients, recipient self-reliance, quick-result projects, inclusion of Chinese equipment, training for recipients, and non-special treatment of Chinese experts (Hong, 2012:138; Mawdsley, 2012:155). As with Western aid principles, China frequently does not meet these aspirational goals, but it does articulate an alternative set of principles that guide Chinese aid. On key values, China offers a real alternative to DAC aid: mutual benefit rather than charity, tying aid, prioritizing economic growth over poverty reduction, emphasizing non-interference without conditionality, and accepting (sometimes requiring) non-transparency (see Brautigam, 2011). All of these differences reflect China's strong commitment to state sovereignty and non-interference in the domestic affairs of other states, a persistent (though declining) suspicion of humanitarianism as imperialism, and the belief that developing states can follow the Chinese model of economic development.

While these are important differences, Chinese aid and DAC aid should not be viewed in terms of binary opposition. DAC donors actually vary a great deal in their aid programs, often exhibiting greater variability than between the major DAC and non-DAC donors (Mawdsley, 2012). To take one example, to reduce indebtedness of recipient states, DAC has sought to have grants comprise 86% of foreign aid (Manning, 2006), although there are reports of DAC aid ranging from 67.1% to over 95% in the form of grants (Manning, 2006:378). The proportion of China's aid provided as grants varies from just 25% (Ogawa, 2018:76) to 41% (Hirono, 2019:212). As a further caveat, however, Mahmoud suggests that China has been far more generous in terms of loan forgiveness and debt relief (2010:199). Brautigam (2011) suggests that China generally prefers providing loans coupled with debt relief to double the positive effects of the initiative: the first for providing the loan for development, the second for forgiving the debt. Once again, there may be double incentives at work: China seeking to maximize soft power gains from foreign aid while also attempting to reduce indebtedness of low-income countries.

Western donors also tie aid, charge interest on loans, insist on support for foreign policy goals, and often prioritize donor over recipient interests (Kapoor, 2008; Brautigam, 2011). Fleck and Kilby (2010) have shown that recipient need as the basis for US aid has been steadily declining for decades. Nor does China tie all of its aid; it provides grants and debt relief. Jacques notes that China has built schools and hospitals, and trained large numbers of professionals in Africa to assist with development initiatives (2012:414). Dreher et al. find that China targets its development assistance mostly toward poorer states, demonstrating that China bases its aid allocation, at least in part, on recipient need (2018:191). As part of its military modernization, China has developed a hospital ship that it can dispatch in cases of humanitarian emergencies and is increasingly participating in humanitarian peacekeeping missions (Murray, 2012:53). By most accounts, China is increasingly adopting some of the key humanitarian principles of DAC without officially endorsing DAC principles or agreeing to be bound by the DAC monitoring mechanisms.

In addition, China's strong emphasis on non-interference and non-conditionality is not ironclad. China has long required aid recipients or development partners to recognize the one-China policy. China is also finding its commitment to no-strings-attached aid and non-interference difficult to maintain in cases where domestic developments threaten the security of Chinese workers and investments (Wenping, 2010:158; Murray, 2012:61). In short, it is important not to overstate the differences between DAC and China's foreign aid programs and to recognize that, as China has become one of the world's leading donors, it has faced some of the same pushback as Western countries have, and it is adapting its policies in ways that more closely resemble DAC norms, albeit framed in terms that resonate with Chinese cultural norms. In addition, there may be aspects of Chinese foreign aid that DAC donors would do well to emulate, namely, the creation of longer term partnerships, centralized coordination and planning, investment in agriculture and infrastructure, and willingness to transfer technology (Mahmoud, 2010:210).

There is evidence of growing interconnections between China and DAC donors. In addition to participating in the Paris Declaration and the Accra Agenda, China and DAC have created the China-DAC study group. China participated in the GPEDC in 2011, which was created following the Fourth

High-Level Forum on Aid Effectiveness. To date, China has been participating as a development partner and has assessed 17 interventions through the monitoring process, totalling close to $900 million US (OECD, 2016). In some areas, China has not been a full participant, such as reporting on transparency and untied aid, nor is the AIIB listed as a partnering multilateral development bank (OECD, 2016). At the time of this writing, it is not clear to what extent China participated in the 2018 monitoring round, so this will bear further examination in the coming years. Increasing participation in the GPEDC may indicate a greater convergence between China and DAC norms on aid and development assistance. In its 2014 White Paper on Foreign Aid, China strongly indicated a commitment to international cooperation: prioritizing support for "development assistance programs of multilateral organizations" and "participation in international exchanges for development cooperation", and holding joint training programs through various organizations, such as the United Nations Development Programme (UNDP), UNICEF, Food and Agriculture Organization (FAO), United Nations Industrial Development Organization (UNIDO), United Nations Office for the Coordination of Humanitarian Affairs (UNOCHA), as well as the IMF and World Bank (PRC, 2014).

Ultimately, the alternatives that China offers to Western forms of foreign aid are relatively minor and offer only subtle variations on the foundational ideologies of international order. Both China and DAC donors are committed to science and technology, free trade, enhancing state power, and nation-state sovereignty, albeit with greater contestation over humanitarianism. Indeed, it would be difficult to find meaningful differences in terms of the modernist and economic liberal foundations of DAC and Chinese foreign aid programs. The primary differences then reflect a slightly different set of prioritizations among realist, nation-statist, and humanitarian principles, although this is fluid as well. A similar trajectory is evident in China's approach to disasters, which I take up in the next section.

China and Disaster Assistance

Although China has a relatively long history of providing foreign aid (development assistance in particular) in the post–World War II period, its involvement in international disaster relief and risk reduction is quite recent. It is only in the last two decades (since 2003) that China has strongly invested in its disaster management capacity, which mirrors the basic standards and institutions of other developed states. China's management of disaster is now highly modernized, centralized and coordinated by the state, acting in combination with a sizable Red Cross Society that specializes in post-disaster assistance. China's Red Cross Society is actually among the oldest in the world, established in 1904, and is among the few institutions to persist throughout the pre-, mid-, and post-revolutionary periods (Krebs, 2015), although it did face periods of near obsolescence (Reeves, 2014). China now even facilitates the (highly regulated) participation of non-governmental actors and private corporations in disaster recovery. While there are still uniquely Chinese features of its disaster management system that reflect the unique division of responsibility between the central state and the five different levels of government departments (Zhang et al., 2013), the Communist party and the state, and the various ministries and army, it does not differ significantly from emergency systems in other states. Hirono (2013) writes that, at present, China's domestic disaster response is now devoid of the anti-Western approach to humanitarian disaster relief characteristic of earlier periods.

China's development of a modern disaster management system was in response to a series of disasters – the 1976 earthquake, the 2003 severe acute respiratory syndrome (SARS) epidemic, and more recently the 2008 Sichuan earthquake. Each of these events profoundly altered China's domestic disaster response capability and encouraged the creation of mechanisms that closely resemble international standards. Increased focus on disaster risk reduction in the 1990s internationally and the 2004 Indian Ocean tsunami seemed to galvanize Chinese focus on disaster management and participation in disaster diplomacy and assistance.

Like the United States, China is a disaster-prone country; it experiences frequent earthquakes, droughts, floods, landslides, storms, and extreme temperature events that affect, on average, over 300 million people per year (Chen, 2012:130). In addition, as Suttmeier (2012) notes, China also experiences frequent technological disasters as it has rapidly modernized with insufficient infrastructure to ensure compliance with existing laws or to respond to technological disasters. According to the Centre for Research on the Epidemiology of Disasters (CRED) database, in the past decade, China has experienced over 314 disaster events, with over 103,000 lives lost (with the majority from the 2008 earthquake), and over 1 billion affected. In addition, disasters have caused hundreds of billions in damage (Suttmeier, 2012). While the 1976 earthquake had a profound impact on the Chinese state, the most notable disasters in recent Chinese history include the 2003 SARS crisis and the 2008 Sichuan earthquake, which claimed close to 88,000 lives and was a watershed moment in Chinese disaster response because it demonstrated that the national emergency management system (NEMS) put in place after the SARS crisis was not yet sufficiently developed to prevent, prepare for, and effectively respond to disasters. Not all regions have fully implemented the requirements of the NEMS, and the system is still not fully "professionalized". The demonstrated weakness of the Chinese state in the case of 2008 earthquake, despite its growing economic prosperity and international influence, led to renewed investment in emergency response capability.

Of course, any discussion of China's current disaster management systems must recognize that China has a long history of disaster awareness and management and of connecting disasters to government legitimacy. In the pre-modern period, disasters were often interpreted as signalling a loss of the "mandate of heaven" for political elites (Suttmeier, 2012:109). Consequently, China was among the earliest societies to attempt to understand, predict, and prevent disasters such as floods and earthquakes (see Chen, 2012). Flood control systems in China were first established around 2100 BCE and became a crucial tool of legitimation for imperial rulers (Hirono, 2013:207; Krebs, 2015:4). It is generally accepted that a seismograph-type device was first developed in China in the second century AD (Sleeswyk and Sivin, 1983). The Southern Song dynasty had well-developed institutions for dealing with disaster and famine as early the twelfth century (Chen, 2012), a practice that continued well into the twentieth century (Fuller, 2015). The combination of Western colonization and civil war essentially destroyed long-standing practices of both state- and community-led involvement in disaster relief (Fuller, 2015).

The communist revolution under Mao introduced substantial changes in disaster management. One of the lasting implications was Mao's introduction of the *hukou* (household registration) system, which restricted migration as an adaptive mechanism to disaster (Fuller, 2015:160). Mao's reforms also embraced a fully modernist account of disasters. For Mao, nature "was an enemy to be overcome, an adversary to be brought to heel" (Dikotter, 2011:174). Prior to this, Chinese society understood disasters as a type of divine retribution, under the concept of the "mandate of heaven". Under this belief system, disasters could signal that imperial rulers had lost the mandate of heaven and thus their political legitimacy. The modernist shift of the late nineteenth and early twentieth century upended the Confucian cosmology that conflated religious and political authority throughout China's history, ushering in, among other changes, a sovereign nation-state order (Phillips, 2011), which, as Fuller notes, disregarded the long history of community involvement in disaster relief. The modernist orientation that displaced the traditional belief system and its interpretation of disasters, combined with the heavy involvement of Western states in providing post-disaster assistance in the early twentieth century, cemented the idea that China needed to adopt some Western ideas and institutions pertaining to disaster management, provided they did not run counter to traditional Chinese thought (Krebs, 2014). Suttmeier summarizes China's modern approach to disaster management by describing it as an approach based on importing international norms and technologies; studying, altering, and deploying them; while preserving "values and traditions of its extant institutions" (2012:110).

Starting with the reforms under Mao, China placed great emphasis on "conquering nature", which included developing greater understanding of the natural causes of earthquakes. Understanding seismic activity was well advanced in China in the 1970s, and the ability to forecast earthquakes formed part of the ideological foundation for the Cultural Revolution (Geller et al., 1997). There are reports that both the 1975 and 1976 earthquakes were "predicted with enough certainty" that certain regions were able to take measures to reduce casualties (Adams, 1976; UN Global Programme, 1996; Chen, 2012:136). The Chinese state's historical involvement in disaster management has meant that more recent initiatives to reform disaster management in China has not required a radical rethinking of the appropriate role of the state or of disasters. Despite China's purported advances in earthquake science, however, the 1976 earthquake claimed between 350,000 and 650,000 lives and demonstrated the inadequacy of the state's disaster management capacity. Although some counties were thoroughly prepared and were spared major loss of life, the general view is that there was insufficient local capacity to prepare for and respond to the earthquake. In response, the Chinese government began to formulate a basic emergency response system.

The SARS outbreak in 2003 had a jolting effect on the Chinese government, resulting in a drastically overhauled emergency response system. The government implemented the Overall National Plan for Responses to Public Emergencies, a multi-hazard planning document meant to cover emergencies ranging from pandemics to earthquakes (Lu and Xue, 2016:415). According to Suttmeier, this initiative created four committees tasked with overseeing the response to distinct types of disaster under a single oversight committee (2012:113). As Lu and Xue note, this plan created a "new institutional structure" for dealing with disasters that also allocated responsibility based on the scale of the disaster, with local actors having primary responsibility and provincial/national governments responding if the disaster exceeded local capacity (2016:418). The 2003 reforms demonstrated the need for the Chinese state to overhaul its emergency management, although the overall impact of these reforms was mixed. Since 2003, the Chinese state has continued to alter its emergency response, with a trend toward greater incorporation of international norms of emergency response, including establishing the ability of the president to declare a state of emergency in cases of natural disaster to mobilize the full resources of the state (Lu and Xue, 2018). Lu and Xue point out that this reform process did not entail wholesale adoption of foreign practices; the dominant forms of emergency management internationally are consistent with disaster management practices established in much earlier periods of Chinese history.

The piecemeal reform of China's emergency response continued throughout the 2000s in response to both domestic challenges and engagement with international institutions. One of the shortcomings revealed in the 2003 SARS case was "low degree of professionalization", poor communication, and an absence of a centralized decision making body (Lu and Xue, 2016:424). To address these shortcomings, in 2005, the Chinese government created the National Committee for Disaster Reduction (NCDR) and developed "Regulations on the Participation of the PLA in Emergency Rescue and Disaster Relief" (Yu, 2012:141), which brought the military into emergency rescue. The 2005 regulations facilitated the rapid mobilization of the army to respond to natural disaster following the Sichuan earthquake (Murray, 2012:55), although the 2008 event revealed ongoing gaps in China's disaster response system – despite a relatively effective response after the earthquake.

The Chinese government implemented a raft of consequential changes after the 2008 earthquake. Lu and Xu report that "emergency search and rescue teams were established in most counties and cities" and government departments at all levels were required to develop emergency plans (2014:432). In addition, the Chinese government invested a great deal in geospatial technology to assist with rescue operations (Lu and Xu, 2014) and implemented a state-wide "color-coded emergency alarm system" as well as an emergency hotline. In addition, the government implemented regulations and statutes governing the use and distribution of emergency relief (Yu, 2012:141). In 2009, China passed the Emergency Response Law, specifying which entities had primary responsibility for emergency response. As is the case in most other states, primary responsibility was allocated to local authorities, with the national government supplementing local efforts in certain emergencies (Yu, 2012). More recently, in 2018, the Chinese government further enhanced its disaster response capability and followed the established practice of many Western states by placing all emergency response under a single ministry (Economist, 2018).

In their reform efforts, the Chinese government has drawn heavily on international resources. According to Suttmeier, throughout their various reforms, the Chinese government studied emergency management systems in a wide range of countries and participated in "international forums devoted to disaster prevention, mitigation and crisis management" (2012:118). The Chinese government turned to the ADB to evaluate the Chinese response to the 2008 earthquake and implemented a number of reforms from that report (Suttmeier, 2012:118). The Chinese government has also worked with various UN agencies to improve its emergency response capability (Suttmeier, 2012). Zhang et al. argue that international factors profoundly influenced the trajectory of China's disaster management system by refocusing on disaster prevention rather than simply increasing rescue capacity and by developing a "standard, systematic and legal" framework for disaster management (2013:2220). Zhang et al. also argue that the international emphasis on disaster risk reduction emphasized technological development and community capacity building, which the Chinese state has incorporated into its disaster management systems and into its negotiating position on further international disaster risk reduction frameworks. Among the disaster management norms that China has implemented as a result of greater international engagement are emphasis on prevention and preparation, increased professionalization and clear allocation of responsibility, openness to international aid, use of armed forces in disaster response, and heavy investment in technological solutions (Chen, 2012).

Other consequential developments to emerge during the establishment and development of emergency response include growing recognition of civil society actors and basic principles guiding the provision of disaster aid. One unexpected development from the 2008 earthquake was the rapid mobilization of civil society actors to assist with the rescue and recovery efforts. The Chinese government has proceeded cautiously in response, supporting civil society involvement and private actor (corporate) participation, while regulating these actors under the premise of ensuring that disaster assistance is provided to all, regardless of race and nationality

(Yu, 2012:140). The Chinese government has also implemented principles governing donations for disaster relief and preventing corruption. Under these guidelines, donations are to be voluntary; spent on disaster relief and rehabilitation; and distributed legally and in a standardized, non-discriminatory way (Yu, 2012:140). At the same time, the government has banned some civil society groups from participating in disaster assistance and works to ensure that donations are directed through officially approved channels, such as the Red Cross (Suttmeier, 2012:120).

In a wide range of areas, the Chinese state is adopting international norms around disaster management, from the centralization of state-led emergency response systems combined with enhanced local capacity to the professionalization of the disaster management system and the loosening of regulations permitting the involvement of civil society actors, private corporations, and international actors. The government has also begun to expand the use of disaster insurance, which is a staple of disaster management in developed states (Suttmeier, 2012). As China further develops its domestic disaster response systems, it does so in ways that more closely resemble and reproduce disaster response systems globally – in turn, further enhancing its ability to contribute its personnel and knowledge to international disaster operations. China's development of disaster and emergency response institutions suggests that the prevailing values are not significantly different than those that shape disaster and emergency response elsewhere.

The various initiatives implemented over the past two decades have had a significant impact. Comparing the 2008 and 2013 earthquakes in the Sichuan region, Lu and Xu found that the 2013 response was swifter and more effective, in large part due to changes implemented after the 2008 disaster (2014:425). At the same time, China's rapid program of economic development and modernization has increased the material capacity of the state to provide disaster preparation, rescue, and relief. China now has technologically advanced forces that can transport relief supplies and engage in sophisticated rescue and relief operations. China's development of an effective disaster response capability signals (1) a growing expectation on the part of citizens that the Chinese state will provide such public goods and (2) acceptance by the Chinese state that this is a basic functional expectation of sovereign states, particularly those that aspire to great power status. The consolidation of emergency response in 2018 under its own ministry signals, among other things, an increased prioritization of disaster response and increased institutional isomorphism with disaster rescue and relief agencies in other states.

As China has transformed its domestic disaster management capability, its response to international offers of aid has also transformed over time toward increased openness. Historically, China has not requested or accepted disaster aid from other countries – which is consistent with the view of humanitarianism as a form of Western imperialism. Mao essentially banned any relief by non-state agencies and refused all foreign assistance, even from other communist states (Krebs, 2014:13). Subsequently, two developments contributed to increased Chinese receptiveness to foreign offers of disaster aid: (1) a changing understanding of humanitarian aid that was less hostile to humanitarian assistance and Western actors, and (2) appreciation of participation in international humanitarian assistance as a basic expectation of a leading power – even as a recipient. As China started to open up in 1979, its attitude to foreign aid began to change (Soderberg, 2010:122). Beginning in 1980, the Chinese state began to accept some forms of assistance, mostly cash and basic necessities, after natural disasters (Hirono, 2013:214). Until 2008, however, the Chinese state did not permit the presence of foreign nationals in disaster response. Yu (2012) reports that the Sichuan earthquake of 2008 was the first instance of China accepting foreign aid. As might be expected in the absence of an established set of policies for accepting international offers of aid, the response in 2008 was not particularly efficient. That said, the government responded quickly to offers of aid and clarified (1) the types of aid it needed and (2) the criteria on which it would accept aid, which did not differ significantly from that prioritized by the United States: cash; basic life necessities; and specialized, highly technical assistance. Chinese authorities prioritized time efficiency and geographical proximity for search and rescue teams; consequently, search and rescue teams from Russia, South Korea, Singapore, and Japan were among the only foreign personnel permitted to the affected area. In response to other offers of aid, the Chinese government called for cash donations, food, and tents.

The most significant obstacle to the acceptance of foreign aid was logistical – the remoteness of the affected location made it difficult to get foreign aid to the region. The Chinese government struggled to get Chinese disaster responders to the region, let alone foreign search and rescue teams. China was hesitant to permit foreign military forces, particularly from certain states, to contribute. While Japan was the first country to provide disaster assistance in the form of search and rescue teams, Chinese officials refused to permit Japanese "self-defence force" planes to deliver aid (Yu, 2012:141).

According to Hirono, China's reversal in 2008 was due to Chinese efforts to demonstrate its status as a "responsible great power", especially prior to the 2008 Olympic Games (2013:214). In subsequent disasters, China has not requested international assistance, but there may be several reasons for this beyond a renewed hostility to foreign humanitarian assistance (Hirono, 2013). China has invested heavily in its domestic emergency response, so it is better equipped to cope with disasters. In addition, China has not experienced a disaster as devastating as the 2008 earthquake – the 2010 earthquake that killed close to 3,000 people did not garner the same level of international attention. While many countries offered condolences and offers of assistance, China did not seek international assistance. This trajectory should be understood not as a return to China's aforementioned hostility to foreign assistance but as China taking on the status of great power that can handle most domestic disasters without foreign assistance. The increased willingness on the part of China to accept international disaster relief signals engagement with the existing international order and the institutions that underpin that order. It also speaks to China's growing confidence as a regional and global power;

while China is increasing its own domestic capacity and therefore seldom requires international aid, its openness to foreign offers of aid signals a drastic shift from China's previous rejection of all humanitarian aid, which it characterized as imperialist. Based on these shifts, I would expect that, in the event of a future major disaster, China will respond as the United States did in the case of Hurricane Katrina: acceptance of some aid from close allies and development partners, as well as technologically sophisticated neighbours, combined with targeted aid refusals for specific states who oppose China's foreign policy goals.

China and the Provision of International Disaster Assistance

We see a similar trajectory in China with regard to the provision of international humanitarian aid. As noted, China has historically focused more heavily on development assistance rather than humanitarian assistance – although the Chinese state funded disaster relief activities through the Chinese Red Cross from its initial founding in the early 1900s. Reeves (2014) notes the Chinese Red Cross provided disaster relief to over 140 countries through the Cold War period (Reeves, 2014). On the whole, however, China has been a relatively minor player in disaster relief, with the DAC countries most active in this realm. In the early 2000s, China began to follow the lead of the DACs. Hirono argues that the legitimacy of the Chinese state is increasingly tied to its role as a major power, which includes participating in international humanitarian assistance (2013:27). In 2001, China created an international rescue team that is comprised of military personnel trained to participate in "international disaster relief and rescue" (Yu, 2012:153). This initial first step, which was used quite sparingly at first, with five deployments over 12 years, signalled an important shift in China's approach to disaster aid (Yu, 2012:153). In 2011, the Chinese government announced "a new stage" for China's foreign aid programs that would increase its focus on humanitarian assistance (Hirono, 2013:202). In recent years, China has invested a significant amount of resources in building its humanitarian response capability.

This shift is significant, given China's historical attitude toward humanitarian assistance. As both Hanna Krebs and Miro Hirano note, China's approach to humanitarianism is heavily shaped by its culture and historical legacy. China's government was slow to engage with international disaster relief, in part due to a desire to engage in areas of strength, such as economic development and infrastructure construction, as well as its perceived need to improve its own domestic disaster response (Krebs, 2014). Krebs also notes that, beginning in the 1990s, the opposition to humanitarianism as a feature of capitalist imperialism began to decline and that more recently China has adopted a strategy of multilateral engagement, which has involved participating in international disaster relief. China contributed after the Indian Ocean tsunami, after Cyclone Nargis in Myanmar, and after the Gorkah earthquake in Nepal (Krebs, 2015). China is also now a member of the International Search and Rescue Advisory

Group and sends medical teams with the World Health Organization (WHO) to assist in disaster-affected areas (Krebs, 2015). China has expanded its humanitarian capacity to include epidemic outbreaks, as in the case of the Ebola outbreak in 2015 (Krebs, 2016), and has sent its earthquake search and rescue corps abroad after several major earthquakes in Haiti, New Zealand, Japan, and Nepal (McElroy, 2017).

China's growing expertise in disaster relief, which has emerged from investing in domestic disaster management, combined with its appetite to be seen and behave as a major world power has altered historical Chinese aversion to international humanitarian assistance. The 2004 Indian Ocean tsunami seems to have initiated a change in China's approach to disaster relief. While Western states mobilized quickly in response to the 2004 Indian Ocean tsunami, China's own response – while sizable historically for China – dwarfed in comparison. Glosny (2006) notes that China provided a total of $153 million in aid, $83 million in governmental aid and $70 million in non-governmental aid, with a number of medical volunteers. Thirteen DAC countries offered more governmental aid than China, with regional competitors, the United States and Japan, offering significantly more (Telford and Cosgrove, 2007). The United States offered close to $1 billion in aid, in addition to a massive military disaster assistance response that included an "aircraft carrier, twenty five ships, 94 aircraft and numerous helicopters" (Glosny, 2006:50). Japan offered over $600 million in governmental aid, seven times that offered by China. The United States' response clearly demonstrated the military and economic superiority of the United States and the effectiveness of that aid in influencing states in the region. Christensen notes that the massive US response to the tsunami significantly improved perceptions of the US in the region and that China recognized the value of disaster assistance in enhancing its soft power (2006:114).

Since then, China has invested in international disaster assistance, made possible in part by its massive investment in its military capabilities. Murray notes that, since 2008, the Chinese state has increased its levels of humanitarian assistance, created a United States Agency for International Development (USAID)–type bureaucracy, established a "Chinese Peace Corps", and established a health diplomacy program that saw 15,000 Chinese doctors dispatched to 47 African states (Murray, 2012:54). In its most recent package of administrative reforms, the Chinese government "expanded the responsibility of its new aid agency and culture ministry" as part of its efforts to expand Chinese soft power (Economist, 2018). This shift is noticeable outside the field of disaster relief as well – China has also become a significant contributor to UN peacekeeping missions, providing more troops than any other permanent five (P5) state (Krebs, 2016:179).

The contradictions, however, between realist and humanitarian values concerning foreign aid continue to play out in the case of China. Following the Typhoon Haiyan disaster in 2013 that killed over 7,000 people in the Philippines, affected 16% of the population, and caused over $10 billion in damage (CRED, 2015), China pledged a rather small aid package – initially just $100,000 in disaster aid. In comparison, the United States offered $20 million in relief plus a wide range of experts, personnel, equipment, and transportation for relief operations. Japan gave $10 million and a search and rescue team, and even Taiwan provided $200,000 in aid. Association of Southeast Asian Nations (ASEAN) members such as Singapore and Indonesia also provided $200,000 and $2 million, respectively. Many observers suggested China had offered such a meagre aid package in response to ongoing disagreements between the two countries regarding disputed islands in the South China Sea, specifically the Philippine decision to take its dispute with China to the UN Arbitral Tribunal (Baltrusaitis, 2013). By using disaster assistance to send a message over a geopolitical dispute, China violated one of the behavioural expectations of great powers engaging in humanitarian assistance. Not surprisingly, this elicited criticism internationally; perhaps more surprising was the domestic opposition to the paucity of the Chinese offer (Krebs, 2016:175). In the end, the Chinese government shifted toward the normative expectation, increasing its aid to $2 million and providing a disaster relief team, emergency medical teams, and one of its hospital ships (Associated Press, 2013).

This episode is highly illustrative of both an identity shift and a normative shift in China. The identity shift reflects China's new status as a major regional power – both neighbouring states and key segments of the domestic population expected China to act in line with the behavioural expectations of major powers by providing a robust aid package commensurate with its economic ability. In this instance, China increased its offer of aid, and it appears that China is becoming more involved in providing humanitarian disaster assistance, at least within the region. In 2015, China offered $500 million in disaster relief to Nepal following a devastating earthquake. In 2017, the Chinese government partnered with the UN to provide $12 million in disaster assistance to Nepal, Bangladesh, and Pakistan (UNDP, 2017). That same year, China worked with the UNDP to provide $5 million for countries affected by hurricanes Irma and Maria (UNDP, 2017). Despite this shift, in most disasters, China is not the dominant disaster aid donor, even in the Asia-Pacific region. Its offer of $300,000 to Taiwan after the 2016 earthquake was dwarfed by Japan's pledge of $1 million (Wilson, 2016). Even China's massive relief effort for Nepal in 2015 totalling $500 million was only half of India's $1 billion aid package (BBC, 2015). While the general trend for China is toward normative convergence for great powers in terms of participating in international disaster relief, at present it seems that China faces a number of constraints on becoming a global leader in disaster relief: longstanding opposition to "charitable" humanitarian assistance, concern about domestic opposition to generous international aid packages, and a preference to invest its limited resources in mutually beneficial development aid.

The second shift concerns the growing import of humanitarian values relative to realist considerations. The current social norm is for great powers to provide humanitarian assistance after disasters regardless of the state of bilateral relations between the countries. As demonstrated in Chapter 6, the United States consistently provides disaster assistance, even to states openly hostile toward US foreign policy, such as Iran and Cuba. The 2013 Philippine incident demonstrated the tendency of the Chinese government to tie disaster aid to geopolitical concerns – as well as a growing recognition that such overt realist policies violated international and even domestic norms. This trend is evident in China's treatment of Taiwan as well. China blocked or delayed Taiwan's access to international assistance following the 1999 earthquake and the 2003 SARS epidemic to ensure international recognition of the one-China policy (Wilson, 2016). More recently, however, this appears to be changing. In response to the 2016 earthquake, China offered aid and did not obstruct the UN or other countries from providing aid to Taiwan (Wilson, 2016). This shift could reflect improved relations with Taiwan or a new strategy to achieve its foreign policy goals concerning Taiwan; perhaps China's leadership is prioritizing humanitarian norms of relieving suffering regardless of foreign policy objectives. China's increased involvement in disaster relief, increasing cooperation with multilateral institutions in the provision of this relief, and less overt use of disaster aid to achieve strategic objectives suggest normative convergence with existing disaster relief norms. At the very least, they suggest that concerns that a rising China will completely undermine humanitarian norms are exaggerated. At the same time, it is clear that China does not view "Western-style" humanitarian aid as a foundational norm of international order. Such a conclusion is further supported by China's approach to the management of humanitarian migration.

Managing Disaster Migration

In recent decades, China has been a source country for refugees and migrants, as well as a destination country for migrants from other countries in the region. In addition, China has experienced large-scale internal migration and spends much of its energy managing internal movements. Despite this (or perhaps because of) extensive experience with refugees, China has not played a significant role in the international management of migration and demonstrates limited normative compliance with international norms on forced migration.

Although China was an early signatory to the Refugee Convention; has been a member of the United Nations High Commissioner for Refugees (UNHCR) Executive Committee since 1958; and has taken in a large number of humanitarian migrants from the region, namely, Vietnam, Myanmar, and North Korea, its orientation toward refugee protection is ambiguous, at best. China does not have a resettlement program and officially resettles very few refugees annually. While China houses a large number of North Korean refugees, it does not permit the UNHCR or other international non-governmental organizations (INGOs) access to these refugees, has a repatriation agreement with North Korea, and often refoules large numbers back to North Korea (Margesson et al., 2007:3). China is still a significant refugee-producing state, and Chinese authorities often ask other states to refoule Chinese refugees and resists efforts by UNHCR to ensure protection for refugees in China (Song, 2018). Given the Chinese emphasis on non-interference and state sovereignty, it often views UNHCR as a threat to these principles (Song, 2018). Historically, China has provided little funding to UNHCR or the International Organization for Migration (IOM), signalling a relatively low interest in international migration management and humanitarian protection. According to Peterson, the treatment of asylum seekers in China at the current time is understood almost entirely in terms of realpolitik and strategic calculations rather than adherence to international law (2018:287).

There are indications of change, however; as with other international initiatives, China has become more involved in refugee protection. In 2012, the Chinese government, for the first time, included "an article on the treatment of refugees" in its "Law on Exit-Entry Administration" (Song, 2018:160). Starting in 2011, China began to increase its contribution to UNHCR and more recently has been at the forefront of trying to solve the Rohingya refugee crisis (Song, 2018). UNHCR reports that China's contribution to UNHCR increased from $2.5 million in 2016 to over $11 million in 2017 (plus $8 million in private Chinese donors), placing China in the top 20 donors for the first time. By way of comparison, China's 2017 contribution pales in comparison to the United States' contribution of $1.5 billion (UNHCR, 2018), evidence that China is either (1) not overly committed to refugee protection and/or (2) still not comparable to the United States in terms using economic clout as soft power. In 2018, China has only committed just over $1 million to UNHCR, suggesting that 2017 may have been a temporary blip in China's contribution.

In addition to reflecting China's insistence on non-interference and suspicion of "humanitarian" interventions, China's hesitant engagement with the refugee protection regime reflects a different understanding of how to address migration. According to Tan (2017), China sees migration as fundamentally driven by economic incentives, and the solution rooted in economic development rather than humanitarian Initiatives and resettlement. Even in the context of North Korean refugees, China views them as economic migrants rather than politically persecuted (Margesson et al., 2007:10). Rather than highlighting resettlement or UNHCR contributions, the Chinese government, in cooperation with UNHCR, has indicated that it sees the development initiatives under BRI as a way to address root causes of refugee displacement and to support developing states that host large numbers of refugees (Tan, 2017).

Not surprisingly then, China has not played a significant role in disaster migration initiatives, such as the Nansen initiative and the Protection Agenda, nor in the Global Compact on Refugees and Migration. On the whole, the evidence suggests that China is not particularly committed to the existing mechanisms of refugee protection or to expanding those protections to environmental- or disaster-induced migration. Of course, the United States has become a laggard in these areas as well, reversing course on the Global Compact and drastically reducing its refugee intake from over 108,00 in 2016 to under 8,500 in 2018 (UNHCR, 2018). It has also slashed its support for UNHCR, from $1.4 billion in 2017 to just over $350 million in 2018 (UNHCR, 2018). Given the Trump administration's decision to undermine the global refugee regime, this would present China an opportunity to increase its soft power and take leadership on an issue that will be of increasing importance. At present, however, it appears that neither China nor the United States are taking global leadership on managing humanitarian migration – prioritizing instead state sovereignty and restriction of migration.

As in other areas, China brings a unique perspective on handling refugee movements that prioritizes state sovereignty and economic development over the humanitarian impulse to offer protection. In negotiations over the Global Compact, the Chinese ambassador to the UN stressed the need to "protect national migration systems and policies" rather than establishing "one size fits all" solutions (Zhaoxu, 2016). In addition, China places more emphasis on addressing root causes, such as conflict and economic development, rather than temporary solutions, with enhanced responsibility for developed states to bring this about (Zhaoxu, 2016; Song, 2018). In the discussion of the Rohingya crisis at the Security Council, the Chinese representative stressed the need to address root causes through "dialogue, increasing development aid… and promoting ongoing development" (Wu, 2017). Wu also stressed that the international community should "avoid interference in internal affairs of the countries in question". The emphasis on root causes fits with China's increased involvement in peacekeeping to prevent renewed hostilities and its massive investment in "mutually beneficial" economic development. Consequently, it seems unlikely that China will become a major resettlement state for displaced persons or that it will sign agreements that impose universal policies on all states, even as it becomes more involved in international institutions that govern migration and refugee movements. This is further supported by China's work with the IOM and European Union (EU) to enhance its border management to reduce irregular migration. As with many other regions of the world, China is engaging with the international migration regime, not to enhance the rights of migrants and refugees but to ensure regular and orderly migration to and within China.

Conclusion

The overall impact of China's approach to foreign aid and disaster management, and its strength as an alternative to Western-led institutions, is indicative of a potentially meaningful shift in international order – but one that primarily reduces the relative weight of what I have termed humanitarian values as a fundamental ideology of international order. There are indications that developing states see the Chinese model as preferable because it is not conditional on domestic political reforms and because China facilitates the rapid implementation of large-scale infrastructure projects (Wenping, 2010). Mamoud notes that most governmental actors, policy makers, and workers in Africa continue to view Chinese investment as a positive (2010:198). The rise of such an alternative may force Western donors to alter their foreign aid and investment programs, presumably by imposing fewer conditions and protections for those negatively affected by these initiatives. Should DAC donors pursue such a course, it would signal a notable change in the international order concerning both the status of humanitarian beliefs as well as the states establishing dominant norms of interaction.

The ultimate success of China's alternative view of international order will depend, however, on whether expectations match results. As noted earlier, one reason why developing states see China as a viable option for development aid is the failure of existing development initiatives to produce promised outcomes: namely, increased prosperity, political stability, rapid modernization, enhanced international standing, and reduced vulnerability to natural hazards. As China becomes an increasingly important actor in development aid and disaster assistance, it too will face scrutiny over the impact of its foreign policy initiatives.

Although it is far too early to assess the impact of China's alternative approach to foreign aid and its impact on international order, there are early indications of internal tensions and contradictions. Murray and others note that China's claim to mutually beneficial development aid does not always match the reality on the ground, with many recipients of Chinese development assistance claiming that it is "exploitative and damaging" (Murray, 2012:50). Many African recipients of development assistance and investment/aid from China are concerned that many Chinese firms employ primarily Chinese workers, thus limiting the contribution to the local economy, and exhibit discriminatory attitudes toward non-Chinese (Jacques, 2012:429–30). In some instances, Chinese officials attempting to promote Chinese initiatives have been met by protestors accusing the Chinese of "being worse than the old colonial and imperialist powers" (Melber, 2010:227). Busse et al. note that the record of African countries receiving aid and investment from China is mixed: increased imports of Chinese goods may have a negative effect on African countries because they may displace local production and block access to the global market for certain goods (2016:255). In the absence of strengthened governance in African states, with which China is not particularly concerned, Busse et al. suggest that African states may be negatively affected by engaging with China (2016:256). Despite a number of successes, Mahmoud observes that China has also financed a number of "disastrous and embarrassing failures" similar to those of Western states (2010:186). Mahmoud writes that

China's aid and development initiatives have produced few "traces of human development", brought an influx of Chinese workers that have displaced African workers, and have weakened regional African blocs (2010:199), a similar effect to the bilateral aid and trade deals of the United States and EU (Melber, 2010:217). For Melber, China's aid and trade policies are not actually a meaningful alternative to those of the West, as China behaves like all "trade-driven industrial powers integrated into the world system" (2010:226). In short, it is already becoming difficult to sustain the claim that China's aid initiatives are as mutually beneficial as claimed or that they differ significantly from those from the West in terms of the values and goals they pursue.

Ultimately, China's vision of international order does not differ dramatically from that of the existing order supported (until recently) by the United States. China promotes modernist, realist, nation-statist, and economically liberal goals and values, shorn of the commitment to humanitarian protection that ameliorates some of the most problematic tensions that arise from the ideological blend that defines the existing international order. China's reluctance to embrace humanitarian values more fully reflects its longstanding commitment to state sovereignty and opposition to Western forms of humanitarianism that legitimize interference and perpetuate dependence. And yet one of the key takeaways from this chapter is that the trend in China has been toward normative convergence – China's current approach to development, foreign aid, international disaster assistance, and even migration is starkly different than it was just two decades ago and more strongly resembles the approach of the United States and other Western states. As China becomes wealthier and increasingly invested in the existing international order, it may find that greater investment in Western-style humanitarianism provides significant order-sustaining benefits. While we should not expect China to completely abandon its commitment to mutually beneficial development, non-interference, and state sovereignty, this analysis does suggest that, provided China continues to experience economic growth and political stability, we should expect China to expand its humanitarian programs to ameliorate the negative effects of sustaining the current international order.

In this context, the greater threat to the existing international order may actually arise from developments within Western states: declining trust in scientific findings, a rising populism that supports economic protectionism, renewed nationalism, unabashed prioritization of the national interest, and a declining commitment to humanitarian values. Many of these populist and nationalist movements are responding to the perceived failures of the existing international order to deliver on the various contradictory promises inherent to their ideological foundations. The rapid pace with which the Trump administration has been undermining modernist, economically liberal, and humanitarian ideologies of international order suggests that those concerned with preserving the existing international order may have been unduly occupied with the rise of

China. These accounts are insufficiently attentive to the contradictory nature of the existing international order and the failure of that order to deliver on its promises, as well as the degree of ideological division within Western states. Paradoxically, the rise of China may be one development that helps preserve key components of the existing international order.

From the perspective of disaster management and the impending impact of global climate change, neither development inspires confidence. China's commitment to rapid and resource-driven economic development on a global scale, combined with its vociferous commitment to nation-state sovereignty and its ability to build and project power, suggests that China is not promoting an alternative vision of international order with human vulnerability to natural hazards as a priority. This is further evidenced by China's positions during negotiations on global climate change agreements. China promotes a differentiated solution that sees developed countries bear the principal responsibility for dealing with climate change and minimizes its own obligations under any global agreement. Along with India and the United States, China opposed establishing any "legally binding obligations" in the Paris Agreement (Rajamani, 2016:498). Dimitrov (2016) notes that, during negotiations, China opposed measures that would require transparency in meeting agreed targets as well as any form of independent and external reviews. Along with other "like-minded developing countries", China endorsed "weak qualitative goals" rather than strong, quantifiable measures, such as fixed emission reductions, and opposed establishing a commission that would assess "the impacts of a 1.5 degree rise" (Dimitrov, 2016:3). In short, while China often takes a leading role in opposing the entrenched position of developed states to minimize their responsibility to cover the cost of mitigating climate change, on other matters it is aligned with developed states in preventing strong, legally binding requirements; external oversight; and transparency. Not unlike most developed states, China insists on protecting its national interest and in ensuring that it can pursue a national policy of development without external oversight.

While China has made significant strides in developing its domestic emergency response and disaster management systems, it is facing severe environmental stress that is likely to worsen over the coming decades. China is increasingly affected by pollution, environmental degradation, and climate change associated its own rapid economic modernization as well as historic greenhouse gas emissions by Western states. Already a disaster-prone state, current developments suggest China's exposure to disasters is likely to increase in the years ahead. Murray reports that acid rain now affects two-thirds of China's arable land; over 25% of its territory is now desert; it has experienced increased flooding, drought, and severity of storms; and many of China's key coastal areas are threatened by sea level rise (2012:67). As a result of these environmental changes, the Chinese people and social, political, and economic institutions are increasingly vulnerable to natural hazards. To the extent that China will be able to play a major role in international disaster management given its current domestic situation, its weakening of humanitarian protections as a balm for the most negative aspects of the current order suggest that those affected and displaced by disaster will have fewer avenues for accessing protection.

The current international order is facing a series of tensions and contradictions that threaten to disrupt that very order. Worsening natural disasters are just one set of developments that expose these contradictions and facilitate support for radical alternatives. To date, however, the radical alternatives that have most successfully capitalized on the readily apparent contradictions of the existing international order have not been those most concerned with reducing human vulnerability and environmental degradation. Rather, the more successful movements in Western states (as indicated by electoral success) thus far are those that have asserted nation-statist and realist ideologies of order to challenge the humanitarian and economically liberal aspects of the current order. In the absence of an alternative that prioritizes vulnerable humans, the impending age of disasters is more likely (1) to become a reality and (2) to produce political orders that further reduce belief in the value of global cooperation to address tensions and contradictions of the current international order.

References

Adams, R.D., 1976. The Haicheng, China, earthquake of 4 February 1975: The first successfully predicted major earthquake. *Earthquake Engineering & Structural Dynamics*, 4(5), pp. 423–37.

Allison, G., 2017. *Destined for War: Can America and China Escape Thucydides' Trap?* Boston: Houghton Mifflin Harcourt.

Associated Press, 2013. After criticism due to slow response, China sending 51 medical staff to Philippines. *Associated Press*. November 20, 2013.

Baltrusaitis, D., 2013. China's revealing Typhoon Haiyan response. *The Diplomat*. https://thediplomat.com/2013/11/chinas-revealing-typhoon-haiyan-response/.

Bartke, W., 1989. *The Economic Aid of the PR China to Developing and Socialist Countries*, London: K.G. Saur.

BBC, 2015. Nepal Earthquake: India and China pledge millions in aid. *BBC*. https://www.bbc.com/news/world-asia-33266422.

Bijian, Z., 2005. China's "peaceful rise" to great-power status. *Foreign Affairs*, 84, pp. 18–25.

Blanchard, J.M.F., & Flint, C., 2017. The geopolitics of China's maritime silk road initiative. *Geopolitics*, 22(2), pp. 223–245.

Brautigam, D., 2011. Aid with "Chinese characteristics": Chinese foreign aid and development finance meet the OECD-DAC aid regime. *Journal of International Development*, 23(5), pp. 752–64.

Breslin, S., 2009. Understanding China's regional rise: Interpretations, identities and implications. *International Affairs*, 85(4), pp. 817–35.

Brooks, S.G., & Wohlforth, W.C., 2016. The rise and fall of the great powers in the twenty-first century: China's rise and the fate of America's global position. *International Security*, 40(3), pp. 7–53.

Busse, M., Erdogan, C., & Mühlen, H., 2016. China's impact on Africa – The role of trade, FDI and aid. *Kyklos*, 69(2), pp. 228–62.

Chen, G., 2012. China's management of natural disasters. In J.H. Chung (ed.). *China's Crisis Management*, London; New York: Routledge.

Cheng, Z., & Smyth, R., 2016. Why give it away when you need it for yourself? Understanding public support for foreign aid in China. *The Journal of Development Studies*, 52(1), pp. 53–71.

Christensen, T.J., 2006. Fostering stability or creating a monster?: The rise of China and U.S. policy toward East Asia. *International Security*, 31(1), pp. 81–126.

Cox, M., 2012. Power shifts, economic change and the decline of the West? *International Relations*, 26(264), pp. 369–388.

CRED, 2015. EM-DAT the International Disaster Database. International Disaster Database. https://www.emdat.be/.

Dikotter, F. 2011. *Mao's Great Famine: The History of China's Most Devastating Catastrophe, 1958–1962*, London: Bloomsbury Press.

Dimitrov, R.S., 2016. The Paris Agreement on climate change: Behind closed doors. *Global Environmental Politics*, 16(3), pp. 1–11.

Dreher, A., Fuchs, A., Parks, B., Strange, A.M., & Tierney, M.J., 2018. Apples and dragon fruits: The determinants of aid and other forms of state financing from China to Africa. *International Studies Quarterly*, 62(1), pp. 182–94.

Economist, The, 2018. Party time: Administrative reform. *The Economist*, p. 44.

Fleck, R.K., & Kilby, C., 2010. Changing aid regimes?: U.S. foreign aid from the Cold War to the war on terror. *Journal of Development Economics*, 91(2), pp. 185–97.

Friedberg, A.L., 2005. The future of U.S.-China relations: Is conflict inevitable? *International Security*, 30(2), pp. 7–45.

Fuller, P., 2015. Changing disaster relief regimes in China: An analysis using four famines between 1876 and 1962. *Disasters*, 39(2), pp. 146–62.

Geller, R.J., Jackson, D., Kagan, Y., & Mulargia, F., 1997. Earthquakes cannot be predicted. *Science*, 275(5306), pp. 1616–17.

Gilpin, R., 1983. *War and Change in World Politics*, Cambridge: Cambridge University Press.

Glosny, M.A., 2006. Heading toward a win–win future? Recent developments in China's policy toward Southeast Asia. *Asian Security*, 2(1), pp. 24–57.

Hamrin, C.L., 1986. Domestic components and China's evolving three world theory. In L.C. Harris & R.L. Vorden (eds.). *China and the Third World: Champion or Challenger?* Dover: Auburn House.

Hirono, M., 2013. Three legacies of humanitarianism in China. *Disasters*, 37(S2), pp. S202–S220.

Holsti, K.J., 2004. *Taming the Sovereigns: Institutional Change in World Politics*, Cambridge: Cambridge University Press.

Hong, Z., 2012. China's evolving aid landscape: Crossing the river by feeling the stones. In Chaturvedi, S., T. Fues, & E. Sidiropoulos (eds.). *Development Cooperation and Emerging Powers: New Partners or Old Patterns*. London: Zed Books.

Huang, Y., 2016. Understanding China's Belt & Road Initiative: Motivation, framework and assessment. *China Economic Review*, 40, pp. 314–21.

Ikenberry, J., 2008. The rise of China and the Future of the West: Can the Liberal System Survive? *Foreign Affairs*, 87(1), pp. 23–37.

Ikenberry, J., 2014. Introduction: Power, order and change in world politics. In J. Ikenberry (ed.). *Power, Order and Change in World Politics*, Cambridge: Cambridge University Press.

Jacques, M., 2012. *When China Rules the World: The End of the Western World and the Birth of a New Global Order*, New York: Penguin Books.
Johnston, A.I., 2003. Is China a status quo power? *International Security*, 27(4), pp. 5–56.
Kapoor, I., 2008. *The Postcolonial Politics of Development (Postcolonial Politics)*, London: Routledge.
Kennedy, P., 1987. *The Rise and Fall of the Great Powers: Economic and Military Conflict from 1500 to 2000*, New York: Random House.
Krebs, H. 2014. Responsibility, legitimacy, morality: Chinese humanitarianism in historical perspective. HPG Working Paper, London, Overseas Development Institute (ODI) September 2014.
Krebs, H., 2015. What you need to understand about Chinese humanitarian aid. *Overseas Development Institute* (ODI). ODI Newsletter. https://www.odi.org/comment/10066-what-you-need-understand-about-chinese-humanitarian-aid.
Krebs, H., 2016. The changing role of China in international humanitarian cooperation: Challenges and opportunities. In V. M. Heins, K. Koddenbrock, and C. Unrau (eds.), *Humanitarianism and Challenges of Cooperation*. London: Routledge.
Kurlantzick, J., 2006. China's Latin leap forward. *World Policy Journal*, 23(3), pp. 33–41.
Kurlantzick, J., 2007. *Charm Offensive: How China's Soft Power Is Transforming the World*, New Heaven, CT: Yale University Press.
Layne, C., 2018. The US–Chinese power shift and the end of the Pax Americana. *International Affairs*, 94(1), pp. 89–111.
Levy, J.S., 1987. Declining power and the preventive motivation for war. *World Politics*, 40(1), pp. 82–107.
Lu, X., & Xue, L., 2016. Managing the unexpected: Sense-making in the Chinese emergency management system. *Public Administration*, 94(2), pp. 414–29.
Lu, Y., & Xu, J., 2014. The progress of emergency response and rescue in China: A comparative analysis of Wenchuan and Lushan earthquakes. *Natural Hazards*, 74(2), pp. 421–44.
Mahmoud, Y.M., 2010. Chinese foreign aid: The tale of a silent enterprise. In J.S. Sörensen (ed.), *Challenging the Aid Paradigm Western Currents and Asian Alternatives*. New York: Palgrave MacMillan.
Manning, R., 2006. Will "emerging donors" change the face of international co-operation? *Development Policy Review*, 24(4), pp. 371–85.
Margesson, R., Chanlett-Avery, E., & Bruno, A., 2007. *North Korean Refugees in China and Human Rights Issues*, Washington, DC.
Mawdsley, E., 2012. *From Recipients to Donors: Emerging Powers and the Changing Development Landscape*, London: Zed Books.
McElroy, A., 2017. China reaffirms disaster risk cooperation commitment-UNISDR, Geneva. https://www.unisdr.org/archive/53104.
Melber, H., 2010. China in Africa: Any impact on development and aid. In J. Sorenson (ed.), *Challenging the Aid Paradigm: Western Currents and Asian Alternatives*. New York: Palgrave MacMillan.
Minkova, K.V., 2018. The economic roots of the Cold War: The IMF, ITO and other economic issues in post-war Soviet-American relations. *Policy Journal of Global Initiatives*, 12(1), pp. 18–31.
Murray, D., 2012. The Chinese dragon. In D. Murray & D. Brown (eds.). *Multipolarity in the 21st Century: A New World Order*, London: Routledge.

Nolke, A., Ten Brink, T., Claar, S., & May, C., 2014. Domestic structures, foreign economic policies and global economic order: Implications from the rise of large emerging countries. *European Journal of International Relations*, 21(3), pp. 538–567.

Nordin, A., & Weissmann, M., 2018. Will Trump make China great again? The Belt and Road Initiative and international order. *International Affairs*, 94(2), pp. 231–49.

OECD, 2016. Making development co-operation more effective: 2016 progress report, Paris: OECD Publishing. http://www.oecd-ilibrary.org/development/making-development-co-operation-more-effective_9789264266261-en.

Ogawa, H., 2018. The superficial success of the Development Assistance Committee: Emerging donors and the revival of economic statecraft. In M. Tadokoro, S. Egashira, & K. Yamamoto (eds.), *Emerging Risks in a World of Heterogeneity*, Singapore: Springer, pp. 67–85.

Organski, A.F., 1958. *World Politics*, New York: Knopf.

Paul, T.V., 2018. Assessing change in world politics. *International Studies Review*, 20(2), pp. 177–185.

Peterson, G., 2018. Forced migration, refugees and China's entry into the 'Family of Nations', 1861–1949. *Journal of Refugee Studies*, 31(3), pp. 274–291.

Phillips, A., 2011. *War, Religion and Empire*, Cambridge: Cambridge University Press.

Pollard, R.A., 1985. Economic security and the origins of the Cold War, Bretton Woods, the Marshall Plan, and American rearmament, 1944–50. *Diplomatic History*, 9(3), pp. 271–89.

PRC, 2014. White paper on China's foreign aid, Beijing. http://www.china.org.cn/government/whitepaper/node_7209074.htm.

Rajamani, L., 2016. Ambition and differentiation in the 2015 Paris Agreement: Interpretive possibilities and underlying politics. *International and Comparative Law Quarterly*, 65(2), pp. 493–514.

Reeves, C., 2014. The Red Cross Society of China: Past, present and future. In J. Ryan, L. Chen, & T. Saich (eds.). *Philanthropy for Health in China*. Bloomington: Indiana University Press.

Rolland, N., 2017. China's "Belt and Road Initiative": Underwhelming or game-changer? *The Washington Quarterly*, 40(1), pp. 127–42.

Rosenau, J.N., 1990. *Turbulence in World Politics: A Theory of Change and Continuity*, Princeton, NJ: Princeton University Press.

Sleeswyk, A.W., & Sivin, N., 1983. Dragons and toads: The Chinese seismoscope of AD 132. *Chinese Science*, 6, pp. 1–19.

Soderberg, M., 2010. Challenges or complements for the West: Is there an "Asian" model of aid emerging? In J.S. Sörensen (ed.), *Challenging the Aid Paradigm*, New York: Palgrave Macmillan.

Song, L., 2018. China and the international refugee protection regime: Past, present, and potentials. *Refugee Survey Quarterly*, 37(2), pp. 139–61.

Staiger, R.W., & Sykes, A.O., 2010. "Currency manipulation" and world trade. *World Trade Review*, 9(4), pp. 583–627.

Stephen, M.D., 2014. Rising powers, global capitalism and liberal global governance: A historical materialist account of the BRICs challenge. *European Journal of International Relations*, 20(4), pp. 912–38.

Suttmeier, R.P., 2012. China's management of environmental crises: Risks, recreancy, and response. In J.H. Chung (ed.). *China's Crisis Management*. London: Routledge.

Tan, V., 2017. China can play key role in solving refugee crises – UNHCR chief, Geneva. http://www.unhcr.org/news/latest/2017/6/593946b64/china-play-key-role-solving-refugee-crises-unhcr-chief.html.

Telford, J., & Cosgrave, J., 2007. The International humanitarian system and the 2004 Indian Ocean earthquake and tsunami. *Disasters*, 31(1), pp. 1–28.

UNHCR, 2018. UNHCR Donor Profiles, Geneva. http://reporting.unhcr.org/sites/default/files/UNHCR 2017 Total Contributions.pdf.

United Nations Development Programme, 2017. China-UNDP Global Cooperation 2017. http://www.cn.undp.org/content/china/en/home/library/south-south-cooperation/china--undp-global-cooperation-2017-highlights-.html.

United Nations Global Programme on the Integration of Public Administration and the Science of Disasters, 1996. Great Tangshan Earthquake: UN Global Programme and UNEPPA Study of a Case of Practical Integration of Disaster Science and Public Administration in Qinglong County, China. Global Watch. http://www.globalwatch.org/ungp/qinglong.htm.

Walz, J., & Ramachandran, V., 2011. *Brave New World A Literature Review of Emerging Donors and the Changing Nature of Foreign Assistance*. Center for Global Development, pp. 1–28. http://www.cgdev.org/content/publications/detail/1425691.

Wenping, H., 2010. China's aid to Africa: Policy evolution, characteristics and its role. In J.S. Sörensen (ed.). *Challenging the Aid Paradigm Western Currents and Asian Alternatives*, New York: Palgrave MacMillan.

Wilson, A.W., 2016. We're no compatriots of yours. *Foreign Policy*. February 12, 2016.

Woods, N., 2008. Whose aid? Whose influence? China, emerging donors and the silent revolution in development assistance. *International Affairs*, 84(6), pp. 1205–21.

Wu, H., 2017. Statement by Ambassador WU Haito at the Security Council Briefing by the UNHCR, New York. http://www.china-un.org/eng/hyyfy/t1511072.htm

Yu, H., 2017. Motivation behind China's "One Belt, One Road" initiatives and establishment of the Asian Infrastructure Investment Bank. *Journal of Contemporary China*, 26(105), pp. 353–68.

Yu, P.K., 2012. I*nternational Governance and Regimes: A Chinese Perspective*, London: Routledge.

Zhang, X., Yi, L., & Zhao, D., 2013. Community-based disaster management: A review of progress in China. *Natural Hazards*, 65(3), pp. 2215–39.

Zhaoxu, M., 2016. Enhancing International Cooperation and Promoting Global Governance in Migration, Geneva. http://www.china-un.ch/eng/dbtzyhd/t1421948.htm.

CONCLUSION

Dire warnings about the potential impact of disasters in the coming years and decades are now commonplace and widespread. The insurance industry; climate scientists; political leaders at the national, regional, and local levels; civil society organizations; as well as many affected communities are drawing attention to the need for immediate, coordinated action. Even oil and gas companies are actively promoting global mechanisms to reduce greenhouse gas emissions to stave off the disasters associated with climate change (Dudley, 2015). Apart from weak international declarations and relatively small-scale local actions, these warnings seem to have largely gone unheeded, particularly at the national and international levels.

The argument developed in this book suggests that the current responses to disaster reflect an understanding of disaster shaped by ideologies of international order and the desire of powerful actors to maintain the current international order without significant alteration. The availability of multiple framings of disaster within the heterogenous international order not only produces tension and instability; it also provides space for actors to frame disasters in multiple ways in order to sustain that order. This process of framing and contesting the meaning of disaster, as well as other developments, such as migration and climate change, is, however, an important source of incremental change in international order.

Looking ahead, the ideologies of international order that will become more prominent in shaping disaster management and those that will be altered depend on three factors. The first is whether actors successfully frame disasters as a failure to follow a particular ideology rather than a failure of that ideology. The second is whether actors successfully promote an alternative understanding of disaster and disaster management. Third, and as Chapter 8 demonstrated, it will depend on the distribution of power internationally.

These factors are evident in current contestation over disaster management in the United States. President Trump's threat to withhold disaster assistance to both Puerto Rico (hurricane) and California (wildfires) amply demonstrates how political elites frame disasters to support or alter policy. For Trump, these disasters are symptomatic of poor local governance and the moral hazard of providing disaster assistance rather than the product of human-caused climate change or a straightforward humanitarian emergency. That President Trump is contesting well-established understandings of disaster should come as no surprise, as he has called into question several of the foundational ideologies of international order. The impact of Trump's presidency highlights one of the key claims developed in this book – change in international order arises from ideological contestation as much as, and in combination with, changing distribution of power.

The level of discontent with the current international order seems to have reached a seminal point. For many, including defenders of the current international order, it is an order in crisis (e.g. Koivisto and Dunne, 2010; Ikenberry et al., 2018). States that played a prominent part in constructing this order (the United States and the United Kingdom) have seen the rise to power of leaders who challenge beliefs foundational to this order. Rising and regional powers such as China and Russia have also called for a significant reorientation emphasizing nation-statist and realist principles, while some of the key remaining defenders of this order (e.g. Germany, Canada, France) have decided or have been forced to abandon modest, progressive alterations of this order in favour of shoring up the status quo. Based on the analysis presented here, continual modification of an existing order is to be expected; widespread dissatisfaction and significant alteration (particularly among its architects and beneficiaries) is less common. The internal tensions and external contradictions that arise within a heterogenous international order necessarily lead to unfulfilled expectations. There will be significant variation, however, in the level of dissatisfaction with an existing order based on the extent to which one (individually or as a group) is exposed to these tensions and contradictions, which often occurs in the aftermath of large-scale disaster.

When and how changes in the underlying foundation of international order take place is of significant interest to scholars of international relations (IR). Disagreement between the dominant accounts revolves around the three factors that I highlight in this book: ideological failure, ideological competition, and changing distribution of power. In this conclusion, I will show that the heterogenous account contributes to existing dominant accounts in the field of IR, namely, realism and liberalism. These two perspectives are often depicted as offering contrasting accounts of the impact of the distribution of power on the international order as well as the nature of the current crisis and the future of the international order. In contrast, I suggest both suffer from profoundly important shortcomings, misdiagnose the source of crisis in the current international order, and ultimately rely on a heterogenous account of international order and change to support their claims. I do not view such accounts as competing with the account offered here; rather, I view them as complimentary, albeit insufficiently comprehensive in their treatment of international order. These accounts are clearly quite helpful in understanding change within a single ideological framework that underpins the existing order or the relative weight placed on a particular belief system, and I have relied on many of these works. Their single foundation analysis, however, inaccurately identifies belief systems in binary terms (realism versus liberalism) rather than complexly interrelated (complimentary and competitive). This leads to some confusion about the interconnectedness of realism and liberalism. For instance, the realist scholar John Mearsheimer (2019) claims that states in the post–World War II period adopted liberal principles for realist purposes, while liberal journalist Kaplan depicts the "architects of the post war order" as realists "in the service of liberalism" (2018:35). While both acknowledge the complex interplay between realist and liberal beliefs, their homogenous account of international order requires simplification and a clear ideological hierarchy. The heterogenous account suggests actors hold and act on multiple belief systems that are not necessarily hierarchical. It also draws greater attention to the influence of other ideological foundations, particularly as changes and/or tensions in other belief systems may influence change within the two ideologies they address. In the remainder of this conclusion, I highlight how the account in this book contributes to the two dominant perspectives on international order.

Realist Perspective on International Order[1]

Realist explanations of international order emphasize distribution of power as the primary determining factor shaping international order (see Morganthau, 1948; Waltz, 1979). Crises or instability in that order emerge when the principles underpinning that order no longer reflect the distribution of power, although, as I will show, realist thinkers often identify other sources of instability and crises as well. John Mearsheimer, who is the most prominent realist currently writing on international order, states that distribution of power determines the set of principles that will define an international order. He concludes that, in bipolar and multipolar systems, international order will be realist, and only under unipolar systems can the order be shaped along liberal (or other ideological) lines.[2] Mearsheimer writes that, since the end of World War II, the United States has presided over two distinct international orders – one characterized by bipolarity (and realist principles) and another unipolar (and liberal principles). The ideology of the major power only matters in a unipolar system, with a unipole holding a universalist ideology that is able to impose its rules and values. In other distributions of power, the ideologies of the great powers do not matter as they engage in a competitive system in which realist considerations predominate (Mearsheimer, 2019:6). Great powers may adopt liberal values and institutions in a realist international order, but Mearsheimer claims that they do so out of realist considerations of maximizing power and will abandon liberal principles when they conflict with their interests (2019:7). From this realist perspective, change to international order results from changing distribution of power – although, for Mearsheimer, a change from bipolar to multipolar would still be a realist order, and thus unchanged in the primary values and goals that states pursue (2019:11).

According to Mearsheimer, unipolar international orders in which the hegemon imposes a universalist ideology, such as the liberal international order, contain additional mechanisms of change, namely, internal inconsistencies within liberal beliefs, combined with persistent challenges from competing belief systems (2019:11). Mearsheimer argues that the liberal international order declined because "it contained the seeds of its own ruin", by which he implies that the promises of liberalism proved illusory or false (2019:23). In addition, he suggests that liberalism's diminishment of sovereignty clashed with continuing belief in nationalism (2019:24). These mechanisms of change are consistent with the argument developed in this book.

The heterogenous account of international order shares with realist accounts an emphasis on the importance of power in shaping the goals and values that define an international order. In addition, both stress that instability and change may emerge from internal ideological inconsistencies as well as the persistence of other belief systems that challenge the influence of a dominant ideology. The key differences on these specific points are that a heterogenous account develops a more expansive conceptualization of power and does not restrict the importance of ideology to unipolar systems only, or ideological inconsistency to liberalism alone.

In other areas, there are significant differences. Many IR scholars, including realist and liberal scholars, identify democracy as the key ideological foundation of the current international order (Edinger, 2018). Mearsheimer similarly sees democracy promotion as the key motivating factor behind most post–Cold War initiatives, such as North Atlantic Treaty Organization (NATO) and European Union (EU) enlargement, the war in Iraq, and engagement with China. There is clearly substantial evidence to show that political liberalism and democracy promotion became an important ideological component of the post–Cold War international order. However, I suggest we should be careful not to overemphasize the influence of this belief system. When I first set out on this project, I had initially hypothesized that political liberalism (of which democracy was a key feature) constituted a foundational ideology of international order. However, the absence of democracy promotion in key regimes that govern disaster, climate, migration, and disaster diplomacy, which are the key areas of study in this book, suggests it was not a particularly significant consideration in the liberal international order as claimed by realists and liberals alike. Democratization efforts did become more common in the post–Cold

War period, but there is compelling evidence to show that it was not international in scope and was secondary to other values and goals.

Key architects of a "new world order" in the post–Cold War period described it as based on "peaceful settlement of disputes, solidarity against aggression, reduced and controlled arsenals and just treatment of all peoples" – with no mention of democracy (Freedman, 1991:196). While democracy promotion may have been one potential avenue to achieve these overarching values and goals, it was not foundational. Examining the United States' commitment to democracy promotion in the post–Cold War period, Kaplan astutely observes that US support for democracy was "indifferent, at best" (2018:57), and focused only on "critical parts of the world" – namely, Europe (2018:30). Citing Nixon, Kaplan notes that many Americans did not (and do not) view democracy as the appropriate form of governments for the peoples of "Asia, Africa, and Latin America" (2018:58). Many other areas of the world were either considered not ready for democracy and freedom, too important for US interests to destabilize friendly dictators, and/or not sufficiently crucial to the international order to bother with promoting political freedom. Democracy promotion was a regional rather than international consideration. The secondary nature of democracy promotion is evident through other programs as well. In his work on the UN's electoral assistance programs, Schroeder observes that democratizing states drove most of these programs, with many Western liberal democracies lukewarm about the idea of increasing UN funding for electoral assistance (2013:215). These same states directly opposed further democratizing international institutions, such as the Security Council, that would undermine the influence of the most powerful states (2013:216). Zanotti goes even further to argue that democracy promotion in the post–Cold War period was subsumed under a larger emphasis on "good governance" that translated into an effort to build particular types of states – the regulatory state, rather than the democratic state (2005:471). Rather than advancing political liberalism, international efforts to foster good governance actually instilled a neo-liberal form of economic liberalism, often at the expense of democracy promotion.

Other post–Cold War developments further call into question the centrality of democracy promotion as the key ideological driver of the liberal international order. The ongoing support for dictatorships by the United States in the post–Cold War period (e.g. Egypt, Saudi Arabia, Jordan, Kuwait, United Arab Emirates [UAE], Bahrain, Ethiopia, Cameroon, Chad, Niger, Qatar, Philippines, Thailand) to maintain US military bases and force projection show that the US government consistently prioritized realist objectives over the goal of democracy promotion or had only a weak commitment to democracy promotion that was largely subsumed under strategic considerations (Calder, 2008; Cooley, 2008). The global expansion of economic liberalism through trade agreements and international institutions, while partially legitimized by a claim to eventual and inevitable political liberalization, often occurred without any explicit political liberal conditionality and often negatively affected transitions to democracy (Li and Reuveny, 2003). The prosecution of the war on terror, ostensibly based on the spread of democracy, saw the United States support several dictators in addition to promoting anti-terror policies domestically and internationally that were essentially anti-democratic (Whitaker, 2007). I raise these counter-examples not to dispute the claim that democracy promotion was an increasingly important belief underlying the post–Cold War international order (it clearly was); rather, I demonstrate that it was limited regionally and that the continued prevalence of realist, statist, and economically liberal objectives and values seriously constrained the stated goal of democracy promotion. Mearsheimer's (2019) tendency to depict US foreign policy blunders as the exclusive product of democracy promotion (and liberalism generally) rather than the pursuit of great power politics (along with other pursuits) demonstrates the problem of relying on a single ideological factor explanation.

Other notable areas of disagreement between these two accounts concern ideology. Mearsheimer's (2019) argument depicts ideology and power as independent from one another, as signified by the claim that ideology matters only in a unipolar order. In contrast, I argue that all orders rest on ideological foundations, including multi- and bipolar orders in which realist principles carry greater moral authority. I have stressed that the values and goals promoted by powerful states as the foundations of an international order are constitutive of their power and, by extension, a potential source of weakness. At times, Mearsheimer seems to agree with such a claim, noting that the combination of liberalism's "intrinsic flaws" and the challenge from a competing belief system (nationalism) undermined US power. These arguments are more consistent with a heterogenous account of order than with a realist account.

A more significant challenge to the realist argument emerges from the claim that dominant belief systems are constitutive of prevailing conceptions of power and the legitimate use of power. In short, material power still matters, but the foundational ideologies of order shape which forms of power matter (Buzan and Lawson, 2015), as well as when and how that power is exercised in the conduct of international relations. Recognizing humanitarian assistance or scientific expertise as a form of power rests on belief systems that identify these as relevant endeavours. At the same time, belief that states may legitimately employ military power to achieve domestic and international objectives without restriction declined due to the rise of humanitarian and economic liberal beliefs promoted by civil society organizations alongside militarily powerful states. Understanding what counts as power, and when and how it may be applied, necessitates an examination of underlying belief systems rather than simply focusing on distribution of power between states.

On the issue of change, Mearsheimer (2019) presents change as structurally determined.[3] The clearest expression of this is his claim that a shift in the distribution of power from unipolar to bi- or multipolar necessarily changes international order. My account provides two correctives to these claims. The first is that all ideologies are intrinsically flawed, including realism; consequently, all orders are unstable and subject to contestation and change. Mearsheimer imputes a level of stability and coherence to realist principles within bi- and multipolar international orders that is unwarranted – as though significant changes have not occurred during these periods. The rise and partial diminishment of imperialist ideology that European states ought to treat each other differently than the non-European "other" arose in a multipolar system, as did the early rise of economic liberalism – both of which profoundly shaped international order and the behaviour of states. The flurry of institution building around the management of disasters that this book has traced (in addition to institution building in a host of other areas) during the multipolar pre–World War I period and during the bipolar Cold War period itself demonstrates the weakness of such a claim and the need to be attentive to other foundational belief systems.

This leads to the second corrective: change in international order is not determined by changing either distribution of power or ideological contestation; international orders are resilient and subject to incremental change that exhibits a high level of continuity with the existing order rather than abrupt and categorical change. The end of the Cold War did not abruptly usher in a new world order that differed dramatically from what preceded it. Mearsheimer and I agree that, at present, realist and nation-statist principles are becoming of greater import, at the expense of humanitarianism, human rights, and political liberalism. We both agree that economic liberalism (in the form of free trade) will remain a dominant feature of the international order, with potential challenges emerging from protectionist policies. The primary difference is that, whereas Mearsheimer sees these developments as an inevitable outcome of the changing distribution of power and failure of liberalism, my account suggests they are contingent, based on political agency and the ability of actors to frame material developments. My account also suggests that the emerging international order is likely to exhibit a great deal of continuity with the present order rather than a substantial break. In short, my work in this book suggests that scholars of international order continue to examine the distribution and operation of power in shaping international order, but also the values and goals that constitute relevant actors, conceptions of power, and the legitimate application of power.

Liberal Perspective on International Order[4]

Liberal theory offers a different perspective on both the importance of distribution of power and how ideological contestation affects change. Liberal theory, as represented by Ikenberry's work, places much greater emphasis on the ideational foundation of international order than does realism. Consequently, liberal theory is perhaps more attuned to the potential impact of ideational crises to affect international order while insisting on the influence of ideas in shaping that order regardless of changes in material power distribution.

For Ikenberry, belief systems are the defining feature of an international order, with liberal order gradually becoming more influential and overcoming a number of challenges and crises, including competition from ideological alternatives such as communism and fascism as well as "internal" tensions such as the Great Depression and the failure of Bretton Woods (Ikenberry, 2018). Thus, the story of liberalism is one of overcoming crises and challenges, due to its ability to deliver material benefits globally and to its ideological adaptability. He emphasizes that liberalism has had a positive global impact in the form of "wealth creation, physical security, and hope for justice" (Ikenberry, 2010:513). The impressive record of achievement of this "liberal" order includes sustained economic growth across seven decades, a nine-fold increase in the number of democratic governments, and a reduction of the number of people living in poverty by 4 billion (Kaplan, 2018:4). For most liberals, there is more than sufficient empirical evidence to show that liberal order has produce material, observable, and measurable improvement around the world.

Despite these strengths, Ikenberry concedes that the liberal order is in crisis – a product of the combination of material developments and ideological tensions. Ikenberry also concedes that changes in the distribution of power reduced the capability of liberal states to "defend and support", and "address problems" inherent in the order (2018:18). A case in point is the 2008 economic recession that severely undermined faith in the US-led international order and in its ability to "manage the tensions of the capitalist system" (2010:511). In addition, the neo-liberal turn has seriously damaged the viability of liberal international order, as Ikenberry claims it is not the strain of liberalism adhered to by most people; as protection for workers have weakened, the influence of corporations has grown, and income inequality has expanded (2018:23). The current crisis is the product of redistribution of power, material developments inconsistent with liberal claims, and readily available ideological alternatives, which is consistent with a heterogenous account of international order.

An important tension emerges in the liberal argument concerning the discrepancy between the positive material outcomes associated with economic liberalism and the current crisis of liberal order. If liberalism has dramatically improved life on the whole for most people on the planet, why is that order in crisis (if one discounts the structural determinism of distribution of power, as Ikenberry and other liberals do)? Ikenberry develops two claims to address this tension: the first is to differentiate a crisis of authority (Polanyi crisis) from a crisis of social purpose (Carr crisis) (2018:10). While the former represents a loss of faith in US leadership while still holding faith in liberalism generally, the latter signals a loss of faith in the ideology of liberalism itself. Ikenberry stresses that we are currently witnessing a crisis of US authority, not a crisis of social purpose associated with liberal thought. In this way, the distribution of power influences international order – it shapes perception of authority of great powers in leading the international order; however, it does not determine the values that will underpin international order. For this reason, Ikenberry claims that liberalism can persist beyond the relative decline of the United States and the crisis of leadership to which this has contributed (2018:9).

Ikenberry's second claim is that the current crisis is not due to a failure of liberalism but to its success. Under US hegemonic leadership, the United States facilitated the rapid expansion of liberal principles around the globe that in turn produced "economic and security interdependence" to an extent that was unimaginable for the original architects of this order (2010:520). The shift of liberal order from a regional order during the Cold War to a global order after the Cold War produced a number of tensions and inconsistencies that require alteration. The United States has failed to implement these necessary alterations and may no longer have the power to do so on its own. Consequently, for Ikenberry, the material developments that have contributed to the current crisis are not due to the failure of liberalism to deliver on its promised outcomes but are due to its rapid success in delivering on these promises and the inability of existing institutions to keep pace with these successes and expectations of a global liberal order.

There are a number of areas of overlap between the liberal theory of international order and the heterogenous account. Both argue that ideology is foundational to international order and shapes the goals and values toward which relevant actors direct resources and construct policies. These accounts reject the structural determination of distribution of power and share an account of change due to crises that emerge from multiple sources: change in material circumstances (including distribution of power), ideological instability produced by internal inconsistencies and unfulfilled promises, and competition from alternative belief systems.

On a number of central issues, however, the heterogenous account of international order departs significantly from liberal theory. First and foremost, I contend that describing the current international order as a liberal international order is empirically inaccurate and normatively misguided. Significant aspects of the so-called liberal international order are profoundly illiberal, while liberalism has always operated alongside other foundational ideologies that are complexly interrelated. This is not to suggest that all foundational ideologies are equally influential at all times; their relative influence waxes and wanes in line with distribution of power, material developments, and extent of ideological contestation. During the post–Cold War period, it is certainly accurate to suggest that liberalism, in both its economic and political variants, gained in influence relative to the Cold War period – although it was never dominant, and certainly not hegemonic. To describe this period as a liberal international order neglects important changes that occurred within economic liberalism, overstates the importance of political liberalism, and denies the continuing relevance of belief systems that now appear to be gaining in influence. It also mistakenly veers toward ideology by reasserting the progressive teleology of liberalism as inevitably triumphant once liberalism had defeated all challengers (see Fukuyama, 1992).

This relates to the second important point of departure – the uniquely adaptive nature of liberalism. For Ikenberry (2010), liberalism's success reflects its unique ability to reform and adjust in line with material developments. The responsive nature of liberalism – responsive to the market and the will of the people – allows it to accommodate novel developments and experiences, a feature that Ikenberry claims is unique to liberalism. My heterogenous account argues that liberalism's purported flexibility is not simply the product of its internal logic but also of its co-existence and dependence upon other ideologies. What some may see as liberalism's internal adaptability may, in some cases, be better understood as strategic framing within a heterogenous ideological environment. Thus, actors may frame failures reasonably attributable to liberalism in realist terms and attribute these failures to a rival power. Similarly, adaptations within the liberal order do not emerge entirely from within the logic of liberalism itself but by borrowing from other belief systems.

I have shown that most other ideologies are similarly flexible and adaptable. Nation-statism has continued to adapt and evolve over time as functional expectations of states and conceptions of "national communities" have changed. Realist beliefs have adapted to new forms of technology, distributions of power, economic interconnectedness, and state forms. The flexibility of these belief systems is what has led to their persistent influence and recent resurgence. Tensions and inconsistencies are inherent in any belief system, particularly under rapidly changing material circumstances; however, most ideologies are sufficiently flexible to accommodate a fair amount of cracks in the foundation and to permit ideological alterations. Ikenberry (2018) claims that alternative ideologies to liberal international order, which he identifies as "political realism, authoritarian nationalism, Social Darwinism, revolutionary socialism, and post-colonialism", are incapable of reform and thus no competitor to liberalism. While I have not addressed some of the ideologies he lists here, I have shown other ideologies to be capable of reform and adaptability. Competing ideologies have co-existed alongside economic liberalism throughout the period often referred to as a "liberal economic order", serving at times to sustain and bolster economic liberalism even as they introduced other tensions and inconsistencies. For instance, the efficient provision of public goods by the state (a key principle of nation-statism) operates as an important legitimizing principle for other ideologies of order. Ideologies considered defeated by liberalism, such as fascism and authoritarianism, have adapted to perceived failures of liberal democracy – such as deadlock, shutdowns, and inefficiency – to gain currency. Indeed, fascism's rise in the early twentieth century was rooted in the perceived inability of democratic states to produce results and ensure that trains run on time. Recognizing the flexibility of all ideologies better explains how these ideologies appear to be resurgent – they have proven adaptable to the perceived failures of liberalism.

The heterogenous account calls into question Ikenberry's (2010, 2018) distinction between the Polanyi and Carr crises. Crises of authority are not easily separable from crises of social purpose, particularly when the appeal of the social purpose is so strongly connected with a particular set of authorities. This distinction is insufficiently attentive to a crisis of liberalism itself. Notwithstanding the myriad successes of economic liberalism, there are major tensions that have undermined faith in economic liberalism: growing inequality, environmental degradation, climate change, and a pervasive sense of loss of communal identity and values under universalist identity claims. All of these tensions are on display in the case of disasters. While liberal principles have survived previous crises, many of which appeared worse than the current crisis (Ikenberry, 2018:22), the range of issues on which the liberal order has failed to deliver is lengthy – and of existential importance.

The distribution of power during previous crises mentioned by Ikenberry (2018) largely favoured liberal states, and they expended great resources to ensure continuity of that order (with modifications). This is no longer the case. The end of the Cold War eliminated the strategic necessity of expending vast resources to maintain this order and gave rise to "excessive optimism and complacency" (Kaplan, 2018:89). In addition, the more recent diffusion of power undermines the ability of one state or group of states to prop up a crisis-ridden order, and political will on the part of liberal states to address international crises of order through massive expenditures of resources appears to be on the wane.

Final Thoughts

The attraction of liberal and realist accounts of international order is that they provide a parsimonious account of the current order and future trajectories. Realists provide a deterministic account in which distribution of power will essentially determine the contours of the future order, which Mearsheimer (2019) sees as a multipolar system founded on realist principles. Liberal theorists such as Ikenberry (2010, 2019) are less deterministic and place greater emphasis on developments within liberal states themselves. The future of liberal order depends on liberal states maintaining their commitment to liberal principles as these states provide the best hope for continuing the liberal order, not through hegemonic imposition but by demonstrating the universal appeal of liberalism itself. Ultimately, Ikenberry envisions a future order that is largely liberal, based on the inherent attractiveness of liberalism, its success in producing positive material outcomes, and the lack of serious competitors.

The heterogenous account rejects binary categorizations of international order and thus does not offer a prediction whether the international order in the next decade will be liberal or realist. In fact, it rejects such a binary account and suggests that it will be both and, even more complicated than that, it is likely to be an amalgam of existing ideologies of order, with slight modifications that reflect shifting distribution of power and ideas. If more radical ideologies of international order, such as deep ecology[5] or some variant of Marxism/socialism, gain traction within existing or rising powers, perhaps due to a series of disasters, we may witness profound ruptures within the existing order.

This does not mean that the heterogenous account does not offer insight into what the next few years or decades may look like. As I have stressed throughout this book, the foundational ideologies of international order are deeply engrained at the local, national, and international levels. Absent major material disruptions, any changes over the next decade or two are likely to be incremental alterations of the existing order rather than a categorical departure. Even a relatively significant redistribution of power among states is unlikely to usher in drastic changes, although this depends greatly on which actors become more powerful and the ideologies of international order that resonate within these societies. Understanding potential changes to international order requires close engagement with the belief systems that shape the dominant powers and not simply the relative distribution of power.

In focusing on ideological foundations and practices around disaster management, this book has demonstrated that there are already indications of which ideologies are likely to become more and less prominent. Nation-statism and realism appear to be increasing in influence, economic liberalism and modernism remain strongly influential with minor alterations, while humanitarianism and human rights are likely to decline in influence. Chapter 8 on China demonstrates, however, that humanitarianism (if not human rights) may become more prominent as China becomes more powerful and sees the value of humanitarian ideology in patching up the cracks in modernism, economic liberalism, realism, and nation-statism. In turn, this may ameliorate the stronger variants of realism and nation-statism that currently appear to be ascendant and soften opposition to the more human rights–oriented strain of humanitarian ideology.

In a book such as this about disasters and the trend toward more frequent and extreme climate events, however, it is problematic to base future expectations on current experience. The reality is that the international community is likely to experience several major material disruptions in the coming decades that could dramatically alter the foundations of international order. To date, the great powers have constructed several regimes designed to manage disasters as part of the project of building and maintaining the modern international order. This book has argued that these regimes are straining under the weight of material changes and the failed promises of these foundational ideologies. Climate change has dramatically increased the likelihood of a "Lisbon" type process that could pave the way for a profound alteration in the foundational ideologies of international order. Whether that leads to an international order based on deep ecological principles that eschew the current dominant ideologies of international order or to an order founded on hyper-realist, nation-statist beliefs will depend on a number of factors: the entities that emerge as most powerful, the belief systems they hold, and the extent to which the developments that led to such a process can be attributed to existing ideologies of international order. If the analysis offered in this book is at all a useful guide, there will be no shortage of evidence to hold each of the currently dominant ideologies of international order to account.

Notes

1 There are multiple realist perspectives on international order; see Brooks (1997) for an overview of competing realist perspectives. Few explicitly theorize the concept of international order as thoroughly as those associated with hegemonic stability, a strand of realist thought applicable to Mearsheimer's (among many others) current writings on international order, so I focus here on that body of work, recognizing that this ignores a wide range of realist thought.
2 It is not clear that Mearsheimer views realism as an ideology in this context, on par with liberalism or communism.
3 Not all realists provide such a determinist account; for instance, Schweller and Pu (2011) show that emerging powers in a multipolar world can adopt a variety of positions vis-à-vis the existing order, including supporting the order.
4 Just as realist thought on international order is more diverse than represented in the brief sketch above, liberal thought is also highly diverse. Considering the diversity within liberal IR theory (see Zacher and Matthew, 1995), as well as areas of overlap with constructivist and even English School theorizing (see Evans and Wilson, 1992), reducing a discussion of a liberal account of international order to a single perspective misses a great deal of diverse thought on international order.
5 Some examples include ecological approaches that draw on indigenous political theory such as James Tully's work (1995, 2008, 2018), that would envision new relationships between humanity and the environment, economic production, and reconciliation between human societies, eco-socialist movements (Pepper, 1993) and political ecological approaches (Bryant and Bailey, 1997) that would disrupt nation-state authority, capitalism, and modernism.

References

Brooks, S., 1997. Dueling realisms. *International Organization*, 51(3), pp. 445–77.

Bryant, R., & S. Bailey. 1997. *Third World Political Ecology*, London, Routledge.

Buzan, B., & Lawson, G., 2015. *The Global Transformation: History, Modernity and the Making of International Relations*, Cambridge: Cambridge University Press.

Calder, K., 2008. *Embattled Garrisons: Comparative Base Politics and American Globalism*, Princeton, NJ: Princeton University Press.

Cooley, A., 2008. *Base Politics: Democratic Change and the US Military Overseas*, Ithaca, NY: Cornell University Press.

Dudley, B., 2015. Why are oil and gas companies calling for more action on climate change? *Reuters*. November 12, 2015. https://www.reuters.com/article/us-climatechange-summit-dudley/why-are-oil-and-gas-companies-calling-for-more-action-on-climate-change-idUSKCN0T103T20151112

Edinger, H., 2018. Bored enough yet?: International order, ideology, and what it takes to get history started again. *International Affairs*, 94(1), pp. 173–83.
Evans, T., & Wilson, P., 1992. Regime theory and the English school of international relations: A comparison. *Millennium: Journal of International Studies*, 21(3), pp. 329–51.
Freedman, L., 1991. The Gulf War and the new world order. *Survival*, 33(3), pp. 195–209.
Fukuyama, F., 1992. *The End of History and the Last Man*, New York: Free Press.
Ikenberry, J., 2010. The liberal international order and its discontents. *Millennium: Journal of International Studies*, 38(3), pp. 509–21.
Ikenberry, J., 2018. The end of liberal international order? *International Affairs*, 94(1), pp. 7–23.
Ikenberry, J., Parmar, I., & Stokes, D., 2018. Introduction: Ordering the world? Liberal internationalism in theory and practice. *International Affairs*, 94(1), pp. 1–5.
Kaplan, R., 2018. *The Jungle Grows Back: America and Our Imperiled World*, New York: Alfred Knopf.
Koivisto, M., & Dunne, T., 2010. Crisis, what crisis?: Liberal order building and world order conventions. *Millennium: Journal of International Studies*, 38(3), pp. 615–40.
Li, Q., & Reuveny, R., 2003. Economic globalization and democracy: An empirical analysis. *British Journal of Political Science*, 33(1), pp. 29–54.
Mearsheimer, J., 2019. Bound to Fail: The rise and fall of liberal international order. *International Security*, 43(4), pp. 7–50.
Morganthau, H., 1948. *Politics among Nations: The Struggle for Power and Peace*, New York: Knopf.
Pepper, D., 1993. *Eco-Socialism: From Deep Ecology to Social Justice*, London: Routledge.
Schroeder, M., 2013. The politics of change: The evolution of UN electoral services, 1989–2006. *Global Governance*, 19(2), pp. 207–26.
Schweller, R., & Pu, X., 2011. After unipolarity: China's visions of international order in an era of U.S. decline. *International Security*, 36(1), pp. 41–72.
Tully, J., 1995. *Strange Multiplicity: Constitutionalism in an Age of Diversity*. Cambridge: Cambridge University Press.
Tully, J., 2008. *Public Philosophy in a New Key: Volume 1, Democracy and Civic Freedom*, Cambridge: Cambridge University Press.
Tully, J., 2018. Reconciliation here on Earth. In M. Asch, J. Borrows, & J. Tully (eds.). *Resurgence and Reconciliation: Indigenous-Settler Relations and Earth Teachings*, Toronto: University of Toronto Press.
Waltz, K., 1979. *Theory of International Politics*, Boston: McGraw-Hill.
Whitaker, B.E., 2007. Exporting the Patriot Act? Democracy and the "war on terror" in the third world. *Third World Quarterly*, 28(5), pp. 1017–32.
Zacher, M., & Matthew, R., 1995. Liberal international theory: Common threads, divergent strands. In C.W. Kegley Jr (ed.). *Controversies in International Relations Theory: Realism and the Neoliberal Challenge*, New York: St Martin's Press.
Zanotti, L., 2005. Governmentalizing the post–Cold War international regime: The UN debate on democratization and good governance. *Alternatives*, 30(4), pp. 461–87.

INDEX

Note: Page numbers followed by n refer to notes.

For Product Safety Concerns and Information please contact our EU representative GPSR@taylorandfrancis.com
Taylor & Francis Verlag GmbH, Kaufingerstraße 24, 80331 München, Germany

www.ingramcontent.com/pod-product-compliance
Lightning Source LLC
Chambersburg PA
CBHW071251180926
51613CB00001B/14
9780367200367